Recent Advances in Hydro- and Biohydrometallurgy

Recent Advances in Hydro- and Biohydrometallurgy

Special Issue Editor

Kostas A. Komnitsas

MDPI • Basel • Beijing • Wuhan • Barcelona • Belgrade

MDPI

Special Issue Editor
Kostas A. Komnitsas
Technical University of Crete (TUC)
Greece

Editorial Office
MDPI
St. Alban-Anlage 66
4052 Basel, Switzerland

This is a reprint of articles from the Special Issue published online in the open access journal *Minerals* (ISSN 2075-163X) from 2018 to 2019 (available at: https://www.mdpi.com/journal/minerals/special_issues/Hydro_Biohydrometallurgy)

For citation purposes, cite each article independently as indicated on the article page online and as indicated below:

LastName, A.A.; LastName, B.B.; LastName, C.C. Article Title. *Journal Name* **Year**, *Article Number*, Page Range.

ISBN 978-3-03921-299-6 (Pbk)
ISBN 978-3-03921-300-9 (PDF)

Contents

About the Special Issue Editor

Kostas A. Komnitsas, Prof., holds a PhD degree in hydrometallurgy and is the director of the Laboratories of (i) Waste Management and Soil Decontamination, (ii) Ceramics and Glass Technology, and (iii) Ore Beneficiation in the School of Mineral Resources Engineering of Technical University Crete, Greece. He is an expert in the fields of hydro- and biohydrometallurgy, waste management and valorization, new materials, soil decontamination, environmental risk assessment, and LCA studies. Prof. Komnitsas has been appointed as a national representative/expert by the Greek General Secretariat of Research and Technology for research issues to the European Commission during the 6th Framework Programme (FP6) for the Action Integration and Strengthening of the European Research Area (2002–2004), for International Cooperation (2006–2009), and for the Action Societal Challenge 'Climate action, environment, resource efficiency and raw materials' (HORIZON 2020) (2013–today). Prof. Komnitsas has been appointed, so far, more than 80 times as evaluator of research proposals for the European Commission and National Research Foundations of several countries, including the US, Russia, Switzerland, Italy, Romania, Kazakhstan, Cyprus, Poland, Armenia, Serbia, etc. He has conducted more than 50 research projects (funded by national bodies and the EC, including life projects, previous FPs, and H2020). He has published more than 100 papers in peer-reviewed journals that have been cited over 2400 times, and his Scopus h-index is 28.

Preface to "Recent Advances in Hydro- and Biohydrometallurgy"

Securing reliable and continuous access to raw materials and extraction of metals are important priorities in almost all countries in order to meet industrial needs, enable high-tech applications, maintain quality of life, and guarantee millions of jobs. Today hydro- and biohydrometallurgical processes are intensely investigated to solve bottlenecks in the raw materials supply; recover critical, base, and precious metals from low-grade ores and various types of wastes; and provide environmental solutions for various industrial problems.

The roots of hydrometallurgy can be traced back to the era of alchemists, while modern hydrometallurgy dates back to 1887, when two important processes were invented, namely the cyanidation process for the treatment of gold ores and the Bayer process for the treatment of bauxites and the production of alumina. On the other hand, there is evidence that bioleaching was used in the Rio Tinto area in Spain prior to Roman occupation for the recovery of copper, as well as in China some 2000 years ago. Modern commercial biohydrometallurgical applications for the processing of ores commenced in the 1950s, focusing on the bioleaching of copper. Since then, biohydrometallurgy has been used for the treatment of various primary and secondary raw materials and the recovery of several metals, including good, copper, and rare-earth elements (REEs). It must be underlined that the critical role of bio- and hydrometallurgy in achieving sustainable development in various industrial sectors was identified more than 30 years ago.

Kostas A. Komnitsas
Special Issue Editor

minerals

MDPI

Editorial

Editorial for Special Issue "Recent Advances in Hydro- and Biohydrometallurgy"

Kostas A. Komnitsas

Technical University of Crete, School of Mineral Resources Engineering, 73100 Chania, Greece; komni@mred.tuc.gr; Tel.: +302821037686

Received: 5 July 2019; Accepted: 10 July 2019; Published: 11 July 2019

Securing reliable and continuous access to raw materials and extraction of metals are important priorities in almost all countries in order to meet industrial needs, enable high-tech applications, maintain quality of life, and guarantee millions of jobs. Today, hydro- and biohydrometallurgical processes are intensely investigated to solve bottlenecks in the raw materials supply, recover critical base and precious metals from low-grade ores and various types of wastes, and also provide environmental solutions for various industrial problems [1–4].

The roots of hydrometallurgy are traced back to the period of the alchemists, while modern hydrometallurgy dates back to 1887, when two important processes were invented, namely the cyanidation process for the treatment of gold ores and the Bayer process for the treatment of bauxites and the production of alumina [5]. On the other hand, there is evidence that bioleaching has been used in the Rio Tinto area in Spain prior to Roman occupation for the recovery of copper, as well as in China some two thousand years ago [6]. Modem commercial biohydrometallurgical applications for the processing of ores commenced in the 1950s, focusing on bioleaching of copper [7]. Since then, biohydrometallurgy has been used for the treatment of various primary and secondary raw materials and the recovery of several metals, including gold, copper, and rare earth elements (REEs) [8–11]. It has to be underlined that the critical role of bio- and hydrometallurgy in achieving sustainable development in various industrial sectors has been identified more than 30 years ago [12].

This Special Issue of *Minerals* presents recent selective studies, carried out in different countries, that highlight advances in the fields of hydro- and biohydrometallurgy. It aims to attract the interest of the readers, and especially of young scientists and students in this fascinating scientific discipline.

Three of the studies investigated leaching of laterites. Mystrioti et al. [13] investigated the efficiency of stirred reactor hydrochloric acid leaching for the treatment of a low-grade saprolitic laterite. The leaching was carried out at 30% pulp density, by applying a counter-current mode of operation in order to better simulate industrial-scale operations and maintain Fe dissolution at low levels. This mode of operation was very efficient in terms of minimizing Fe dissolution, which was maintained at 0.6%, but had a negative effect on Ni and Co extraction, which was 55% and 63%, respectively, probably due to the passivation of ore grain surfaces by secondary iron precipitation products. The treatment of PLS (pregnant leach solution) involved a precipitation step for the removal of trivalent metals, Fe, Al, and Cr with the use of $Mg(OH)_2$. Komnitsas et al. [14] investigated the efficiency of column leaching of low-grade limonitic laterites with the use of H_2SO_4 for the extraction of Ni and Co. Parameters studied were acid concentration (0.5 M or 1.5 M) and addition of 20 or 30 g/L of sodium sulfite (Na_2SO_3). The experimental results showed that (i) Ni and Co extractions increased with the increase of H_2SO_4 concentration and reached 60.2% and 59.0%, respectively, after 33 days of leaching with the use of 1.5 M H_2SO_4, and (ii) addition of 20 g/L Na_2SO_3 in the leaching solution resulted in higher extractions for both metals (73.5% for Ni and 84.1% for Co, respectively). Finally, the extractions of Fe, Mg, Al, and Ca were quite low, namely, 7.9, 40.2, 23.3, and 51.0%, respectively. Miettinen et al. [15] investigated iron control during atmospheric acid leaching of two laterite types, a limonite and a silicate, in order to decrease acid consumption and iron dissolution. The process involved direct acid leaching of the limonitic laterite followed by simultaneous iron precipitation as jarosite after the addition of the silicate laterite for pH neutralization. The combined leaching and precipitation process reduced acid consumption and iron concentration in the pregnant leach solution (PLS). The acid consumption, which during the direct atmospheric leaching was approximately 0.7 kg H_2SO_4 per kg of laterite was reduced during the combined process to 0.42 kg H_2SO_4 per kg of laterite. In addition, Fe concentration in the PLS decreased from 10 g/L to approximately 2–3 g/L, resulting in significant savings compared with the conventional process.

Salinas et al. [16] investigated copper extraction from a typical porphyry copper sulfide deposit from Antofagasta, Chile, using chloride–ferrous leaching. They carried out large-scale column leaching tests using 50 kg of agglomerated ore that was first cured for 14 days and then leached for 90 days. The highest Cu extraction, 50.23%, was achieved at 32.9 °C with the addition of 0.6 kg of H_2SO_4, 0.525 kg of NaCl, and 0.5 kg of $FeSO_4$ per ton of ore. The effect of agglomeration, curing, and temperature on the leaching kinetics of Cu was also assessed.

Hernandez et al. [17] studied leaching of chalcopyrite ore containing 1.6 wt% Cu in a nitrate-acid–seawater system. The parameters studied were water quality (pure water and seawater), temperature (25–70 °C), reagent concentration, nitrate type ($NaNO_3$ or KNO_3), and leaching duration. Results showed that up to 80 wt% of Cu can be extracted during leaching at 45 °C in 7 days. In the absence of nitrates, under the same leaching conditions, only 28 wt% of Cu was extracted. The Cu extraction increased to 97.2 wt% with the use of 1 M H_2SO_4 and 1 M $NaNO_3$ when the temperature increased to 70 °C. The main disadvantage of this approach was the production of NOx gases that should be controlled in industrial operations.

In another study, Hernandez et al. [18] investigated leaching of chalcopyrite in acid-nitrate–chloride media using mini-columns. The effect of ore pretreatment, involving agglomeration and curing, as well as of several factors, namely addition of nitrate as $NaNO_3$ (11.7 and 23.3 kg/ton), chloride as NaCl (2.1 and 19.8 kg/ton), curing time (20 and 30 days), and temperature (25 and 45 °C) was also evaluated. The maximum Cu extraction, 58.6%, after 30 days of curing at 45 °C, was obtained during leaching with the addition of 23.3 kg of $NaNO_3$/ton and 19.8 kg of NaCl per ton of ore. Copper extraction from the pretreated ore reached 63% during leaching at pH 1 and 25 °C with the use of a solution containing 6.3 g/L of $NaNO_3$ and 20 g/L of NaCl.

Castillo et al. [19] investigated the effect of NaCl on the leaching of white metal from a Teniente converter in NaCl-H_2SO_4 media and proposed a simplified two-stage mechanism. Parameters studied involved the concentration of ferric ion (1–10 g/L), NaCl (30–210 g/L), and H_2SO_4 (10–50 g/L). The results showed that the addition of NaCl increased the dissolution of Cu from 55% to nearly 90%, whereas the effect of sulfuric acid was only minor. The positive effect of NaCl is mainly related to the action of chloro-complexes oxidizing agents in relation to the Cu^{+2}/Cu^+ couple. Leaching of Cu takes place in two stages involving (i) transformation of chalcocite into covellite and production of Cu^{2+} ions and (ii) reaction of covellite for the generation of Cu^{2+} ions and elemental sulfur.

Xu et al. [20] investigated the galvanic effect of pyrite and arsenopyrite during leaching of gold ores in sulfuric acid, ferric ion, and HQ0211 bacterial strain solutions with the use of electrochemical testing (open-circuit potential, linear sweep voltammetry, Tafel, and electrochemical impedance spectroscopy, EIS) and frontier orbital calculations. The results indicated that (i) the linear sweep voltammetry curve and Tafel curve of the galvanic pair were similar to those of arsenopyrite, (ii) the corrosion behavior of the galvanic pair was consistent with that of arsenopyrite, and (iii) the galvanic effect promoted the corrosion of arsenopyrite by simultaneously increasing the cathode and anode currents and reducing oxidation resistance. The frontier orbital calculation explained the principle of the galvanic effect of pyrite and arsenopyrite from the view of quantum mechanics.

Three papers studied leaching of marine nodules in the presence of reducing agents to extract Mn. In the first study [21], the surface optimization methodology was used to assess the effect of three independent variables, namely time, particle size, and sulfuric acid concentration, on Mn extraction during leaching with H_2SO_4 in the presence of foundry slag. In a second study [22], the effect of magnetite-rich tailings produced from slag flotation during leaching at room temperature (25 °C) was explored. Other factors studied included MnO_2/Fe_2O_3 ratio in solution and agitation speed. The highest Mn extraction, 77%, was obtained at MnO_2/Fe_2O_3 ratio 0.5, 1 mol/L H_2SO_4, particle size −47 + 38 μm, and leaching time 40 min. Finally, in their third study [23], the authors optimized the main operating parameters through factorial experimental design. It is mentioned that the generation of Fe^{2+} and Fe^{3+} improved Mn extraction that reached 73% within 5 to 20 min.

Khaing et al. [24] explored the factors that affect bioleaching of gold ores in the presence of iodide-oxidizing bacteria. The factors studied, in order to maximize gold dissolution, included concentration of nutrients and iodide, initial cell number, incubation temperature, and shaking speed. The culture medium contained marine broth, potassium iodide, and gold ore. The main findings of the study were (i) gold contained in the ore was almost completely dissolved in the culture solution, incubated at 30 °C and 35 °C, (ii) the pH and redox potential of the culture solution were 7.7–8.4 and 472–547 mV, (iii) gold leaching rate in iodine–iodide solution was much faster compared with that of the conventional cyanidation process, and (iv) iodine can be recovered after leaching.

Makinen et al. [25] investigated the efficiency of a two-step sequential leaching process, involving bioleaching and chemical leaching, to treat apatite ores containing P and U impurities. The first leaching step, at pH ≥ 2, Eh +650 mV and Fe^{3+} concentration ≥1.0 g/L, enabled 89% extraction of U in 3 days. After solid–liquid separation, the second leaching step at pH ≤ 1.5 enabled the recovery of phosphorus from the solid leach residue. It is mentioned that despite the high leaching degree for P (98%), the duration of the process was quite long (28 days).

In Brazil, lateritic deposits are often associated with rare earth element enriched phosphate minerals such as monazite. Given that monazite is highly refractory, rare earth elements (REEs) extraction is difficult and normally involves high-temperature digestion with concentrated NaOH and/or H_2SO_4. Nancucheo et al. [26] assessed the effect of bioreductive dissolution of ferric iron minerals associated with monazite in stirred reactors using Acidithiobacillus (A.) species in pH and temperature-controlled tests. The results indicated that under aerobic conditions, A. thiooxidans at extremely low pH can enhance substantially the solubilization of iron from ferric iron minerals.

Finally, Makinen et al. [27] investigated a robust and simple heap leaching approach for the recovery of Zn and Cu from municipal solid waste incineration bottom ash (MSWI BA). Also, they studied the effect of autotrophic and acidophilic bioleaching microorganisms. Leaching yields for Zn and Cu varied between 18–53% and 6–44%, respectively. The main contaminants present in MSWI BA, namely Fe and Al, were easily liberated by sulfuric acid leaching, lowering the quality of PLS and imposing limitations for the industrial utilization of the process.

Author Contributions: K.K. wrote this editorial.

Acknowledgments: The guest editor would like to sincerely thank all authors, reviewers and the editorial staff of *Minerals* for their efforts and devotion to successfully complete this Special Issue.

Conflicts of Interest: The author declares no conflict of interest.

References

1. Panda, S.; Akcil, A.; Pradhan, N.; Deveci, H. Current scenario of chalcopyrite bioleaching: A review on the recent advances to its heap-leach technology. *Bioresour. Technol.* **2015**, *196*, 694–706. [CrossRef] [PubMed]
2. Musariri, B.; Akdogan, G.; Dorfling, C.; Bradshaw, S. Evaluating organic acids as alternative leaching reagents for metal recovery from lithium ion batteries. *Miner. Eng.* **2019**, *137*, 108–117. [CrossRef]
3. Ndlovu, S. Biohydrometallurgy for sustainable development in the African minerals industry. *Hydrometallurgy* **2008**, *91*, 20–27. [CrossRef]
4. Komnitsas, K.; Petrakis, E.; Bartzas, G.; Karmali, V. Column leaching of low-grade saprolitic laterites and valorization of leaching residues. *Sci. Total Environ.* **2019**, *665*, 347–357. [CrossRef] [PubMed]
5. Habashi, F. A short history of hydrometallurgy. *Hydrometallurgy* **2005**, *79*, 15–22. [CrossRef]
6. Holmes, D.S. Review of International Biohydrometallurgy Symposium, Frankfurt, 2007. *Hydrometallurgy* **2008**, *92*, 69–72. [CrossRef]

7. Brierley, J.A.; Brierley, C.L. Present and future commercial applications of biohydrometallurgy. *Process. Metall.* **1999**, *9*, 81–89. [CrossRef]

8. Komnitsas, C.; Pooley, F.D. Bacterial oxidation of a refractory gold sulphide concentrate from Olympias, Greece. *Miner. Eng.* **1990**, *3*, 295–306. [CrossRef]

9. Pradhan, N.; Nathsarma, K.C.; Srinivasa Rao, K.; Sukla, L.B.; Mishra, B.K. Heap bioleaching of chalcopyrite: A review. *Miner. Eng.* **2008**, *21*, 355–365. [CrossRef]

10. Chen, S.; Yang, Y.; Liu, C.; Dong, F.; Liu, B. Column bioleaching copper and its kinetics of waste printed circuit boards (WPCBs) by Acidithiobacillus ferrooxidans. *Chemosphere* **2015**, *141*, 162–168. [CrossRef]

11. Baniasadi, M.; Vakilchap, F.; Bahaloo-Horeh, N.; Mousavi, S.M.; Farnaud, S. Advances in bioleaching as a sustainable method for metal recovery from e-waste: A review. *J. Ind. Eng. Chem.* **2019**, *76*, 75–90. [CrossRef]

12. Conard, B.R. The role of hydrometallurgy in achieving sustainable development. *Hydrometallurgy* **1992**, *30*, 1–28. [CrossRef]

13. Mystrioti, C.; Papassiopi, N.; Xenidis, A.; Komnitsas, K. Counter-current leaching of low-grade laterites with hydrochloric acid and proposed purification options of pregnant solution. *Minerals* **2018**, *8*, 599. [CrossRef]

14. Komnitsas, K.; Petrakis, E.; Pantelaki, O.; Kritikaki, A. Column leaching of Greek low-grade limonitic laterites. *Minerals* **2018**, *8*, 377. [CrossRef]

15. Miettinen, V.; Mäkinen, J.; Kolehmainen, E.; Kravtsov, T.; Rintala, L. Iron control in atmospheric acid laterite leaching. *Minerals* **2019**, *9*, 404. [CrossRef]

16. Salinas, K.E.; Herreros, O.; Torres, C.M. Leaching of primary copper sulfide ore in chloride-ferrous media. *Minerals* **2018**, *8*, 312. [CrossRef]

17. Hernández, P.C.; Taboada, M.E.; Herreros, O.O.; Graber, T.A.; Ghorbani, Y. Leaching of chalcopyrite in acidified nitrate using seawater-based media. *Minerals* **2018**, *8*, 238. [CrossRef]

18. Hernández, P.C.; Dupont, J.; Herreros, O.O.; Jimenez, Y.P.; Torres, C.M. Accelerating copper leaching from sulfide ores in acid-nitrate-chloride media using agglomeration and curing as pretreatment. *Minerals* **2019**, *9*, 250. [CrossRef]

19. Castillo, J.; Sepúlveda, R.; Araya, G.; Guzmán, D.; Toro, N.; Pérez, K.; Rodríguez, M.; Navarra, A. Leaching of white metal in a NaCl-H_2SO_4 system under environmental conditions. *Minerals* **2019**, *9*, 319. [CrossRef]

20. Xu, J.; Shi, W.; Ma, P.; Lu, L.; Chen, G.; Yang, H. Corrosion behavior of a pyrite and arsenopyrite galvanic pair in the presence of sulfuric acid, ferric ions and HQ0211 bacterial strain. *Minerals* **2019**, *9*, 169. [CrossRef]

21. Toro, N.; Herrera, N.; Castillo, J.; Torres, C.M.; Sepúlveda, R. Initial investigation into the leaching of manganese from nodules at room temperature with the use of sulfuric acid and the addition of foundry slag—Part I. *Minerals* **2018**, *8*, 565. [CrossRef]

22. Toro, N.; Saldaña, M.; Castillo, J.; Higuera, F.; Acosta, R. Leaching of manganese from marine nodules at room temperature with the use of sulfuric acid and the addition of tailings. *Minerals* **2019**, *9*, 289. [CrossRef]

23. Toro, N.; Saldaña, M.; Gálvez, E.; Cánovas, M.; Trigueros, E.; Castillo, J.; Hernández, P.C. Optimization of parameters for the dissolution of Mn from manganese nodules with the use of tailings in an acid medium. *Minerals* **2019**, *9*, 387. [CrossRef]

24. Khaing, S.Y.; Sugai, Y.; Sasaki, K.; Tun, M.M. Consideration of influential factors on bioleaching of gold ore using iodide-oxidizing bacteria. *Minerals* **2019**, *9*, 274. [CrossRef]

25. Mäkinen, J.; Wendling, L.; Lavonen, T.; Kinnunen, P. Sequential bioleaching of phosphorus and uranium. *Minerals* **2019**, *9*, 331. [CrossRef]

26. Mäkinen, J.; Salo, M.; Soini, J.; Kinnunen, P. Laboratory scale investigations on heap (bio)leaching of municipal solid waste incineration bottom ash. *Minerals* **2019**, *9*, 290. [CrossRef]

27. Nancucheo, I.; Oliveira, G.; Lopes, M.; Johnson, D.B. Bioreductive dissolution as a pretreatment for recalcitrant rare-earth phosphate minerals associated with lateritic ores. *Minerals* **2019**, *9*, 136. [CrossRef]

minerals

MDPI

Article

Counter-Current Leaching of Low-Grade Laterites with Hydrochloric Acid and Proposed Purification Options of Pregnant Solution

Christiana Mystrioti [1], Nymphodora Papassiopi [2], Anthimos Xenidis [2] and Konstantinos Komnitsas [1,*]

[1] Department Mineral Resources Engineering, Technical University Crete, 73100 Chania, Greece; chmistrioti@metal.ntua.gr
[2] School of Mining and Metallurgical Engineering, National Technical University of Athens, 15780 Zografos, Greece; papasiop@metal.ntua.gr (N.P.); axen@metal.ntua.gr (A.X.)
* Correspondence: komni@mred.tuc.gr; Tel.: +30-28210-37686

Received: 22 November 2018; Accepted: 14 December 2018; Published: 18 December 2018

Abstract: A hydrochloric acid hydrometallurgical process was evaluated for Ni and Co extraction from a low-grade saprolitic laterite. The main characteristics of the process were (i) the application of a counter-current mode of operation as the main leaching step (CCL), and (ii) the treatment of pregnant leach solution (PLS) with a series of simple precipitation steps. It was found that, during CCL, co-dissolution of Fe was maintained at very low levels, i.e., about 0.6%, which improved the effectiveness of the subsequent PLS purification step. The treatment of PLS involved an initial precipitation step for the removal of trivalent metals, Fe, Al, and Cr, using $Mg(OH)_2$. The process steps that followed aimed at separating Ni and Co from Mn and the alkaline earths Mg and Ca, by a combination of repetitive oxidative precipitation and dissolution steps. Magnesium and calcium remained in the aqueous phase, Mn was removed as a solid residue of Mn(III)–Mn(IV) oxides, while Ni and Co were recovered as a separate aqueous stream. It was found that the overall Ni and Co recoveries were 40% and 38%, respectively. About 45% of Ni and 37% of Co remained in the leach residue, while 15% Ni and 20% Co were lost in the Mn oxides.

Keywords: low-grade saprolitic laterite; counter-current leaching; pregnant leach solution; purification

1. Introduction

Nickel is a metal in high demand for industrial applications such as the production of stainless steel, non-ferrous alloys, alloy steels, plating, foundry, and batteries due to its beneficial properties (strength, corrosion resistance, high ductility, good thermal and electric conductivity, magnetic characteristics, and catalytic properties) [1]. Nickel can be found naturally in lateritic ores which are formed by weathering of ultramafic rocks or in sulfide resources. Due to the decline of sulfide deposits and high-grade laterites, there is a need for adapting or optimizing technologies aiming at increasing metal recovery from low-grade laterites in order to make their treatment economically feasible [2]. It is well known that conventional mineral processing techniques, including flotation heavy-media separation and others, cannot be readily applied to oxide ores such as laterites, unlike sulfides, due to the complex nature of the ores and the fact that nickel is hosted in several mineral phases [3].

Hydrometallurgical processes for nickel and cobalt recovery from low-grade laterite ores can be classified as high-pressure acid leaching (HPAL), atmospheric acid leaching (AL), heap leaching (HL), and biological leaching (BL).

HPAL is often selected for the treatment of low-grade laterite ores, but presents drawbacks such as expensive equipment in order to stand the harsh leaching conditions (typical operating temperature from 250 to 270 °C) and high initial capital cost [4,5].

AL of low-grade laterites is considered as a promising method which efficiently recovers Ni and Co with low cost due to milder leaching conditions. The majority of studies were carried out using sulfuric acid as lixiviant and fewer tested nitric and hydrochloric acid [1,6–8]. More recent studies comparing the performance of HCl and H_2SO_4, under atmospheric conditions indicated that hydrochloric acid is more effective for both nickel and cobalt extraction [4,8]. However, AL using hydrochloric acid as a lixiviant results in high iron and magnesium dissolution which is related to high acid consumption and requires a complex purification treatment.

HL is a well-known process due to the numerous applications for the treatment of copper, uranium, and gold ores, and requires low investment and operating costs. However, Ni and Co recoveries by HL are relatively low compared with leaching in stirred reactors [9,10]. Studies were carried out evaluating the bio-hydrometallurgical HL of laterites using organic acids, citric acid, and oxalic acid, or using heterotrophic micro-organisms which metabolize organic compounds and excrete organic acids, namely carboxylic acids. These microorganisms should be tolerant to metal mixtures [11,12].

The majority of studies aim at optimizing the conditions of atmospheric leaching, in order to maximize Ni and Co extraction, and fewer deal with the purification of pregnant leach solution (PLS), which has low concentrations of Ni (0.8–3.3 g/L) and Co (0.05–0.2 g/L) and high concentrations of the major elements, mainly Fe (13–37 g/L) and Mg (8–40 g/L) [13,14]. The main processes for PLS purification include precipitation, solvent extraction, and ion exchange. Precipitation takes place when the chemical species exceed their limit of solubility by changing the solution pH or temperature, or by adding a reagent [15]. Solvent extraction presents high selectivity and involves mixing of PLS with an organic phase; the resulting emulsion is allowed to separate, and the valuable metals are transferred to the organic phase [11,16,17]. The use of ion-exchange media, such as resins, zeolite, and active carbon, in order to exchange cations or anions from a solution, presents high recoveries and is environmentally friendly, but these media have limited efficiency in mixed stream solutions [15].

The high ferric iron content constitutes the major impurity in the PLS produced by the atmospheric acid leaching of laterites [18]. The removal of iron from PLS involves a difficult and costly step. Moreover, it often results in important Ni and Co losses (5–20%) [19]. Iron removal has been extensively studied in zinc industry in the early 1960s and several techniques have been developed for iron precipitation as jarosite, goethite and hematite [19]. The majority of studies for the purification of sulphate leach liquors involve the addition of limestone or calcium hydroxide, in order to increase the pH of the solution, so that metal hydroxides are formed and precipitate [16,20,21]. Ni and Fe hydroxides precipitate at different pH, however it is difficult to avoid co-precipitation of Ni during the removal of Fe. Nickel co-precipitation takes place due to its adsorption on the surface of iron hydroxides or substitution in the precipitates. The temperature increase affects the stability and the crystallization of the precipitates. For these reasons the process conditions such as pH, temperature, and duration of precipitation have serious impact on the properties of the precipitates and the potential Ni and Co losses [18,22].

Some steps of the purification treatment of the hydrochloric acid leach solutions are yet to be thoroughly investigated. Filippou and Choi [23] evaluated the step of iron precipitation by adding sodium hydroxide or urotropin at 100 °C. However, the formation of various allotropes of FeOOH made the step of solid/liquid separation difficult and resulted in high losses of the other metals. Increasing the pH to 5.5 improved solid/liquid separation, but caused higher losses of Ni and Co [24]. Beukes et al. [24] suggested that nickel and cobalt adsorption on the surface of iron oxides (goethite and hematite) is dependent on the presence of precipitates and not on their quantity. The researchers focused on solvent extraction processes due to the difficulties which occurred upon the precipitation of iron oxides. Many neutral extractants such as tri-*n*-butyl phosphate (TBP), tri-*n*-octylphosphine oxide (TOPO), amines, and mixtures of them were tested, but effective iron removal requires many steps due to the absence of an adequate "salting out agent". In a solution with concentration of 200 g/L $MgCl_2$, the removal of iron (40 g/L) was equal to 99.4% using TBP and methylisobutyl ketone (MIBK) with some losses of Co [11]. Demopoulos et al. [25] tested hydrolytic distillation for the precipitation of ferric iron as Fe_2O_3 and the recovery of Cl^- as HCl. Nickel and cobalt were precipitated as oxides at an operating temperature of ~170 °C. Zhang et al. [2] applied hydrolytic distillation on synthetic laterite-leaching solutions and showed that 95.5% of iron was removed as hematite, 94% of Ni and Co remained in the aqueous phased as soluble chlorides, and excess HCl was distilled with a final recovery in the order of 77.9%. The removal of Mn and Mg constitutes another significant challenge for PLS purification treatment [15,17].

In this study, a counter-current leaching (CCL) scheme was evaluated for the HCl leaching of a saprolitic ore. The CCL mode of operation was adopted in order to maintain the co-dissolution of Fe at low levels and obtain satisfactory extraction of Ni and Co, while operating at high pulp densities close to the industrial practice. The proposed flowchart for the treatment of PLS consists of a series of simple precipitation steps.

2. Materials and Methods

2.1. Investigated Flowchart

A conceptual flowchart illustrating the main steps of the HCl leaching process, which were investigated in the framework of this study, is shown in Figure 1. In this flowchart, the first step concerns the leaching of laterite ore. The next step (precipitation of trivalent metals) involves the removal of iron and other trivalent metals, e.g., Cr(III), Al(III), etc., in the form of hydroxides, using MgO as a neutralization agent and maintaining the pH at relatively acidic conditions, i.e., 3.5. The solution resulting from this treatment step contains all divalent metals, including Ni, Co, Mn, Mg, and Ca. The following step aims primarily at the separation of alkaline earths, Mg and Ca, from the other divalent metals by increasing the pH close to 9. This step is combined with the oxidation of Mn(II) to the tetravalent state, Mn(IV), and, for this reason, is denoted as alkaline oxidation (AO) in the flowchart of Figure 2. Mg and Ca remain in the aqueous phase, while Ni, Co, and Mn are recovered in the form of a solid residue containing $Ni(OH)_2$, $Co(OH)_2$, and MnO_2. The next step, acidic dissolution (AD), involves the acidic treatment of the solid residue, in order to enable the selective dissolution of Ni and Co, given that the high-valence Mn is expected to remain in the solid state.

The aqueous solution of the AO step contains mainly Mg, Ca, and Cl^- anions. Calcium can be removed from these solutions via precipitation as $CaSO_4 \cdot xH_2O$ [26,27]. The final stream containing Mg and Cl can be treated by pyrohydrolysis. By this process, it is possible to regenerate HCl, which is recycled in the leaching step, producing MgO.

As seen in Figure 1, the investigated flowchart involves a series of successive precipitation steps. Initial thermodynamic calculations were carried out using Visual Minteq software [28] to identify the optimum pH ranges, which could allow the selective separation of impurities, through precipitation of metal hydroxides, from the valuable metals. For this reason, the model was run using fixed pH values in the whole pH range between 0 and 14. The following metal hydroxides were considered as possible precipitates: ferrihydrite (Fh), $Al(OH)_3$, $Cr(OH)_3$, $Ni(OH)_2$, $Co(OH)_2$, pyrochroite $(Mn(OH)_2)$ or pyrolusite (MnO_2), brucite $(Mg(OH)_2)$, and portlandite $(Ca(OH)_2)$.

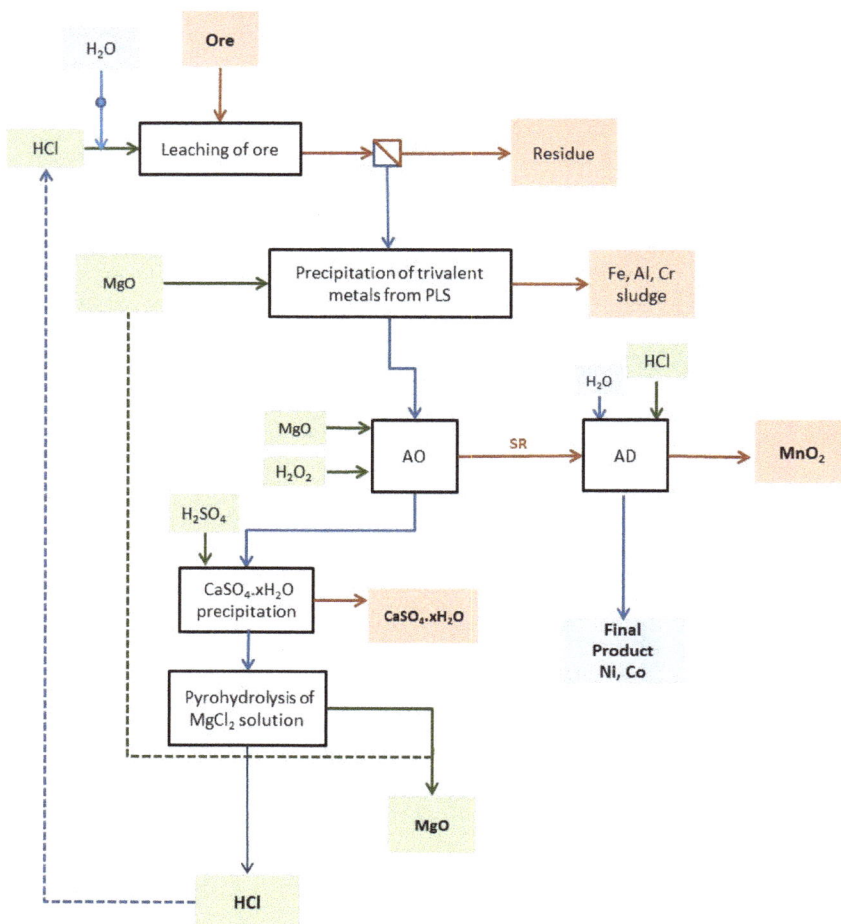

Figure 1. The main steps of the investigated HCl leaching process for the treatment of laterite ores.

2.2. Materials

The saprolitic laterite used in this study was collected from Larco deposits in the area of Kastoria (northern Greece). The sample (50 kg) was homogenized, ground using vibration milling for 10 min, and the particle size of the working sample was under 74 μm.

The chemical composition of the laterite is given in Table 1. For the elemental analysis, the representative subsample of the ore was subjected to digestion with aqua regia, and the resulting solution was analyzed using atomic absorption spectroscopy (AAS). As seen in Table 1, the laterite contained nickel 0.88%, cobalt 0.06%, iron 22.42%, and magnesium 7.58%. The X-ray diffraction (XRD) analysis showed that the ore consisted mainly of quartz, goethite, calcite, and lizardite ($(Mg,Al)_3[(Si,Fe)_2O_5](OH)_4$). Reagents used for the leaching experiments and for the purification of PLS were HCl (Chem-Lab, Zedelgem, Belgium), $Mg(OH)_2$ (Sigma Aldrich, Tokyo, Japan), H_2O_2 (Sigma Aldrich, Steinheim, Germany), and NaOH (Sigma Aldrich, Steinheim, Germany).

2.3. Leaching of Laterite Ore

The atmospheric leaching experiments were conducted using glass reactors of 1000-mL capacity, equipped with a heating mantle connected with a temperature controller, and mechanically agitated with a Heidolph Mixer at 500 rpm. An initial series of tests were carried out using one leaching step and investigating the effect of HCl concentration (1 to 4 M), temperature (65–90 °C), and pulp density (10–30% w/v). A second series of tests were conducted using two leaching steps in a counter-current mode of operation (counter-current leaching, CCL). The experimental procedure for the simulation of CCL operation is shown in Figure 2. It should be noted that it is difficult to simulate the steady-state conditions of a continuous CCL system using batch experiments.

Table 1. Chemical analysis of ore sample.

Main Elements	%
Fe	22.42
Ni	0.88
Mg	7.58
Co	0.06
Ca	2.78
Al	0.48
Cr	0.19
Mn	0.42
Si	0.04
Insoluble [a]	24.27
Loss on ignition [b]	13.22

(a) Solid residue of the aqua regia digestion; (b) weight loss after heating at 1000 °C for 30 minutes.

In the leaching step 1a (Figure 2), 300 g of ore was treated with 1000 mL of 2 M HCl at 80 °C for six hours. Due to the relatively low extraction of Ni and Co during this step, the resulting solid residue was leached again with 1000 mL of 2 M HCl under similar conditions (step 1b). The second acid attack was able to dissolve an additional amount of Ni and Co from the residue, but the solution contained also high concentration of the undesirable iron. To enable re-precipitation of dissolved Fe, the leaching solution from step 1b was used for the treatment of fresh ore (step 1a′). The iron-free solution emanating from step 1a′ was enriched in Ni and Co and constituted the pregnant solution, which was used in the subsequent PLS purification steps. The leach solution LS-1a from step 1a was not considered as representative of the PLS composition of a continuous CCL operation.

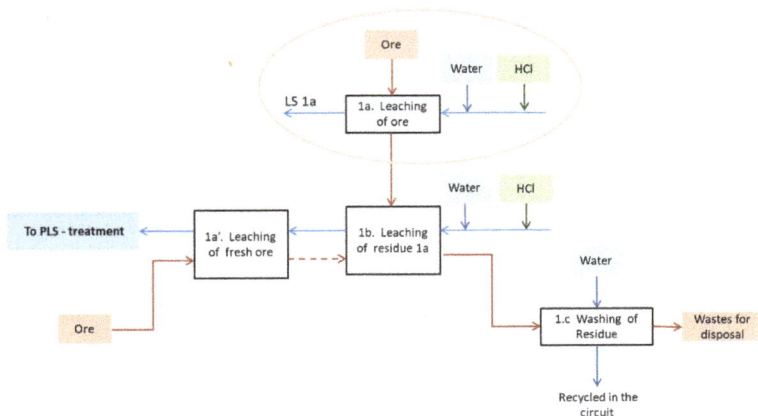

Figure 2. The experimental procedure for the simulation of counter-current leaching (CCL) operation during HCl leaching of the laterite ore.

The solid residues produced from step 1b were washed using a solid/liquid (S/L) ratio of 30% (step 1c). A representative sample of the washed solids was subjected to the EN12457.02 standard leaching test procedure (L/S = 10 L/kg) [29]. The test was carried out in duplicate.

2.4. Treatment of Pregnant Solution

2.4.1. Experiments with the PLS of One-Step Leaching

The initial determination of optimum conditions for the treatment of pregnant solution was carried out using the PLS of the one-step leaching of ore at 10% pulp density, temperature of 80 °C, and 2 M HCl with six hours of treatment time.

The main parameter investigated for the precipitation of trivalent metals step was the final precipitation pH. Four pH values were tested, namely 2.5, 3, 3.5, and 4. The tests were carried out using 50 mL of PLS and adding dropwise a $Mg(OH)_2$ slurry, 10% *w/v*, to obtain the predetermined final pH value. The suspension was agitated for 2 h in an orbital shaker (150 rpm) and, after the end of the experiment, centrifugation (Hettich Universal 320A centrifuge) took place in order to separate the solution from the solids. The supernatant was filtered through a 0.45-μm membrane filter and analyzed for Fe, Ni, Co, Mg, Mn, Cr, Al, and Si using atomic absorption spectroscopy with flame emission.

The AO treatment was then carried out in the solution which was produced after the precipitation of Fe, Al, and Cr at pH 3.5. The main parameter which was investigated during AO was the amount of the added H_2O_2. The experiments were carried out by adding H_2O_2 (1%, 2%, 5%, or 10% *w/v*) to 50 mL of the purified PLS. The pH was increased from 3.5 to 9.0 by adding NaOH solution (1 M). The tests were conducted in 250-mL conical flasks at room temperature, using a magnetic stirrer for agitation, and the total duration was 1.5 h. At the end of treatment, the suspensions were filtered and the aqueous solutions were analyzed for Ni, Co, Mg, Mn, Ca, and Al using atomic absorption spectroscopy with flame emission.

The solid residue, produced by the AO process, was treated by mixing with 50 mL of distilled water and adjusting the pH using 2 M HCl. The effect of pH on the recovery of Ni, Co, Mn, and Mg was investigated by conducting dissolution tests at pH values equal to 1, 3, and 6. The tests were conducted in 100-mL conical flasks at room temperature, using a magnetic stirrer for agitation, and the total duration was 1 h.

In order to enrich the concentrations of Ni and Co and remove the remaining Mg and Mn, the AO and AD steps were repeated twice. The AO was conducted using 5 wt.% H_2O_2 and the AD was carried out decreasing the pH to 0.8. The solid residues from AD-2 were characterized by SEM and energy-dispersive X-ray spectroscopy (EDS) analysis.

2.4.2. Experiments with the PLS of CCL

The parameters investigated during the treatment of PLS from CCL included the final pH of the precipitation of trivalent metals step and the pH of the AD step.

For the pyrohydrolysis treatment, the performance was estimated using the HSC Chemistry software (version 7.6.1). The main objective of using the HSC software was to determine the temperature required in order to recover the HCl in the gas phase and MgO as a solid precipitate.

2.5. Analytical Methods

The aqueous samples were analyzed using AAS with flame emission (Perkin Elmer 2100, Wellesley, MA, USA) and inductively coupled plasma mass spectrometry (ICP-MS; X series Thermo Fisher Scientific, Waltham, MA, USA). Divalent Fe(II) was determined using the phenanthroline method. The morphology of the solids was also examined by scanning electron microscopy (SEM) using a JEOL 6380LV Microscope (JEOL, Tokyo, Japan). Chemical analysis of observed grains was carried out using an Oxford INCA energy-dispersive spectrometer (EDS) (Oxford Instruments, Abingdon, UK) connected to SEM. Centrifugation was carried out in a Hettich Centrifuge, model Universal 320A.

The XRD analysis was performed using a Bruker D8-Focus powder diffractometer (Bruker, Karlsruhe, Germany), with nickel-filtered CuKα radiation (λ = 1.5405 Å).

3. Results

3.1. Leaching Experiments

3.1.1. Initial One-Step Leaching Tests (10% Pulp Density)

The results of the one-step leaching tests are presented in Table 2. High HCl concentration (4 M) at 80 °C and S/L 10% resulted in 98% and 96% extraction of Ni and Co, respectively. Increase of the temperature from 65 to 90 °C had a positive effect on the extraction of both Ni and Co, whereby Ni increased from 75.5% to 94.4% and Co from 76.8% to 88.6%. Higher ratios of S/L resulted in a decrease of extraction percentage of all elements, i.e. Ni, Co, Fe, and to a lesser extent Mg (see Table 2). Based on leaching results, it was deduced that hydrochloric acid in higher ratios of S/L resulted in lower Fe co-dissolution. In order to enrich the PLS in Ni and Co and preserve the low Fe co-dissolution, it was decided to investigate the performance of a CCL scheme. It should be noted that the major part of dissolved Fe, i.e., more than 98%, was in the trivalent state.

3.1.2. Counter-Current Leaching of Ore (30% Pulp Density)

The CCL treatment of ore was carried out using 30% pulp density. The final extraction of metals and the concentrations in the PLS are shown in Tables 3–6. As seen in Table 3, leaching of the ore at high pulp density (30%) resulted in very low dissolution of Fe, 1.62%, but Ni and Co extraction was also low—49.29% and 43.03%, respectively (step 1a). When the residue was treated with fresh HCl (step 1b), extraction of Ni and Co increased and reached 84.06% and 80.52%, respectively, but the second treatment step resulted also in increased dissolution of Fe to 31.64%. The leaching solution from step 1b was used for the treatment of fresh ore (step 1a′, Figure 2). It can be seen that almost all Fe was extracted during step 1b precipitated, indicated by the negative extraction percentage (−29.46%) shown in Table 3. The composition of all leaching solutions is presented in Table 4. As seen in this table, the concentration of Fe dropped substantially from 20.19 (step 1b) to 0.38 g/L (step 1a′). The concentration of Ni increased from 0.917 to 1.46 g/L, while that of Co increased from 0.066 to 0.113 g/L.

From the elemental composition of leaching solutions, the residual acidity was calculated. It can be seen that all the acidity of the solution in step 1a was consumed, mainly from the dissolution of Mg and Ca. The solid residue treated in step 1b contained lower amounts of Ca and Mg and, for this reason, the corresponding minerals did not consume all the acidity of the fresh HCl solution. In the solution of step 1b, the residual acidity was high, i.e., 0.088 M, and this was the reason for the observed relatively high dissolution of Fe in step 1b. When this solution was used for the treatment of fresh ore in step 1a′, the residual acidity was again consumed by Ca and Mg, while the dissolved Fe precipitated.

Table 2. Metal extraction (%) from the saprolitic ore after six hours leaching with HCl.

HCl Molarity M	Temperature °C	Solid/Liquid (S/L) Ratio % w/v	Extraction (%)			
			Ni	Co	Fe	Mg
1	80	10	41.1	40.5	3.0	100.0
2	80	10	79.3	84.3	54.0	100.0
3	80	10	95.9	96.5	93.7	100.0
4	80	10	97.9	96.0	97.5	100.0
2	65	10	75.5	76.8	39.4	100.0
2	90	10	94.4	88.6	56.7	100.0
2	80	20	59.1	54.1	11.9	100.0
2	80	30	45.1	39.8	1.5	77.4

Table 3. Counter-current leaching (CCL) results (acid concentration 2 M, 80 °C, S/L 30%, time 6 h).

Step	pH	Ni (%)	Co (%)	Fe (%)	Mg (%)	Mn (%)
1a Leaching of ore	1.04	49.29	43.03	1.62	75.60	41.04
1b Leaching of 1a step residue with fresh acid		34.77	37.89	30.02	32.60	37.37
1a+1b		84.06	80.92	31.64	108.20	78.41
1a′ Leaching of fresh ore pregnant leach solution (PLS)	1.31	20.59	26.86	-29.46	33.41	27.49
1a′+1b		55.36	64.75	0.56	66.01	64.86

Table 4. Composition of PLS streams obtained by the CCL (steps in Figure 2).

	Step 1a		Step 1b		Step 1a′	
Concentration	g/L	mol/L	g/L	mol/L	g/L	mol/L
Cl^-		2.000		2.000		2.000
Fe^{+3}	1.09	0.019	20.19	0.361	0.380	0.007
Cr^{+3}	0.019	0.000	0.063	0.001	0.017	0.0003
Al^{+3}	0.314	0.012	0.391	0.014	0.526	0.019
Ni^{+2}	1.3	0.022	0.917	0.016	1.460	0.025
Co^{+2}	0.075	0.001	0.066	0.001	0.113	0.002
Mn^{+2}	0.515	0.009	0.469	0.009	0.814	0.015
Mg^{+2}	17.2	0.708	8.1	0.333	15.70	0.646
Ca^{+2}	9.08	0.227	1.32	0.033	10.90	0.273
H^+ (*)		−0.030		0.088		0.001

(*) Calculated based on electroneutrality.

3.1.3. Washing of Leaching Solid Residues—Assessment of Environmental Quality

Washing was applied to the solid residue of step 1b. The composition of wash water is shown in Table 5. As seen in the Table, the pH of the wash water was 1.60, suggesting that the high acidity of the leach solution in step 1b which remained in the moisture of solid residue was not completely removed in the washing step.

Table 5. Composition of wash water (mg/L) obtained after washing of the leach solid residue (step 1c in Figure 2).

pH	Fe	Cr^{+3}	Al^{+3}	Ni^{+2}	Co^{+2}	Mn^{+2}	Mg^{+2}	Ca^{+2}
1.60	1210	6.0	22	55	4	30	520	86

The washed solid residue was subjected to EN12457.02 leaching, and the results are presented in Table 6. The leachate resulting from the mixing of deionized water with the solid residue showed that the leachability of As, Ba, Cd, Cu, Pb, and Se met the criteria established by Directive 2003/33/EC for the acceptance of solid residue in landfills for inert waste. Dissolution of Zn was greater than the limits set for inert waste disposal, but satisfied the non-hazardous waste criterion. It was found that the solid residue contained a high leachable content of Cr and Ni, which amounted to 142.6 mg/kg and 540.4 mg/kg, respectively, and exceeded the limits set for non-hazardous waste disposal. It can also be seen that the residue retained a high level of acidity, since the pH of the leachate was 1.64.

The results suggest that the washing procedure, as applied during the laboratory tests, was not sufficient for the complete removal of acidity and of the leachable part of Cr and Ni. Thus, a multi-stage counter-current washing installation, combined with addition of alkalinity, was required in order to obtain a solid residue appropriate for safe environmental disposal.

3.1.4. Characterization of CCL Solid Residues

XRD Analysis

The dried leach residues obtained after CCL were examined using XRD, and the patterns are shown in Figure 3. In comparison with the raw ore, the peaks of lizardite and calcite present in the ore disappeared after leaching with HCl in the solid residues of steps 1b and 1c. However, even though the peak of calcite also disappeared in the solid residue of step 1a', the peak of lizardite remained. Quartz and weaker goethite peaks remained in the HCl residues from all the steps of CCL process since these minerals are leached poorly in the acidic medium.

Table 6. Results of the EN 12457.02 (L/S = 10 L/kg) leaching test for the washed CCL solid residue and comparison with the existing criteria for the acceptance of waste at inert non-hazardous and hazardous landfills (Directive 2003/33/EC).

	Solution	Sample	Criteria for the Acceptance of Waste at Landfills		
			Inert	Non-Hazardous	Hazardous
pH	1.64				
Electrical conductivity (EC; mS/cm)	15.29				
	mg/L	mg/kg		mg/kg	
As	<0.006	<0.06	0.5	2	25
Ba	0.165	1.659	20	100	300
Cd	0.0006	0.006	0.04	1	5
Cr	14.255	**142.6**	0.5	10	70
Cu	0.130	1.305	2	50	100
Ni	54.04	**540.4**	0.4	10	40
Pb	0.010	0.104	0.5	10	50
Se	0.0026	0.026	0.1	0.5	7
Zn	0.662	6.627	4	50	200

* Directive 2003/33/EC.

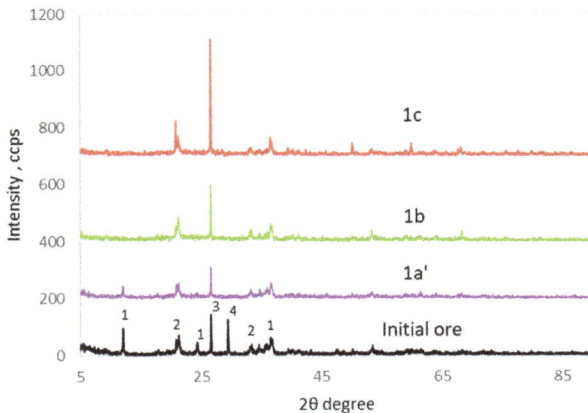

Figure 3. X-ray diffraction (XRD) patterns of leach residues from CCL in comparison with the pattern of the initial ore (1, lizardite $(Mg,Al)_3[(Si,Fe)_2O_5](OH)_4$; 2, goethite (FeOOH); 3, quartz; 4, calcite).

SEM Analyses

The elemental composition of the saprolitic ore, before and after leaching, as derived by EDS analysis, is presented in Figure 4. In the examined grains of solid residue derived from step 1b

(Figure 4b), Ni and Co disappeared and the content of Fe and Mg decreased in comparison with the initial ore (Figure 4a). The presence of Cl anions, related to the composition of leach solution, was also observed in the grains of residue obtained from step 1a' (Figure 4c). The atomic ratio of major elements (O/Mg/Si/Fe) of the grain in Figure 4c was similar to the grain of the initial ore in Figure 4a. However, the morphology of the grain in Figure 4c was very different, and this difference could be attributed to the presence of fine Fe precipitates on the surface of the grain. It should be reminded that approximately 19 g/L Fe was removed from the aqueous phase via precipitation during ore leaching in step 1a'.

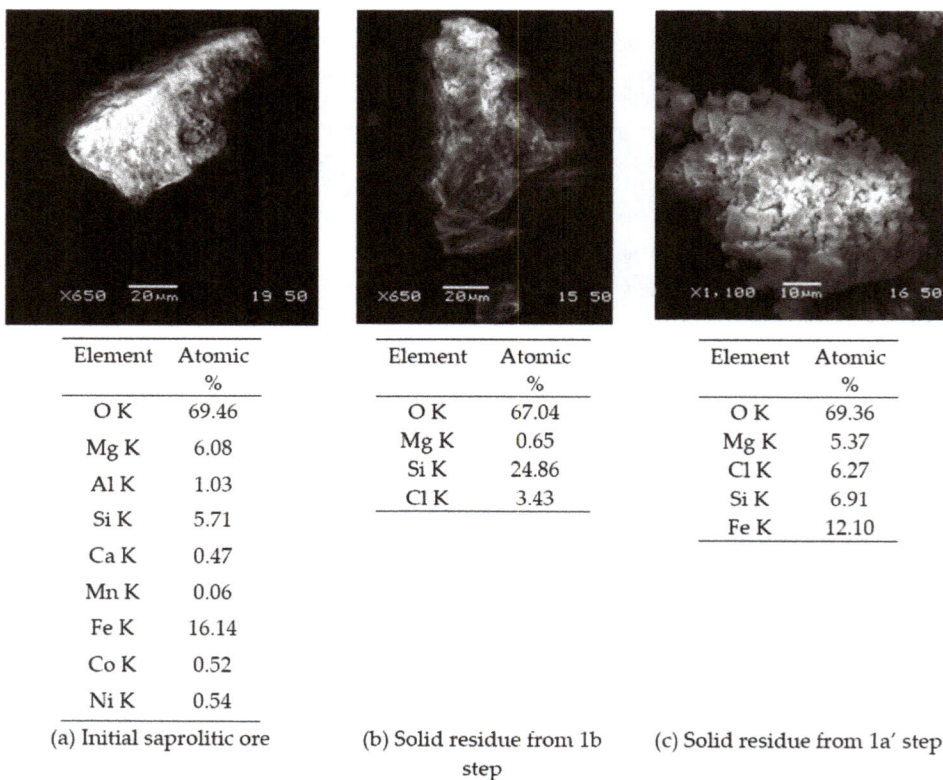

Element	Atomic %
O K	69.46
Mg K	6.08
Al K	1.03
Si K	5.71
Ca K	0.47
Mn K	0.06
Fe K	16.14
Co K	0.52
Ni K	0.54

(a) Initial saprolitic ore

Element	Atomic %
O K	67.04
Mg K	0.65
Si K	24.86
Cl K	3.43

(b) Solid residue from 1b step

Element	Atomic %
O K	69.36
Mg K	5.37
Cl K	6.27
Si K	6.91
Fe K	12.10

(c) Solid residue from 1a' step

Figure 4. SEM microphotographs and energy-dispersive X-ray spectroscopy (EDS) analyses of (**a**) initial grains, (**b**) grains after CCL step 1b, and (**c**) grains after CCL step 1a'.

3.2. Precipitation of Trivalent Metals from Pregnant Leach Solution

3.2.1. Thermodynamic Calculations using Visual Minteq

The thermodynamic calculations using VMinteq were carried out in order to identify the pH ranges, where precipitation of metal hydroxides took place for the separation of impurities from the valuable elements. As shown in Figure 5, precipitation of iron hydroxides (as ferrihydrite) was observed at pH values above 2. The increase in pH above 4 resulted in the precipitation of Cr and Al as $Al(OH)_3$ and $Cr(OH)_3m$ respectively. Ni and Co precipitated as $Ni(OH)_2$ and $Co(OH)_2$ at pH values higher than 7.5, while Mg precipitated as brucite at pH values higher than 9. Precipitation of Ca as portlandite occurred above pH 12. Based on the thermodynamic data of Visual Minteq, the solution that contained 0.073 M of Ca was undersaturated with respect to lime. Also, according to the calculations, if Mn is divalent, it remains in solution in the whole range of pH values (Figure 5a), while, if tetravalent, it precipitates as pyrolusite (MnO_2) in the pH range of 4 to 9 (Figure 5b).

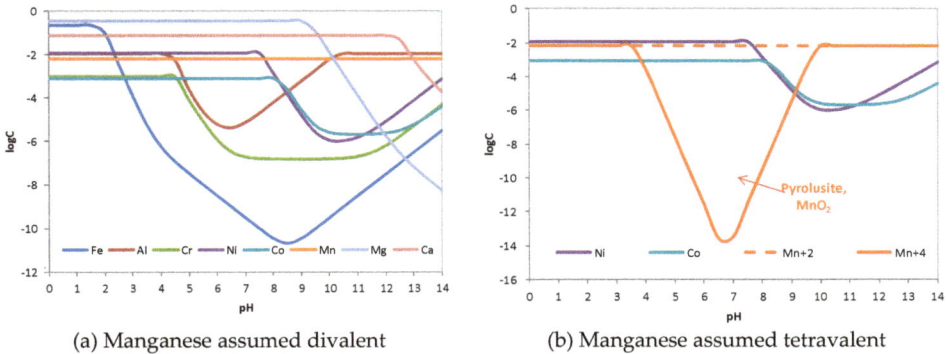

| (a) Manganese assumed divalent | (b) Manganese assumed tetravalent |

Figure 5. Thermodynamic calculations using Visual Minteq for the identification of optimum pH ranges for the selective precipitation of metals in the form of hydroxides.

3.2.2. Experiments with the PLS Obtained after One-Step Leaching (10% Pulp Density)

Precipitation of Trivalent Metals from PLS

The treatment with $Mg(OH)_2$ was conducted at four different target pHs (2.0, 3.0, 3.5, and 4.0). The final volume of solution was increased due to the addition of $Mg(OH)_2$ in the form of pulp 10% w/v. The percentage of metal removal was calculated by taking into consideration the factor of volume increase. As seen in Figure 6, the removal of Fe and Cr was higher than 99% above pH 3. The removal percentage for Al was 97% at pH 3.5 and increased to 99% at pH 4. Co-precipitation (losses) of Ni and Co were estimated to be close to 0.6% and 1.2% for Ni, and 4.5% and 7.6% for Co at pH 3.5 and 4, respectively.

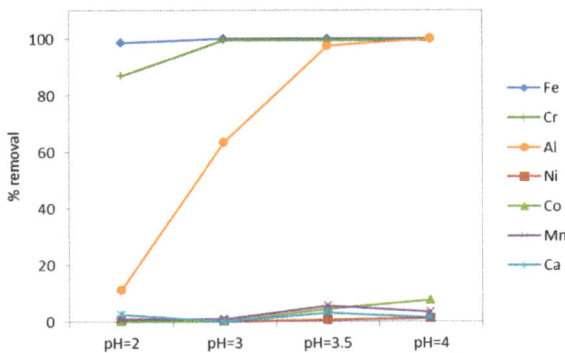

Figure 6. Effect of pH on metal removal during precipitation of trivalent metals from pregnant leach solution (PLS) obtained from one-step leaching test at 10% pulp density.

Alkaline Oxidation (AO)

The AO treatment was applied to the solution produced after the precipitation of Fe with $Mg(OH)_2$ at pH 3.5. The experiments were carried out using different doses of H_2O_2 ranging from 1% to 10% w/v and increasing pH to 9 with the addition of NaOH (1 M). A control experiment without addition of H_2O_2 was also carried out at increased pH 9. In the control experiment, Ni and Co precipitation was higher than 99%, but a relatively high percentage of Mn (12–20%) remained in solution. When

H_2O_2 was added, the precipitation of the three metals, Ni, Co, and Mn, was almost complete. The dose of H_2O_2 had no effect on the final degree of precipitation. It should be noted that the color of the suspensions was different in the absence or presence of H_2O_2. Without H_2O_2, the color was green, while, in the presence of H_2O_2, it turned brown and this can be attributed to the presence of higher-valence Mn oxides.

According to the thermodynamic analysis of the system (see Figure 5a) precipitation of pure Mn^{+2} hydroxide (pyrochroite) was not expected to occur. However, it is known that Ni can form mixed hydroxides with many divalent and trivalent metals, including Mn(II) and Mn(III) [30]. According to Zhou et al., under anaerobic conditions, Mn^{+2} is incorporated into the structure of $Ni(OH)_2$ hydroxide at all Mn/Ni molar ratios [31]. Under aerobic conditions, Mn is oxidized to the Mn^{+3} state and layered double hydroxide (LDH) phases are formed. In the structure of LDH phases (similar to hydrotalcite), the layers have the approximate composition $[Ni^{+2}_{1-x}M^{+3}_x(OH)_2]^{+x}$, and different anions are intercalated in the interlayer space for charge compensation and stability. Tetravalent cations like Mn^{+4} are not easily incorporated in the structure of LDH [30]. If Mn is oxidized in the +4 state, it will probably precipitate as pyrolusite (MnO_2) (see Figure 5b). These mechanisms could explain the removal of Mn together with Ni, either with or without the addition of H_2O_2. If Mn(II) remains in solids, it is impossible to separate it from Ni and Co using the acidic treatment of AO solid residue.

Acid Dissolution (AD) of AO Solid Residue

Preliminary acid dissolution tests carried out at pH 3 indicated that the highest selective dissolution of Ni with respect to Mn was obtained from the solid residues produced at 5% and 10% w/v H_2O_2, without major differences between the two residues. For this reason, subsequent tests were conducted using the AO solid residues obtained after treatment with the use of 5% w/v H_2O_2. The dissolution tests were carried out at different pH values in order to determine the effect of pH on the recovery of metals in the aqueous solution. The results are shown in Figure 7. According to the thermodynamic calculations (Figure 5b), at pH 6, tetravalent Mn(IV) remained in the solids, while Ni and Co hydroxides were completely dissolved. However, the experimental results indicate that all metals remained in the solids at pH 6. As previously discussed, more complex hydroxides with low solubility, containing Ni, Co, and Mn, are probably formed close to neutral pH values. When the pH was decreased from 6 to 1, 95% dissolution of Ni and 89% dissolution of Co were obtained. However, Mn was also re-dissolved up to a percentage of 47%. Moreover, the aqueous solution contained high levels of Mg. A representative composition of AD aqueous stream was Ni 801, Co 51, Mn 194, and Mg 5564 mg/L.

To enable the separation of Mg and Mn from Ni and Co, a treatment scheme involving three successive cycles of AO and AD of the solid residue was examined. The results are presented in Figure 8. As seen in the figure, the removal of Mg and Mn from the final Ni/Co stream was very efficient with the three-stage treatment. However, the recovery of Ni and Co also decreased from 95% and 89% to 70% and 63%, respectively, indicating certain losses during the successive treatment steps.

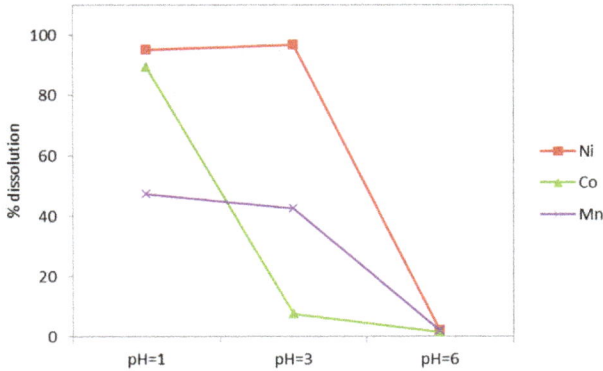

Figure 7. Effect of pH applied at the acidic dissolution (AD) step on dissolution of Ni, Co, and Mn from the solid residue of the alkaline oxidation (AO) step (with the use of H_2O_2 5% w/v).

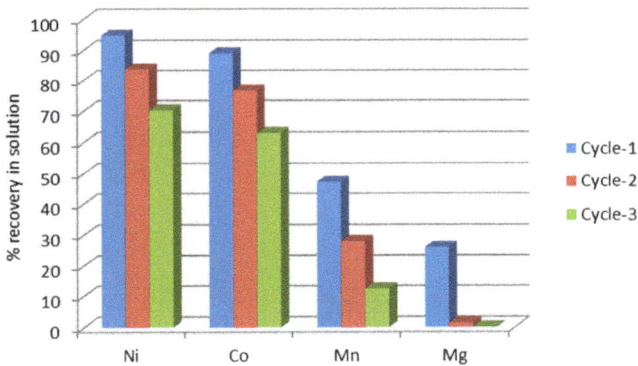

Figure 8. Percent recovery of Ni, Co, Mn, and Mg in the aqueous solution, after three repetitive cycles of AO and AD steps.

3.2.3. Experiments with the PLS of CCL (30% w/v Pulp Density)

Precipitation of Trivalent Metals from CCL Pregnant Solution

The precipitation of trivalent metals from PLS produced by CCL was carried out with $Mg(OH)_2$ at two pH values, 3.5 and 4. The percentage of precipitated metals is presented in Figure 9. At pH 3.5, Fe was 92.9%, and some minor co-precipitation of Ni was noted (4%), while no co-precipitation (losses) of Co was observed. The removal of Cr was also high (98.3%), but the precipitation of Al was very low, only 15.9%. When the precipitation was carried out at pH 4.0 almost complete removal of Al was obtained and, for this reason, subsequent treatment steps were conducted using the PLS solution treated at pH 4.0. It should be mentioned that the filtration of the suspension was difficult due to the colloidal behavior of iron hydroxides; thus, S/L separation was carried out using centrifugation at 9000 rpm for 20 minutes.

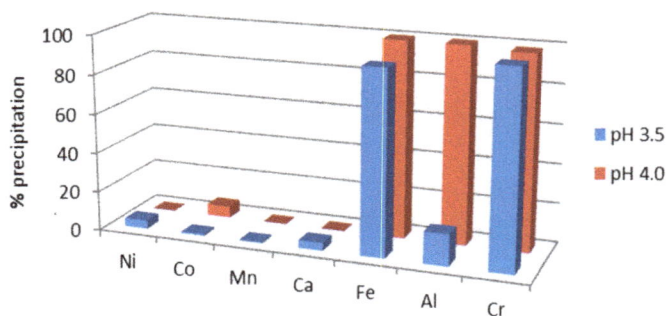

Figure 9. Metal removal (%) from PLS during the precipitation of trivalent metals step.

Separation of Mn, Ca, and Mg from Ni and Co after Two-Stage AO and AD Treatment

In order to separate Mn, Ca, and Mg from the valuable metals, Ni and Co, the PLS obtained after Fe removal was subjected to AO followed by AD of the solid residue in two successive steps. The total mass balance of Ni, Co, Mn, Ca, and Mg, during the successive PLS purification steps, is shown in Table 7 and Figure 10. To reduce losses of Ni and Co in the MnO_2 solids, the acidic dissolution of AO solids in this series of tests was conducted at pH equal to 0.8.

As seen from the results in Table 7 and Figure 10, the recovery of Ni in the final product was equal to 71.6% and that of Co to 61.3%. The product stream contained a relatively high amount of Mg, Mn, and Ca. It was, thus, concluded that the optimization of Mg, Mn, and Ca separation steps seems to be crucial for the improvement of the quality of the final product.

Table 7. Metal distribution (in mg) among streams remaining in the circuit, Fe sludge, MnO_2 precipitates, and final Ni/Co product stream. AO—alkaline oxidation.

Product Stream	Ni	Co	Mn	Mg	Ca
PLS before Fe removal	1460	113.3	814	15700	8340
Precipitation of trivalent metals at pH 4.0					
Mg added for Fe removal				820	
Removed from the circuit with Fe sludge	0.0	6.71	0.0	0.0	0.0
Aqueous stream from AO steps	2.04	1.01	0.55	15626	8135
Dissolution of AO solids at pH 0.8					
Removed from the circuit with MnO_2	412.85	36.17	517.07	91	10.45
Ni/Co product stream	1045.11	69.41	296.37	803	195.0

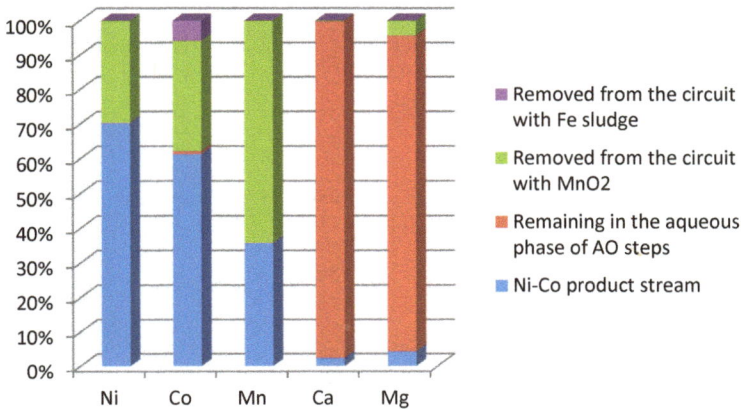

Figure 10. Metal distribution (%) among the streams remaining in the circuit, Fe sludge, MnO_2 precipitates, and final Ni/Co product stream.

The MnO_2 solid residues produced from the AO/AD process were studied using SEM/EDS analysis and the results are presented in Figure 11. The solids were rich in Mn, but also retained a relatively high content of Ni, close to 2% in the first and 4% in the second AO/AD stage. They also contained Mg, whose content decreased from 5% in the first to 2% in the second AO/AD stage.

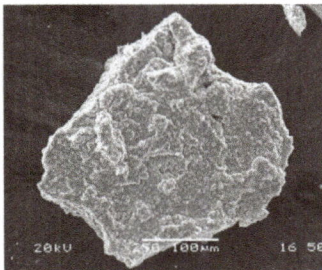

Element	Atomic %
O K	65.21
Mg K	4.74
Al K	1.68
Cl K	14.90
Mn K	8.53
Ni K	1.94

(a) Solid residue of first AO/AD stage

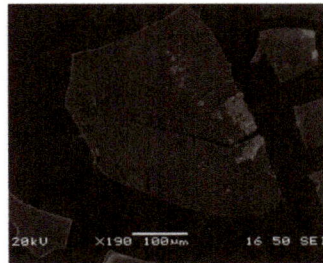

Element	Atomic %
O K	61.41
Mg K	1.96
Al K	3.62
Cl K	16.80
Mn K	12.08
Ni K	4.13

(b) Solid residue of second AO/AD stage

Figure 11. SEM and EDS analyses of MnO_2 solid residue obtained from the first and second AO/AD stages.

Removal of Ca from PLS

As seen in Table 8, the aqueous stream of the alkaline oxidation steps constituted basically a solution of $MgCl_2$ and $CaCl_2$ with molar concentrations approximately equal to 0.65 and 0.20 M. It is known that the presence of Ca adversely affects the performance of pyrohydrolysis step [32].

For this reason, a step for removing Ca from the circuit must be included in the flowchart before the pyrohydrolysis treatment, and an option is to precipitate Ca in the form of gypsum $CaSO_4 \cdot xH_2O$. The residual concentrations of Ca and SO_4 ions in the aqueous solution after calcium sulfate precipitation were estimated using V-Minteq software, assuming that the solution initially contained 0.65 M $MgCl_2$

and 0.20 M $CaCl_2$. An amount of H_2SO_4 stoichiometrically equivalent to Ca (0.20 M) was added in this solution, and calculations were carried out assuming precipitation of gypsum ($CaSO_4 \cdot 2H_2O$). Input data and calculated results are presented in Table 8. As seen in the table, the calculated equilibrium concentrations of Ca and SO_4 were equal to 0.08 M.

Table 8. Input data and calculated results for gypsum precipitation using V-Minteq software.

Component	Initial Composition of Aqueous Phase	Calculated Amount in Equilibrium
	Molar Conc. (M)	Molar Conc. (M)
Mg	0.65	0.65
Ca	0.20	0.08
Cl	1.70	1.70
H$^+$	0.40	0.40
SO$_4{}^{-2}$	0.20	0.08

Pyrohydrolysis of the Aqueous Stream after Calcium Removal

The distribution of species during the thermal process of pyrohydrolysis was estimated based on calculations carried out with the HSC Chemistry software. The composition of the aqueous stream was simulated assuming a mixture containing 0.65 mol of $MgCl_2$, 0.08 mol of $CaSO_4$ $2H_2O$, 0.40 mol of HCl, and 55 mol of H_2O. The release of HCl and the evolution of Mg phases as a function of temperature are shown in Figure 12a. The evolution of Ca phases is shown in Figure 12b. Between 200 and 340 °C, the release of HCl was equal to 0.96 mol, and the residual 0.74 mol of chlorides was bound to the Mg(OH)Cl and $CaCl_2$ compounds. Between 340 and 400 °C, Mg(OH)Cl decomposed and the released HCl increased to 1.52 mol. A small amount of Cl, equal 0.18 mol, was retained in the form of $CaCl_2$, until 770 °C. Above this temperature, all the chlorides fed to the pyrohydrolysis step, i.e., 1.70 mol, were released as HCl. At 800 °C, the solid phases of Mg consisted of oxides MgO (0.643 mol) and traces of $MgSO_4$ (0.007 mol). Calcium was found in the form of $CaSO_4$ (0.072 mol) with a small amount of CaO (0.008 mol).

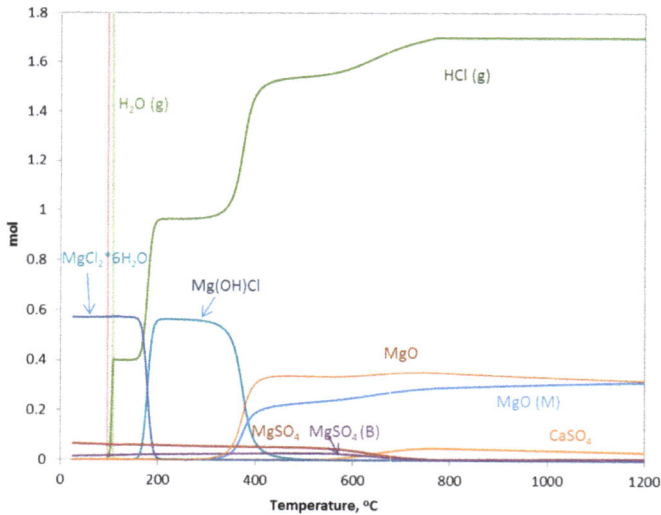

(a)

Figure 12. *Cont.*

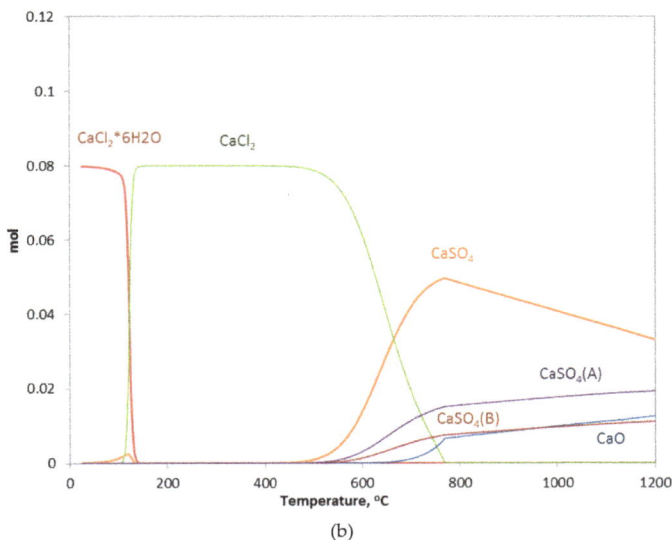

(b)

Figure 12. Produced solid and gas phases from the aqueous stream after calcium sulfate precipitation, increasing the temperature from 25 to 1200 °C, based on HSC equilibrium calculations. (**a**) Release of HCl gas and evolution Mg phases; (**b**) evolution of Ca phases.

4. Discussion

The experimental evaluation of the HCl process, applying one-step leaching of laterite at 10% *w/v* pulp density, indicated that the most important steps determining the final recovery of Ni and Co and the purity of Ni/Co stream are those involved in the combined AO/AD process. Repetitive application of the AO/AD treatment steps significantly improved Mg and Mn separation from the valuable metals. For instance, the Mg/Ni ratio dropped from 6.9 to 0.43 and 0.05, applying one, two, and three cycles of the combined treatment, respectively.

Operation at 10% pulp density requires a high excess of water and is unrealistic under industrial conditions. Much closer to the industrial practice is operation at 30% S/L. The initial experimental results (Table 2) indicated that operating at this higher pulp density had the great advantage of suppressing co-dissolution of Fe from 54% to 1.5%, but the extraction of Ni dropped from 79% to 45% and that of Co from 84% to 40%.

The CCL mode of operation, at 30% pulp density, was evaluated experimentally in order to increase the extraction of Ni and Co and maintain co-dissolution of Fe at low levels. Indeed, during CCL, the dissolution of Fe was very low, as expected, on the order of 0.56%, while the extraction of Ni improved from 45% to 55% and that of Co from 40 to 63%. However, the final recoveries of Ni and Co were low (40% and 38%, respectively) due to additional losses in the subsequent treatment steps. The mass balance of metals, based on the experimental results simulating the CCL operation and the following PLS purification steps, is presented in the flowchart of Figure 13.

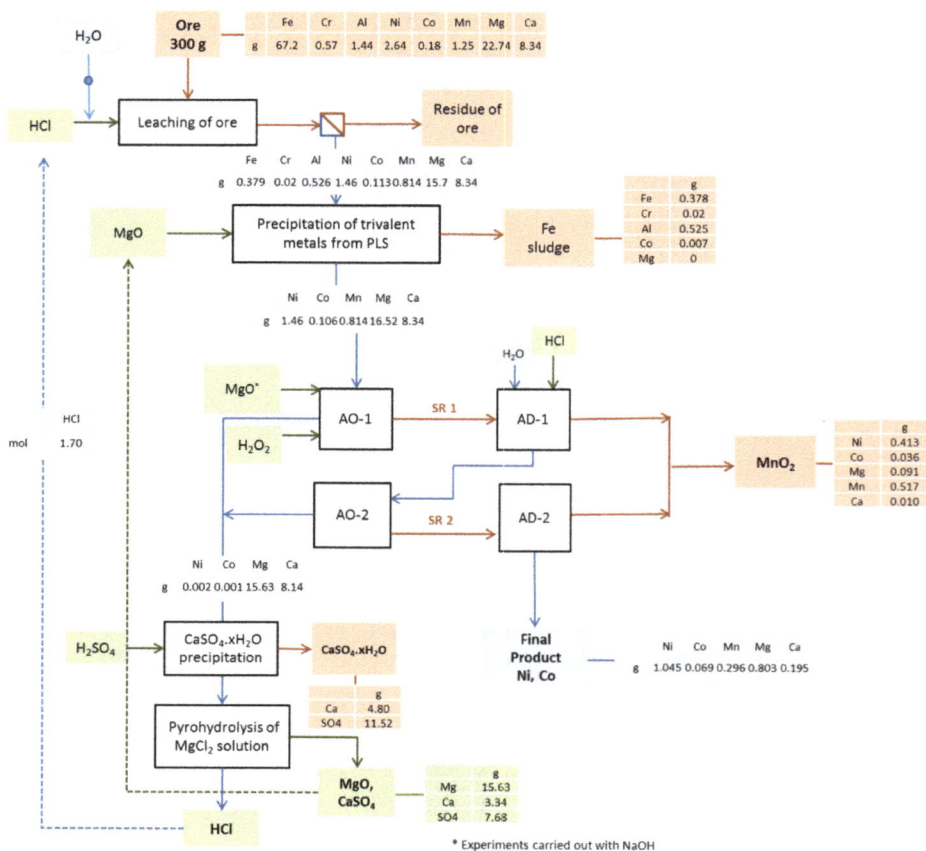

Figure 13. Flowchart simulating CCL operation at 30% pulp density, including mass balance of metals.

From the mass balance, it was calculated that 45% of Ni and 37% of Co remained in the residue produced after CCL. About 15% Ni and 20% Co were also lost in the MnO_2 solids; 64% of the extracted Mn was removed with the MnO_2 solids, while 99% of the extracted Mg was fed in the pyrohydrolysis step for MgO recovery and HCl regeneration. Co-extraction of Ca had a negative effect on MgO recovery and HCl regeneration; thus, Ca had to be removed from the circuit in the form of gypsum before the pyrohydrolysis step. It was estimated that, in this case, 60% of the extracted Ca would be removed as gypsum and 40% would be recovered as $CaSO_4$, together with the MgO solids during the pyrohydrolysis step.

It is evident that the two crucial processes of the flowchart in order to maximize Ni and Co extraction and reduce their losses in the solid residues were the CCL and AO/AD operations.

According to Rice [11], it is not possible to increase Ni concentration in the leach liquor by applying a CCL type of operation due to the passivation caused as a result of the formation of a product layer around the ore particles. Most probably, this is the mechanism explaining the high Ni and Co losses in the residue of CCL. As seen in Table 4, approximately 20 g/L iron precipitated during the leaching of 300 g of fresh ore at step 1a'. It is possible that the iron oxide precipitates were attached on the surface of the ore particles, inhibiting further extraction of Ni and Co. A possible way of overcoming this problem could be to carry out leaching in the presence of an inert abrasive medium, e.g., quartz particles of gravel size, in order to avoid the formation of this passivation layer.

Regarding Ni and Co losses in the Mn solids, they were primarily due to the fact that the oxidation of Mn(II) to the tetravalent state, Mn(IV), was incomplete. For the optimization of this step, it is, thus, necessary to increase the efficiency of Mn(II) oxidation. In addition to H_2O_2, several other oxidants were tested for this purpose, including O_3, an SO_2/O_2 mixture, peroxy-sulfuric acids, hypochlorite, and chlorate [15,33]. The investigated flowchart was based on the use of HCl; it is, thus, better to avoid the use of sulfur-based oxidants. The use of ClO_2 as tested by Park et al. [33] would not alter the composition of aqueous streams and could be an interesting alternative.

Another option for the efficient separation of Mn from Ni and Co comprises the following steps: (i) precipitation of Ni, Co, and Mn as hydroxides, and (ii) leaching of the precipitates with ammonia in the presence of air and CO_2 to selectively dissolve Ni and Co as ammine complexes, while Mn is left in the solid residue as manganese carbonate [34].

5. Conclusions

An atmospheric leaching process based on the use of hydrochloric acid was evaluated for Ni and Co extraction from a low-grade saprolitic laterite. The main advantage of using HCl, instead of other acids like H_2SO_4, is the fact that HCl can effectively extract both Ni and Co from the laterite ore, while H_2SO_4 is primarily efficient for Ni extraction. On the other hand, HCl can be regenerated and recycled in the leaching stage.

The leaching was carried out at 30% pulp density, applying a counter-current mode of operation, in order to better simulate industrial-scale operations and maintain Fe dissolution at low levels. This mode of operation was indeed very efficient regarding Fe dissolution, which was maintained at 0.6%, but proved to have a negative effect on Ni and Co extraction, which was limited to 55% and 63%, respectively, probably due to the passivation of ore minerals by secondary Fe-oxide precipitation products.

The treatment of PLS involved a series of simple precipitation steps. In this series of operations, the most crucial step was found to be the separation of Mn from the valuable Ni and Co metals. The adopted process was based on the oxidation of Mn(II) to Mn(IV) in order to form a stable MnO_2 oxide, which would remain in the solid phase under acidic conditions, while Ni and Co could be selectively recovered in the aqueous phase. However, the oxidation of Mn was not complete, and a relatively high amount of Ni and Co was retained in the solids, reducing the final recovery of the two elements to the levels of 40% for Ni and 38% for Co. At this stage, additional tests were carried out to overcome some of these drawbacks and improve the overall efficiency of the process.

Author Contributions: C.M. performed the literature search, carried out experiments and analytical techniques, analyzed data, and wrote a first draft of the paper. N.P. designed the experiments, critically analyzed results, and wrote the paper. A.X. designed the experiments and reviewed the paper. K.K. designed the experiments, contributed in the analysis of the results, and reviewed the paper.

Acknowledgments: The authors would like to acknowledge the financial support of the European Commission in the frame of Horizon 2020 project "Metal recovery from low-grade ores and wastes", www.metgrowplus.eu, Grant Agreement n. 690088.

Conflicts of Interest: The authors declare no conflicts of interest.

References

1. McDonald, R.G.; Whittington, B.I. Atmospheric acid leaching of nickel laterites review. Part I. Sulphuric acid technologies. *Hydrometallurgy* **2008**, *91*, 35–55. [CrossRef]
2. Zhang, P.; Guo, Q.; Wei, G.; Qu, J.; Qi, T. Precipitation of α-Fe$_2$O$_3$ and recovery of Ni and Co from synthetic laterite leaching solutions. *Hydrometallurgy* **2015**, *153*, 21–29. [CrossRef]
3. Feng, Q.; Zhao, W.; Wen, S. Surface modification of malachite with ethanediamine and its effect on sulfidization flotation. *Appl. Surf. Sci.* **2018**, *436*, 823–831. [CrossRef]
4. Whittington, B.I.; Muir, D. Pressure acid leaching of nickel laterites: A review. *Miner. Process. Extr. Metall. Rev.* **2000**, *21*, 527–599. [CrossRef]

5. McDonald, R.G.; Whittington, B.I. Atmospheric acid leaching of nickel laterites review. Part II. Chloride and bio-technologies. *Hydrometallurgy* **2008**, *91*, 56–69. [CrossRef]

6. Komnitsas, K. Kinetic Study of Laterite Atmospheric Leaching with Sulfuric Acid. Master's Thesis, National Technical University of Athens, Athens, Greece, 1983. (In Greek)

7. Kontopoulos, A. Sulphuric acid leaching of laterites. In *Extraction and Processing Division Congress*; Gaskell, D.R., Ed.; The Minerals, Metals and Materials Society: Warrendale, PA, USA, 2001; pp. 147–163.

8. Halikia, I. Parameters influencing kinetics of nickel extraction from a Greek laterite during leaching with sulphuric acid at atmospheric pressure. *Trans. Inst. Min. Metall. Sect. C* **1991**, *100*, C154–C164.

9. Agatzini-Leonardou, S.; Dimaki, D. Heap leaching of poor nickel laterites by sulphuric acid at ambient temperatures. In *Hydrometallurgy '94*; Chapman and Hall: London, UK, 1994; pp. 193–208.

10. Komnitsas, K.; Petrakis, E.; Pantelaki, O.; Kritikaki, A. Column leaching of Greek low grade limonitic laterites. *Minerals* **2018**, *8*, 377. [CrossRef]

11. Tzeferis, P.G.; Agatzini-Leonardou, S. Leaching of nickel and iron from Greek non-sulphide nickeliferous ores by organic acids. *Hydrometallurgy* **1994**, *36*, 345–360. [CrossRef]

12. Tzeferis, P.G. Leaching of a low grade hematitic laterite ore using fungi and biologically produced acid metabolites. *Int. J. Miner. Process.* **1994**, *42*, 267–283. [CrossRef]

13. Mystrioti, C.; Papassiopi, N.; Xenidis, A.; Komnitsas, K. Comparative Evaluation of Sulfuric and Hydrochloric Acid Atmospheric Leaching for the Treatment of Greek Low Grade Nickel Laterites. In *Extraction 2018, Proceedings of the First Global Conference on Extractive Metallurgy, The Minerals, Metals & Materials Series, Ottawa, QC, Canada, 26–29 August 2018*; Davis, B.R., Moats, M.S., Wang, S., Gregurek, D., Kapusta, J., Battle, T.P., Schlesinger, M.E., Flores, G.R.A., Jak, E., Goodall, G., et al., Eds.; Springer: Cham, Switzerland, 2018; pp. 1753–1764.

14. Rice, N.M. A hydrochloric acid process for nickeliferous laterites. *Miner. Eng.* **2016**, *88*, 28–52. [CrossRef]

15. Baba, A.A.; Ibrahim, L.; Adekola, F.A.; Bale, R.B.; Ghosh, M.K.; Sheik, A.R.; Pradhan, S.R.; Ayanda, O.S.; Folorunsho, I.F. Hydrometallurgical Processing of Manganese Ores: A Review. *J. Miner. Mater. Charact. Eng.* **2014**, *2*, 230–247. [CrossRef]

16. Agatzini-Leonardou, S.; Tsakiridis, P.E.; Oustadakis, P.; Karidakis, T.; Katsiapi, A. Hydrometallurgical process for the separation and recovery of nickel from sulphate heap leach liquor of nickeliferous laterite ores. *Miner. Eng.* **2009**, *22*, 1181–1192. [CrossRef]

17. Meng, L.; Qua, J.; Guo, Q.; Xie, K.; Zhang, P.; Han, L.; Zhang, G.; Qi, T. Recovery of Ni, Co, Mn, and Mg from nickel laterite ores using alkaline oxidation and hydrochloric acid leaching. *Sep. Purif. Technol.* **2015**, *143*, 80–87. [CrossRef]

18. Wang, K.; Li, J.; McDonald, R.G.; Browner, R.E. Iron, aluminium and chromium co-removal from atmospheric nickel laterite leach solutions. *Miner. Eng.* **2018**, *116*, 35–45. [CrossRef]

19. White, D.T.; Gillaspie, J.D. Atmospheric leaching of nickel laterites with iron precipitation. In Proceedings of the Alta Conference Proceedings, Perth, Australia, 23–30 May 2015.

20. Willis, B. Downstream processing options for nickel laterite heap leach liquors. In *ALTA 2007 Nickel/Cobalt 12*; ALTA Metallurgical Services: Melbourne, Australia, 2007; 25p.

21. Readett, D.; Sullivan, J. GME's NiWest Nickel Project-ongoing development of Ni laterite heap leach, sx-ew process. In Proceedings of the ALTA 2015 Conference, Perth, Australia, 23–30 May 2015.

22. Demopoulos, G.P. Aqueous precipitation and crystallization for the production of particulate solids with desired properties. *Hydrometallurgy* **2009**, *96*, 199–214. [CrossRef]

23. Filippou, D.; Choi, Y. A contribution to the study of iron removal from chloride leach solutions. In *Chloride Metallurgy, Proc. 32nd Annual CIM Hydrometallurgical Conference*; Peek, E., van Weert, G., Eds.; CIM: Montreal, QC, USA, 2002; Volume 2, p. 729.

24. Beukes, J.P.; Giesekke, E.W.; Elliott, W. Nickel retention by goethite and hematite. *Miner. Eng.* **2000**, *13*, 1573–1579. [CrossRef]

25. Demopoulos, G.P.; Li, Z.; Becze, L.; Moldoveanu, G.; Cheng, T.; Harris, G.B. New technologies for HCl regeneration in chloride hydrometallurgy. *World Metall. ERZMETALL* **2008**, *61*, 84–93.

26. Feldmann, T.; Demopoulos, G.P. Influence of impurities on crystallization kinetics of calcium sulfate dihydrate and hemihydrate in strong HCl-CaCl$_2$ solutions. *Ind. Eng. Chem. Res.* **2013**, *52*, 6540–6549. [CrossRef]

27. Feldmann, T.; Demopoulos, G.P. Effects of crystal habit modifiers on the morphology of calcium sulfate dihydrate grown in strong CaCl2-HCl solutions. *J. Chem. Technol. Biotechnol.* **2014**, *89*, 1523–1533. [CrossRef]

28. Gustafsson, J.P. Visual MINTEQ, Version 3.0: A Window Version of MINTEQA2, Version 4.0. 2010. Available online: http://www.lwr.kth.se/english/OurSoftware/Vminteq (accessed on 10 April 2018).

29. BSI. *EN 12457.02: Characterization of Waste—Leaching—Compliance Test for Leaching of Granular Waste Materials and Sludges—Part 2: One Stage Batch Test at a Liquid to Solid Ratio of 10 l/kg for Materials with Particle Size below 4 mm (without or with Size Reduction)*; BSI: London, UK, 2002.

30. Kosova, N.; Devyatkina, E.; Kaichev, V. Mixed layered Ni–Mn–Co hydroxides: Crystal structure, electronic state of ions, and thermal decomposition. *J. Power Sources* **2007**, *174*, 735–740. [CrossRef]

31. Zhou, F.; Zhao, X.; van Bommel, A.; Rowe, A.; Dahn, J. Coprecipitation synthesis of $Ni_xMn_{1-x}(OH)_2$ mixed hydroxides. *Chem. Mater.* **2010**, *22*, 1015–1021. [CrossRef]

32. Yildirim, M.; Akarsu, H. Preparation of magnesium oxide (MgO) from dolomite by leach precipitation pyrohydrolysis process. *Physicochem. Probl. Miner. Process.* **2010**, *44*, 257–272.

33. Park, K.H.; Kim, H.I.; Das, R.P. Selective Acid Leaching of Nickel and Cobalt from Precipitated Manganese Hydroxide in the Presence of Chlorine Dioxide. *Hydrometallurgy* **2005**, *78*, 271–277. [CrossRef]

34. Manson, P.G.; Groutsch, J.V.; Mayze, R.S.; White, D.R.S. *Process Development and Plant Design for the Cawse Nickel Project*; ALTA Metallurgy Service: Perth, Australia, 1997.

minerals

MDPI

Article

Column Leaching of Greek Low-Grade Limonitic Laterites

Kostas Komnitsas *, Evangelos Petrakis, Olga Pantelaki and Anna Kritikaki

Technical University of Crete, School of Mineral Resources Engineering, University Campus, Kounoupidiana, 73100 Chania, Greece; vpetraki@mred.tuc.gr (E.P.); olgapan@mred.tuc.gr (O.P.); akritik@mred.tuc.gr (A.K.)
* Correspondence: komni@mred.tuc.gr; Tel.: +30-28210-37686

Received: 15 July 2018; Accepted: 29 August 2018; Published: 31 August 2018

Abstract: In this study, column leaching experiments were carried out to investigate the extraction of Ni and Co from low-grade limonitic laterites from Agios Ioannis mines in central Greece. Tests were carried out in laboratory Plexiglas columns using H_2SO_4 as leaching solution. Parameters determining the efficiency of the process, i.e., acid concentration (0.5 M or 1.5 M) and addition of 20 or 30 g/L of sodium sulfite (Na_2SO_3) in the leaching solution, were also studied. Upflow transport of the leaching solution with the use of peristaltic pumps was carried out, while the pregnant leach solution (PLS) was recycled several times over the entire test duration. The concentration of Ni, Co, Fe, Ca, Al, Mg, and Mn in the PLS was determined by Atomic Absorption Spectroscopy (AAS). The ore and the leaching residues were characterized by different techniques, i.e., X-ray fluorescence (XRF), X-ray diffraction (XRD), Fourier transform infrared (FTIR) spectroscopy, and differential scanning calorimetry and thermogravimetry (DSC/TG). The experimental results showed that (i) Ni and Co extractions increased with the increase of H_2SO_4 concentration—60.2% Ni and 59.0% Co extractions were obtained after 33 days of leaching with 1.5 M H_2SO_4; (ii) addition of 20 g/L Na_2SO_3 in the leaching solution resulted in higher extraction percentages for both metals (73.5% for Ni and 84.1% for Co, respectively), whereas further increase of Na_2SO_3 concentration to 30 g/L only marginally affected Ni and Co extractions; and (iii) when leaching was carried out with 1.5 M H_2SO_4 and 20 g/L Na_2SO_3, its selectivity was improved, as deduced from the ratios Ni/Mg, Ni/Ca and Ni/Al in the PLS; on the other hand, the ratio Ni/Fe dropped as a result of the higher Fe extraction compared with that of Ni.

Keywords: limonitic laterites; column leaching; sulphuric acid; sodium sulfite

---

1. Introduction

Nickel is the fifth most common element on earth and is widely used in many industrial, transport, aerospace, marine, architectural, military, and consumer applications. Its biggest use is in alloying, particularly with chromium and other metals, to produce stainless and heat-resisting steels [1]. It must also be underlined that nickel is considered today as the most important metal by mass in the Li-ion battery cathodes used by electric vehicle manufacturers [2].

Sulphide ores comprise about 30% of the global nickel reserves and result in almost 55% of world metal production [3]. The gradual depletion of high-grade nickel sulphides and the increasing demand for nickel has initiated studies into the exploitation of the huge reserves of nickel laterites [4–6].

It is known that conventional mineral processing techniques cannot be readily applied to laterites due to the complex nature of the ores and the fact that nickel is hosted in several mineral phases [7,8]. The treatment of laterites is mainly carried out by pyrometallurgical techniques to produce ferronickel (FeNi) [9,10]. However, due to the high capital and operating costs of pyrometallurgy, the use of hydrometallurgical techniques, including atmospheric, heap, and high-pressure acid leaching (HPAL), as well as bioleaching, has gained more interest [11–14].

Earlier studies of column leaching indicated that Ni or Co extraction from laterites is strongly dependent on the type of nickel-bearing phases, leaching time, acid type or concentration, and particle size. In several cases, the use of agglomerating was considered to improve permeability of the reaction bed and accelerate leaching of the useful elements [15]. Duyvesteyn et al. [16] suggested that the ore should be crushed and pelletized with the use of sulphuric acid with a concentration of at least 100 g/L to neutralize magnesia. Horizonte, S.G. and Horizonte, D.O. [17] carried out tests in columns with 1 and 4 m height using agglomerates, concentration of H_2SO_4 20 to 200 g/L, and irrigation rate 10 L/(m^2·h). Their results indicated that higher acidity leads to faster kinetics and recoveries of 84% for Ni and 70% for Co after 150 days of leaching. Elliot et al. [18] performed column leaching tests using various Australian nickel laterites, which were agglomerated with the use of water and sulfuric acid in a rotating drum. The leaching solution contained 200 g/L H_2SO_4 while no recycling was involved. Depending on the ore mineralogy, extractions of 10–98% for Ni and 12–97% for Co were obtained after 120 days of leaching. The same study indicated that Ni and Co extractions were increased for laterites containing low goethite and moderate smectite content. Quast et al. [11] studied the behavior of siliceous goethitic ore and the effect of agglomeration on Ni and Co extractions in column studies, with the use of H_2SO_4 for a period of 100 days; also in this case, no recycling of the leaching solution was carried out. The authors mentioned that the feed size influences Ni and Co extractions which were, for the finest size used, 90% and 80%, respectively. In the study by Quaicoe et al. [19], column leaching experiments were performed using agglomerates of saprolitic (SAP) and goethitic (G) laterite ores, 200 g/L H_2SO_4, and an irrigation rate of 8.5 L/(m^2·h). Ni and Co extractions for the G ore were, after 100 days, 62.7% and 55.6%, respectively, while for the SAP ore higher extractions of 90.0% Ni and 72.8% Co were obtained. This is due to the fact that serpentine minerals dissolve more rapidly in H_2SO_4 compared with the more refractory goethite. Other studies have investigated the potential of SO_2 as reductant, also in the presence of Cu^{2+} ions as catalyst, to accelerate the rate of atmospheric acid leaching of laterites [20–24].

Greek laterite deposits are mainly located in three regions, i.e., Evia island and Lokrida in central Greece and Kastoria in northwestern Greece. Evia and Lokrida deposits are of limonitic type, with high iron and low magnesium content, whereas Kastoria laterites are of saprolitic type with high magnesium and low iron content. Greek laterites have been processed for more than 50 years at the Larco S.A. pyrometallurgical plant at Larymna to produce FeNi containing 18–20% Ni [25,26]. The hydrometallurgical treatment of these Greek ores, with atmospheric and pressure H_2SO_4 leaching, was first studied in the mid-1980s at National Technical University of Athens (Athens, Greece) [27–30]. Later, at the same institution, research efforts were focused on heap leaching [31,32]. The developed approach, through laboratory and pilot studies, involved the use of H_2SO_4, purification of the pregnant leach solution (PLS) by chemical precipitation, and, finally, recovery of Ni and Co [33]. No further studies to assess the potential of leaching of Greek laterites were then carried out, while in the last years the quality of the ores has gradually dropped. Thus, the aim of this study is to investigate the potential of column leaching for the treatment of low-grade Greek limonitic laterite and also to assess the effect of the addition of Na_2SO_3 in the leaching solution on the overall efficiency of leaching.

2. Materials and Methods

The ore used in this study (~50 kg) is a low-grade (0.58 wt % Ni) laterite from Agios Ioannis (LAI), Greece, obtained from Larco S.A mines. It was dried and homogenized by the cone and quarter method, and a representative subsample was crushed to minus 16 mm using a jaw crusher. Chemical and mineralogical analyses of the ore, after pulverization with the use of a FRITSCH-BICO pulverizer (Fritsch, Dresden, Germany), and of the leaching residues, after drying overnight at 80 °C, were carried out with the use of (i) a Bruker-AXS S2 Range type X-ray fluorescence energy dispersive spectrometer (XRF-EDS) (Bruker, Karlsruhe, Germany) and (ii) an X-ray diffractometer (XRD)(Bruker AXS (D8 Advance type), Karlsruhe, Germany) (Cu tube, scanning range from 4° to 70° 2θ, step 0.02° and measuring time 0.2 s/step) with the use of the DIFFRACplus EVA v. 2006 software

(Bruker, Karlsruhe, Germany) and the Powder Diffraction File (PDF-2) database(Bruker, Karlsruhe, Germany). A quantity of 2 g of each ore sample and leaching residue was also digested with the use of aqua regia and analyzed with AAS to carry out mass balance calculations [27]. Since the differences between aqua regia digestion and XRF were only minor (±5%), results presented in this paper are those derived by XRF. The functional groups present in laterite and leaching residues were identified through Fourier transform infrared (FTIR) spectroscopy using KBr pellets and a Perkin Elmer Spectrum 1000 spectrometer (Akron, OH, USA); each sample was mixed with KBr at a ratio 1:100 w/w and pressed to obtain a disc. In addition, differential scanning calorimetry and thermogravimetry (DSC/TG) were performed using a Setaram LabSys Evo TG-DTA-DSC analyzer (SETARAM Inc., Cranbury, NJ, USA). The samples were heated in a nitrogen atmosphere from 40 to 1000 °C with a heating rate of 10 °C/min. All analyses were carried out in duplicate and average values are given in this study.

Four leaching tests were carried out in columns to study the effect of H_2SO_4 concentration and the addition of sodium sulfite (Na_2SO_3) in the leaching solution on Ni, Co, and other element extractions. The leaching solutions and the process parameters considered are presented in Table 1. Laboratory Plexiglas columns with a diameter of 5 cm and height of 50 cm were used. The bed consisted of 1000 g laterite (bed height 40 cm) and two layers of 2 cm of silica sand placed at both ends of each column to act as filters. Cotton glass was also added at the bottom of each column to prevent transport of ultra-fine particles and precipitates and subsequent blockage of the columns. The leaching solution used, with an initial volume of 10 L, was either 0.5 M or 1.5 M H_2SO_4. The effect of the addition of 20 or 30 g/L Na_2SO_3 in the solution containing 1.5 M H_2SO_4 was also investigated. It is noted that the conditions used in the present study were the ones which were considered optimum after a series of previous tests investigating leaching of saprolitic and limonitic ores were carried out.

The leaching solution was pumped from plastic vessels with a flowrate of 3 L/day and collected in similar vessels. The solution was pumped upwards using a variable-speed peristaltic pump (Masterflex L/S economy variable-speed drive, Cole-Parmer Instrument Co, Vernon Hills, IL, USA) at a Darcy velocity of 152.9 cm/day. This means that an input flowrate of 2.08 ± 0.1 mL/min was used. This flowrate can be also expressed as ~175.4 L/(m^2·h). It is mentioned that in column leaching studies, velocities greater than 20 cm/day minimize the impact of axial (e.g. longitudinal) diffusion on transport [34]. Hence, the total empty bed contact time (EBCT), defined as the ratio of actual bed length to approach velocity, was 15.7 h [35].

The leaching solution was recycled every 3 days. Recirculation started after the entire feed solution passed through the column (laterite bed). The total duration of the tests was 33 days. At the end of the tests, columns were flushed with distilled water so that the final PLS volume was ~10 L. Samples from the outflow were taken initially every day and at later stages at the end of each cycle to determine the concentration of Ni, Co, Fe, Ca, Al, Mg, and Mn by Atomic Absorption Spectroscopy (AAS). pH and Eh measurements were carried out with the use of a WTW pH 7110 inoLab pH/Eh meter. Acid consumption was determined by calculating the acid strength of the initial and the solution at the end of each cycle by a titrimetric method, as proposed by Quaicoe et al. [19]. The residues were dried at 80 °C for 1 day before characterization with the use of analytical techniques.

All leaching tests were done twice and metal concentration values in PLS given in this paper are averages; it is noted that the difference in concentration for all metals was less than 1% in all cases.

Table 1. Experimental details.

Test Number	Strength of H_2SO_4 Solution (M)	Na_2SO_3 conc. (g/L)	Weight Loss (%)
1	0.5		10.5
2	1.5		8.8
3	1.5	20	8.6
4	1.5	30	8.5

3. Results and Discussion

3.1. Ore Characterization

The product of the crusher was dry sieved using a series of screens with an aperture ratio of 2 and the determined particle size distribution is presented in Figure 1. The results showed that the 80% passing size (d_{80}) of the raw material was 8.8 mm. The chemical composition of each size fraction is shown in Table 2. It is seen that grinding is not selective and does not result in noticeable nickel upgrade in the finer fractions. The XRD pattern of LAI and selected leaching residues is shown in Figure 2. Chemical and mineralogical analyses showed that the ore is of limonitic type (Fe content > 32% and MgO < 10%). The main mineral phases identified in the ore are hematite (Fe_2O_3), goethite ($FeO(OH)$), and quartz (SiO_2), while calcite ($CaCO_3$) and cryptomelane (KMn_8O_{16}) are minor phases. The main Ni-bearing phases of the ore are chromite ($Cr_2O_3 \cdot NiO$), clinochlore ($(Mg,Fe)_5Al(Si_3Al)O_{10}(OH)_8$), and willemseite ($(Ni,Mg)_3Si_4O_{10}(OH)_2$). The mineralogy of the laterites used in the present study is quite similar to the mineralogy of other limonitic laterites originating either from the same area or from other countries and investigated in earlier studies [36,37].

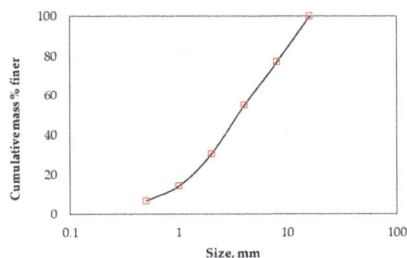

Figure 1. Particle size distribution of laterite from Agios Ioannis (LAI) used in column leaching tests.

Table 2. Chemical composition (% w/w) of LAI size fractions.

Size Fraction (mm)	Ni	Co	Fe₂O₃	SiO₂	Al₂O₃	CaO	MgO	Cr₂O₃	K₂O	TiO₂	MnO	LOI	Sum
+8	0.66	0.031	49.25	27.66	4.79	0.63	2.98	3.65	0.47	0.60	0.44	7.51	98.67
+4	0.49	0.035	51.75	27.05	4.94	0.53	3.11	2.59	0.51	0.54	0.32	7.85	99.72
+2	0.56	0.033	51.40	27.56	4.75	0.55	2.91	2.68	0.56	0.58	0.32	7.37	99.27
+1	0.61	0.031	49.27	28.12	5.00	0.76	3.05	2.52	0.69	0.58	0.32	8.00	98.95
+0.5	0.55	0.032	49.44	28.73	5.01	0.75	2.95	2.29	0.59	0.57	0.31	8.36	99.58
−0.5	0.67	0.028	41.20	32.46	5.87	1.73	4.02	2.66	0.74	0.53	0.37	9.14	99.42
Total	0.58	0.032	49.81	27.98	4.94	0.69	3.07	2.82	0.57	0.57	0.35	7.81	99.22

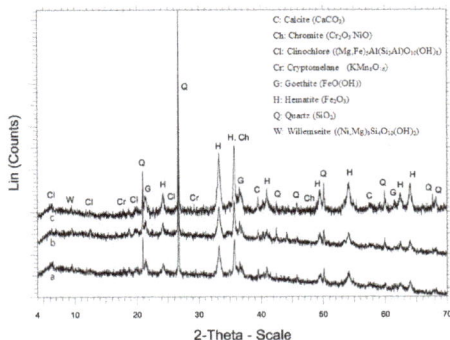

Figure 2. XRD patterns of (**a**) LAI, (**b**) residue after leaching with 1.5 M H₂SO₄, and (**c**) residue after leaching with 1.5 M H₂SO₄ and 20 g/L Na₂SO₃.

The XRD patterns of the leaching residues, resulting from H_2SO_4 leaching with and without the use of Na_2SO_3, are shown in Figure 2b,c. The main difference in these patterns is the higher intensity of hematite peaks, especially when Na_2SO_3 was added in the leaching solution. No gypsum was detected in the residues, as in the case of reactor leaching of the same ore [38], due to the conditions prevailing in columns, i.e., low temperature and high H_2SO_4 concentration [39]; some gypsum may be formed but not enough to be detected by XRD.

The FTIR spectra of LAI and its residues after column leaching with 1.5 M H_2SO_4 and 1.5 M H_2SO_4 with the addition of 20 g/L Na_2SO_3 are shown in Figure 3. In all spectra, three regions can be identified, including several low-, mid-, and higher-frequency weaker bands. The band seen at 466 cm^{-1} in LAI (Figure 3a), which was slightly shifted to 458 cm^{-1} in leaching residues (Figure 3b,c), can be attributed to the bending motions of the Al- and Si-containing phases and the formation of Fe phases. The peak at 784 cm^{-1} in LAI is mainly due to Si–O–Si symmetric stretching of bridging oxygen between SiO_4 tetrahedra. The band seen at 1010 cm^{-1} in LAI, which has been shifted to higher values (1086 cm^{-1}) in the leaching residues (Figure 3b,c), is attributed to asymmetric stretching vibrations of the silicate tetrahedral network [40]. The weak band shown at 1626 cm^{-1} in LAI, which becomes more intense in the leaching residues and especially in the one obtained after leaching with the addition of Na_2SO_3, is probably due to –OH bending vibrations. The small band seen around 2360 cm^{-1} in LAI is mainly associated with the infrared band position of HCO_3^- ions. The band shown between 2892 and 2922 cm^{-1} is due to hydrocarbon stretches. The broad band seen at 3416 cm^{-1} only in leaching residues (Figure 3b,c) corresponds to –OH stretching vibrations [41]. The band seen at 3416 cm^{-1} only in leaching residues may be assigned to Fe^{3+}–OH–Fe^{3+} stretching and deformation vibrations [42,43].

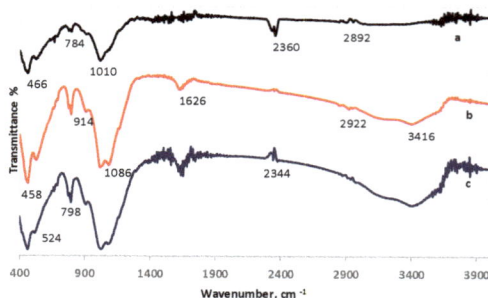

Figure 3. FTIR spectra of (**a**) LAI, (**b**) residue after column leaching with 1.5 M H_2SO_4, and (**c**) residue after column leaching with 1.5 M H_2SO_4 with 20 g/L Na_2SO_3.

The behavior of LAI and its residues during heating was investigated through DSC/TG analysis (Figure 4). The peaks at 110 °C and 140 °C in all samples (Figure 4a–c) are due to loss of free water. The peaks seen between 330 and 350 °C, also in all samples, are associated with the removal of crystalline water and OH^- group from the structure of goethite as well as the formation of hematite [44]. This peak was shifted to a higher temperature (350 °C) in the residue obtained after leaching with the use of Na_2SO_3, probably due to the higher degree of crystallinity of goethite [45]. The peaks at 530 and 570 °C (Figure 4a,b) are due to phase transformations of iron, silica, and calcium compounds; these peaks are almost invisible in the leaching residue obtained after leaching with the addition of Na_2SO_3. The exothermic peak at 800 °C (Figure 4c) is associated with recrystallization of forsterite (Mg_2SiO_4) and transformation of $NiSO_4$ and $CoSO_4$ to NiO and CoO, respectively [45,46]. Concerning TG analysis, the total weight loss of LAI was almost 15% (Figure 4a) and increased to 40% and 63% for the residues obtained after leaching with 1.5 M H_2SO_4 in the absence/presence of 20 g/L Na_2SO_3 (Figure 4b,c, respectively).

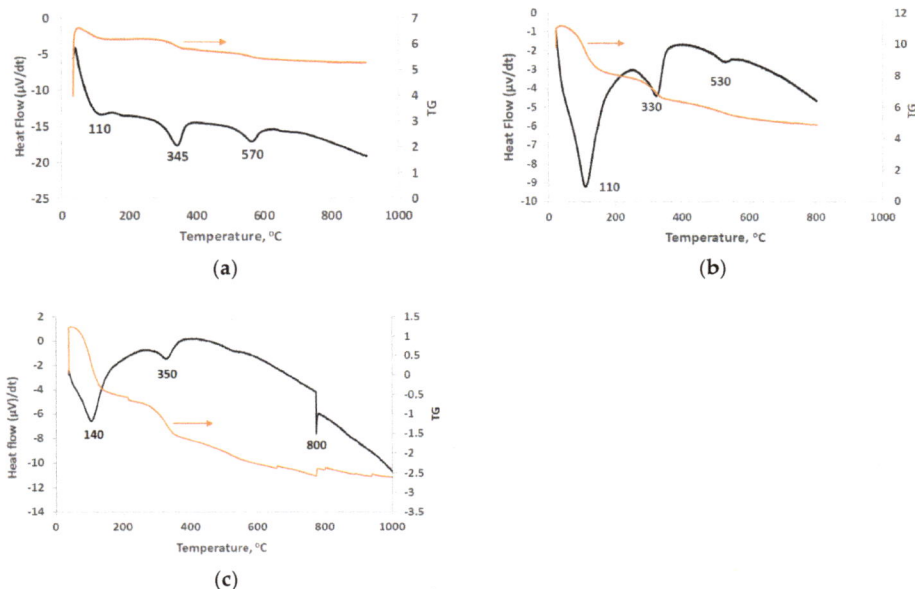

Figure 4. DSC/TG curves of (**a**) LAI, (**b**) residue after leaching with 1.5 M H$_2$SO$_4$, and (**c**) residue after leaching with 1.5 M H$_2$SO$_4$ with 20 g/L Na$_2$SO$_3$.

3.2. Leaching Efficiency

Figures 5–11 show the evolution of metal extractions (Ni, Co, Fe, Mg, Al, Ca, and Mn) versus time (in days) during leaching for the different column tests. It is seen that the extraction percentage for each metal increases with time but the leaching rate differs and depends on the conditions prevailing in each column.

The extraction of Ni and Co when 0.5 M H$_2$SO$_4$ is used is rather low and does not exceed 40 and 55%, respectively; this is also due to the fact that some acid is consumed for the solubilization of Mg and Ca compounds present in the ore. Increase of the acid strength to 1.5 M improves Ni and Co extractions to 60.2% and 59.0% for Ni and Co, respectively. Further increase of the Ni and Co extractions is shown when 20 g/L of Na$_2$SO$_3$ is added to the leaching solution, reaching 73.5% and 84.1%, respectively. Increase of the addition of Na$_2$SO$_3$ to 30 g/L only marginally affects extraction of Ni and Co: Ni extraction increases from 73.5 to 76.0%, whereas Co extraction decreases from 84.1 to 75.5%; this decrease may be due to the formation of neophases and their precipitation in the residues.

The extraction of Ni and Co is increased in the presence of Na$_2$SO$_3$ as a result of the following reactions. In the presence of H$_2$SO$_4$, Na$_2$SO$_3$ reacts with H$^+$ ions to form H$_2$SO$_3$, as shown in Reaction (1). Then, H$_2$SO$_3$ is dissociated according to Reaction (2).

$$Na_2SO_3 + H_2SO_4 = Na_2SO_4 + H_2SO_3 \tag{1}$$

$$H_2SO_3 = H^+ + HSO_3^- = H_2O + SO_2\,(aq) \tag{2}$$

Goethite reacts according to the following reactions ((3) and (4)) and, thus, nickel associated to this mineral phase is extracted.

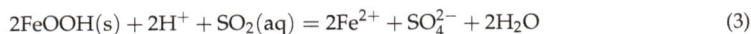

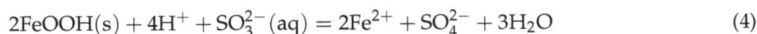

$$2FeOOH(s) + 2H^+ + SO_2(aq) = 2Fe^{2+} + SO_4^{2-} + 2H_2O \tag{3}$$

$$2FeOOH(s) + 4H^+ + SO_3^{2-}(aq) = 2Fe^{2+} + SO_4^{2-} + 3H_2O \tag{4}$$

The $SO_2(aq)$ generated from Reaction (2) will lower the potential of Reactions (3) and (4) and accelerate Fe extraction and Ni liberation from goethite. An in-depth analysis of the leaching chemistry of laterites with the use of H_2SO_4 in the presence of H_2SO_3 is given in a recent paper [24].

It is seen from Reactions (3) and (4) that less acid is required for the initial dissolution of goethite, namely 0.5 or 1 mole of H_2SO_4 per mole of goethite, compared with the consumption of acid without the presence of H_2SO_3, which is 1.5 moles per mole of goethite [47], as shown in Reaction (5).

$$FeOOH(s) + 3H^+ = Fe^{3+} + 2H_2O \qquad (5)$$

Iron extraction is, in general, very low; however, the increase of the acid strength and the addition of Na_2SO_3 result in an increase of Fe extraction which does not exceed 8.2%. It is known that Fe extraction during laterite leaching is the main parameter that defines the selectivity of the process. In this study, the low Fe extraction is due to the room temperature used, the use of columns instead of stirred reactors, the coarser feed size used in comparison to the size used in agitated leaching, and the precipitation of iron compounds during leaching. In the study carried out by Luo et al. [24] and involving leaching of a limonitic laterite with liquid/solid (L/S) ratio 10:1 and 12% (w/w) H_2SO_4 at 90 °C with or without the use of Na_2SO_3, Fe extraction after 6 h was much higher and reached 70% and 55%, respectively. It is known that in the conditions prevailing in our experiments, iron can be removed through the formation of hydronium jarosites and ferrihydrites. The formation of hydronium jarosites is favored at very low pH values, i.e., pH = 0.2, whereas hematite can only be formed when tests are carried out in much higher temperatures [44].

The maximum extractions of the other three elements that affect the selectivity of leaching, namely, Mg, Al, and Ca, do not exceed 40.2, 23.3, and 51.0%, respectively, in the optimum conditions (1.5 M H_2SO_4, 20 g/L Na_2SO_3). The low extraction percentages of these elements during laterite leaching with the use of H_2SO_4 is an advantage of column leaching, since the solubility of metal sulphates in water increases with the increase of temperature; on the other hand, inverse solubilities may be noticed at elevated temperatures. It is known that limonitic laterites have lower content of Ca and Mg and higher content of Al compounds compared with saprolitic ores; this affects the selectivity of the process, especially when leaching is carried out in stirred reactors and higher temperature [38].

pH values were very low, close to zero, in all column leaching tests; the only exception was noted in the use of 0.5 M H_2SO_4, where pH increased with time to 0.36 at the end of the test. On the other hand, final Eh values decreased slightly with the addition of Na_2SO_3 and varied between 410 and 370 mV when 0.5 M H_2SO_4 and 1.5 M H_2SO_4 with 30 g/L Na_2SO_3 were used, respectively.

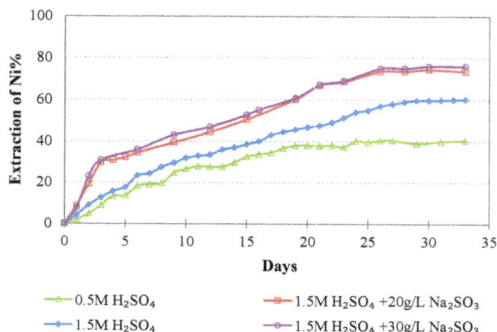

Figure 5. Evolution of % Ni extraction vs time during LAI leaching for different tests.

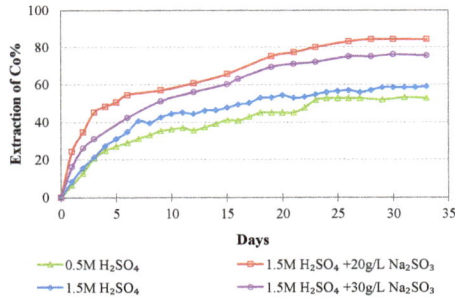

Figure 6. Evolution of % Co extraction vs time during LAI leaching for different tests.

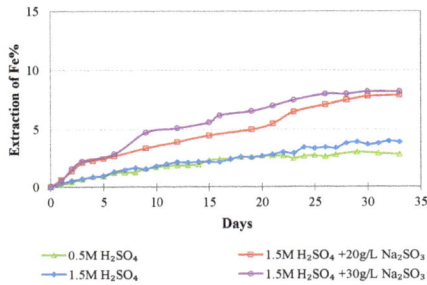

Figure 7. Evolution of % Fe extraction vs time during LAI leaching for different tests.

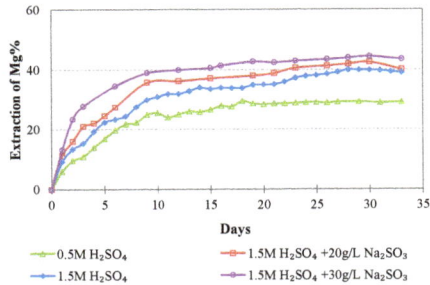

Figure 8. Evolution of % Mg extraction vs time during LAI leaching for different tests.

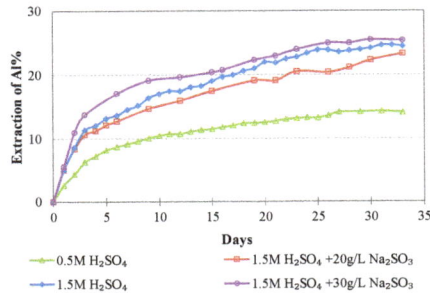

Figure 9. Evolution of % Al extraction vs time during LAI leaching for different tests.

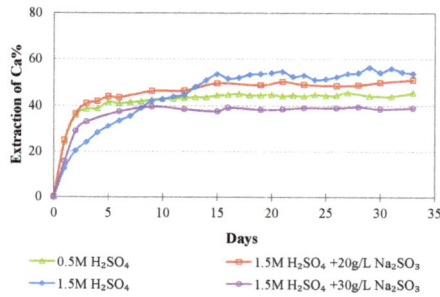

Figure 10. Evolution of % Ca extraction vs time during LAI leaching for different tests.

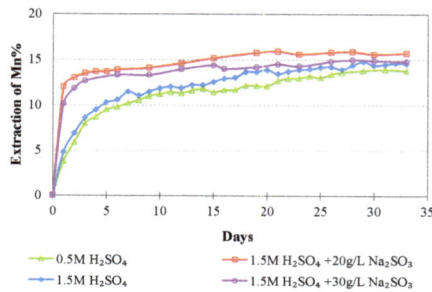

Figure 11. Evolution of % Mn extraction vs time during LAI leaching for different tests.

Table 3 presents the concentration of Ni and other main elements in the PLS in order to assess the selectivity of leaching in each case by comparing the concentration ratios of Ni/Fe, Ni/Mg, Ni/Ca, and Ni/Al. The comparisons done do not include the use of 0.5 M H_2SO_4, which does not result in high Ni and Co extractions, or the use of 30g/L Na_2SO_3, which does not seem to improve the efficiency of leaching compared with the addition of 20 g/L.

It is seen from the results that the addition of 20 g/L Na_2SO_3 in the leaching medium improves the selectivity of leaching compared with the test which involves only the addition of 1.5 M H_2SO_4, as the ratios Ni/Mg, Ni/Ca, and Ni/Al increase by almost 18, 30, and 28%, respectively. The only exception is the ratio Ni/Fe which decreases by almost 40%; this is mainly due to the much higher concentration of Fe in the PLS, which increases from 1.56 to 2.88 g/L, compared with the concentration of Ni, which also increases from 0.40 to 0.45 g/L. It is worth mentioning that the purification of PLS for the recovery of Ni and Co as well as of other useful elements including Mg, has been extensively investigated [48–50].

Table 3. Comparison of pregnant leach solution (PLS) quality and selectivity of each leaching medium.

Test No.	Ni	Co	Fe	Mg	Ca	Al	Ni/Fe	Ni/Mg	Ni/Ca	Ni/Al
	(mg/L)									
1	266	19.0	1105	611	263	422	0.24	0.43	1.01	0.63
2	399	21.5	1556	829	314	741	0.26	0.48	1.27	0.54
3	445	28.0	2878	776	271	645	0.15	0.57	1.64	0.69
4	450	24.7	2922	823	203	687	0.15	0.55	2.22	0.66

Table 4 compares the results of this study with those of other leaching studies pertinent to the leaching of limonitic laterites carried out over the last 25 years, as derived from an extensive literature search. Some data presented in this table were derived after calculations done by the authors of this study.

Minerals **2018**, *8*, 377

Table 4. Comparison of results of various column leaching studies.

Ore Type	Ni %	Weight (kg ore)	Ore Size (mm)	H_2SO_4 (N)	Duration (days)	Flow Rate (L/day)	Extraction (%)			Ni/Fe in PLS	Acid Consumption (kg/tonne ore)	Ref.
							Ni	Fe	Co			
L	0.73	8–22	<6.75	1	90	4.5	80	14.0	65	0.5	380	[31]
L	1.05	6.6	<15	3	10	4.0	60	8.1	45	0.2	284	[32]
SG	1.01		<15 (a)	4	102	2.3	71		45		684	[4,11, 51]
	1.01	5	<2 (a)	4	104	2.3	81		67		569	
	1.25		<0.038 (a)	4	101	2.3	89		80		541	
G	0.96	5	<15 (a)	4	106	2.3	55				453	[19]
G	0.96	5	<2 (a)	4	100	2.3	62.7	48.5	55.6	0.03	525	
L	0.58	1	<16	3	33	3.0	60.2	3.9	59.0	0.26	764.4	This study
L	0.58	1	<16	3 + (ss)	33	3.0	73.5	7.9	84.1	0.15	688.8	

L: Limonitic laterite, SG: Siliceous goethitic laterite, G: Goethitic laterite, (ss): Sodium sulfite, (a): use of agglomerated ore.

Agatzini-Leonardou and Dimaki [31] performed large column leaching tests on low-grade limonitic laterites using 0.5 M H_2SO_4. Extractions of 80% for Ni and 60% for Co were achieved after 90 days of leaching, while acid consumption was around 380 kg H_2SO_4/tonne ore. In another leaching study involving different types of laterites, Agatzini-Leonardou and Zafiratos [32] noted that the high calcite content in the ore reduces the permeability of the heap and adversely affects leaching. The recoveries achieved for the limonitic type after 10 days of leaching reached 60% for Ni and 45% for Co. The Fe/Ni ratio was reduced from 37:1 in the limonitic ore to 5:1 in the PLS. The acid consumption was around 284 kg H_2SO_4/tonne ore. Other studies investigated the leaching behavior of goethitic (G) and siliceous goethitic laterites (SG) [4,11,51] and found that the particle size of the agglomerated ore (<15, <2, and <0.038 mm) had a significant impact on Ni and Co extractions which increased with decreasing feed size, also resulting in reduction of acid consumption. Quite similar results were obtained in the study by Quaicoe et al. [19] who performed column leaching tests using <2 mm agglomerated goethitic ore. Extractions of 62.7% for Ni and 55.6% for Co were achieved after 100 days of leaching. The most interesting results of the present study, which involved leaching of a very poor laterite (0.58% Ni), are the good extractions of Ni and Co, which exceed 70 and 80%, respectively, when Na_2SO_3 is added in the leaching solution; the very low extraction of Fe, which remains below 8%; and the short duration (33 days). It is worth mentioning that the higher acid consumption noted in the present study may be decreased if bigger quantities of ore are leached in larger columns with the use of the same volume of leaching solution; this is an issue that is currently under study.

Leaching results of the present and earlier studies indicated that the variability in the efficiency of column leaching of limonitic laterites depends on the complexity of the ore and the strong bonding of Ni with iron minerals, compared with clay-like and high-magnesium ores (i.e., saprolitic) [52,53]. As mentioned in an excellent recent study [54], a full and comprehensive mineralogical characterization of the ore is necessary, especially at the mineral grain scale, in order to fully understand the overall leach behavior. The same authors also proved that the experimental scale noticeably affects the leaching behavior of ores at ambient conditions and proposed that the use of synchrotron methods is required to ascertain the coordination chemistry of key elements within both the fresh ore and leaching residues, in order to elucidate mineral formation and identify the major leaching mechanisms.

Other crucial factors that affect the economics of the process are the efficient control of Fe and other secondary elements' co-extraction and the possibility to recover saleable byproducts, such as MgO and MnO_2, during PLS purification. Finally, in order to minimize the environmental impacts and improve the environmental footprint of the process, the leaching residues may be alkali activated for the production of inorganic polymers. By taking into account that the residues contain sufficient amounts of SiO_2 and Al_2O_3, the residues may be transformed into solid matrices comprising a Si–O–Al network and exhibiting beneficial physical and chemical properties, including high early strength, low shrinkage and porosity, as well as high fire and corrosion resistance. These new materials may find several applications, mainly in the construction sector [55,56].

4. Conclusions

The results of the present study are considered very promising and confirm the potential of column leaching of very low-grade Greek limonitic laterites with the use of 1.5 M H_2SO_4 and the addition of 20 g/L Na_2SO_3. These types of laterites cannot be treated economically with the use of pyrometallurgical techniques due to the extremely high energy requirements. The experimental results indicated high extractions of Ni and Co, namely, 73.5 and 84.1%, respectively, while the extractions of Fe, Mg, Al, and Ca were quite low, namely, 7.9, 40.2, 23.3, and 51.0%, respectively. The low extractions of the unwanted elements are the result of the ore granulometry and the low temperature used.

Further studies are required though to optimize leaching efficiency with the use of larger columns and ore quantities in order to enrich the quality of the PLS, improve selectivity, and reduce acid consumption. The use of agglomerates and the presence of catalysts to accelerate the leaching rate are also issues which are currently under study. In addition, a full and comprehensive mineralogical characterization of the ore and the leaching residues is necessary in order to fully elucidate the overall leach behavior and identify the major leaching mechanisms.

In order to minimize the environmental impacts and improve the environmental footprint of the process, the leaching residues can be alkali activated for the production of inorganic polymers and their potential use in the construction sector. This subject is also under study.

Author Contributions: K.K. designed the experiments, critically analyzed results, and reviewed the paper. E.P. performed a literature search, carried out experiments and analytical techniques, analyzed data, and wrote a first draft of the paper. O.P. carried out experiments and analyzed data. A.K. carried out analytical techniques and analyzed data.

Acknowledgments: The authors would like to acknowledge the financial support of European Commission in the frame of Horizon 2020 project "Metal recovery from low-grade ores and wastes", www.metgrowplus.eu, Grant Agreement n° 690088.

Conflicts of Interest: The authors declare no conflicts of interest.

References

1. Kursunoglu, S.; Kaya, M. Atmospheric pressure acid leaching of Caldag lateritic nickel ore. *Int. J. Miner. Process.* **2016**, *150*, 1–8. [CrossRef]
2. Zeng, X.; Zhan, C.; Lu, J.; Amine, K. Stabilization of a high capacity and high-power nickel-based cathode for Li-Ion batteries. *Chem* **2018**, *4*, 690–704. [CrossRef]
3. Mohammadreza, F.; Mohammad, N.; Ziaeddin, S.S. Nickel extraction from low grade laterite by agitation leaching at atmospheric pressure. *Int. J. Min. Sci. Technol.* **2014**, *24*, 543–548. [CrossRef]
4. Nosrati, A.; Quast, K.; Xu, D.; Skinner, W.; Robinson, D.J.; Addai-Mensah, J. Agglomeration and column leaching behaviour of nickel laterite ores: Effect of ore mineralogy and particle size distribution. *Hydrometallurgy* **2014**, *146*, 29–39. [CrossRef]
5. MacCarthy, J.; Nosrati, A.; Skinner, W.; Addai-Mensah, J. Atmospheric acid leaching mechanisms and kinetics and rheological studies of a low grade saprolitic nickel laterite ore. *Hydrometallurgy* **2016**, *160*, 26–37. [CrossRef]
6. Dalvi, A.D.; Bacon, W.G.; Osborne, R.C. The past and the future of nickel laterites. In Proceedings of the Prospectors & Developers Association of Canada (PDAC) 2004 International Convention, Trade Show & Investors Exchange, Toronto, ON, Canada, 7–10 March 2004.
7. Girgin, I.; Obut, A.; Ucyildiz, A. Dissolution behaviour of a Turkish lateritic nickel ore. *Miner. Eng.* **2011**, *24*, 603–609. [CrossRef]
8. Onal, M.A.R.; Topkaya, Y.A. Pressure acid leaching of Çaldağ lateritic nickel ore: An alternative to heap leaching. *Hydrometallurgy* **2014**, *142*, 98–107. [CrossRef]
9. Takeda, O.; Lu, X.; Miki, T.; Nakajima, K. Thermodynamic evaluation of elemental distribution in a ferronickel electric furnace for the prospect of recycling pathway of nickel. *Resour. Conserv. Recycl.* **2018**, *133*, 362–368. [CrossRef]
10. Wang, Z.; Chu, M.; Liu, Z.; Wang, H.; Zhao, W.; Gao, L. Preparing Ferro-Nickel Alloy from Low-Grade Laterite Nickel Ore Based on Metallized Reduction–Magnetic Separation. *Metals* **2017**, *7*, 313. [CrossRef]
11. Quast, K.; Xu, D.; Skinner, W.; Nosrati, A.; Hilder, T.; Robinson, D.J.; Addai-Mensah, J. Column leaching of nickel laterite agglomerates: Effect of feed size. *Hydrometallurgy* **2013**, *134*, 144–149. [CrossRef]
12. Oxley, A.; Smith, M.E.; Caceres, O. Why heap leach nickel laterites? *Miner. Eng.* **2016**, *88*, 53–60. [CrossRef]
13. Alibhai, K.A.K.; Dudeney, A.W.L.; Leak, D.J.; Agatzini, S.; Tzeferis, P. Bioleaching and bioprecipitation of nickel and iron from laterites. *FEMS Microbiol. Rev.* **1993**, *11*, 87–95. [CrossRef]
14. du Plessis, C.A.; Slabbert, W.; Hallberg, K.B.; Johnson, D.B. Ferredox: A biohydrometallurgical processing concept for limonitic nickel laterites. *Hydrometallurgy* **2011**, *109*, 221–229. [CrossRef]
15. Watling, H.R.; Elliot, A.D.; Fletcher, H.M.; Robinson, D.J.; Sully, D.M. Ore mineralogy of nickel laterites: Controls on processing characteristics under simulated heap-leach conditions. *Aust. J. Earth Sci.* **2011**, *58*, 725–744. [CrossRef]
16. Duyvesteyn, W.P.C.; Liu, H.; Davis, M.J. Heap Leaching of Nickel Containing Ore. U.S. Patent 6,312,500 B1, 6 November 2001.

17. Horizonte, S.G.; Horizonte, D.O. Process for Extraction of Nickel, Cobalt, and Other Base Metals from Laterite Ores by Using Heap Leaching and Product Containing Nickel, Cobalt, and Other Metals from Laterite Ores. European Patent EP 1790739 A1, 21 June 2007.

18. Elliot, A.; Fletcher, H.; Li, J.; Watling, H.; Robinson, D.J. Heap leaching of nickel laterites—A challenge and an opportunity. In *Hydrometallurgy of Nickel and Cobalt 2009, Proceeding of the 39th Annual Hydrometallurgy Meeting Held in the Conjunction with the 48th Annual Conference of Metallurgists, Sudbury, ON, Canada, 23–26 August 2009*; Budac, J.J., Fraser, R., Mihaylov, I., Papangelakis, V.G., Robinson, D.J., Eds.; Canadian Institute of Mining, Metallurgy and Petroleum: Montreal, QC, Canada, 2009; pp. 537–549.

19. Quaicoe, I.; Nosrati, A.; Skinner, W.; Addai, M.J. Agglomeration and column leaching behaviour of goethitic and saprolitic nickel laterite ores. *Miner. Eng.* **2014**, *65*, 1–8. [CrossRef]

20. Das, G.K.; de Lange, J.A.B. Reductive atmospheric acid leaching of West Australian smectitic nickel laterite in the presence of sulphur dioxide and copper (II). *Hydrometallurgy* **2011**, *105*, 264–269. [CrossRef]

21. Lee, H.Y.; Kim, S.G.; Oh, J.K. Electrochemical leaching of nickel from low-grade laterites. *Hydrometallurgy* **2005**, *77*, 263–268. [CrossRef]

22. Senanayake, G.; Das, G.K. A comparative study of leaching kinetics of limonitic laterite and synthetic iron oxides in sulfuric acid containing sulfur dioxide. *Hydrometallurgy* **2004**, *72*, 59–72. [CrossRef]

23. Youzbashi, A.A.; Dixit, S.G. Leaching of nickel from supported nickel waste catalyst using aqueous sulfur dioxide solution. *Metall. Trans. B* **1991**, *22*, 775–781. [CrossRef]

24. Luo, J.; Li, G.; Rao, M.; Peng, Z.; Zhang, Y.; Jiang, T. Atmospheric leaching characteristics of nickel and iron in limonitic laterite with sulfuric acid in the presence of sodium sulfite. *Miner. Eng.* **2015**, *78*, 38–44. [CrossRef]

25. Zevgolis, E.; Zografidis, C.; Halikia, I. The reducibility of the Greek nickeliferous laterites: A review. *Miner. Process. Extract. Metall.* **2010**, *119*, 9–17. [CrossRef]

26. Bartzas, G.; Komnitsas, K. Life cycle assessment of FeNi production in Greece: A. case study. *Resour. Conserv. Recycl.* **2015**, *105*, 113–122. [CrossRef]

27. Komnitsas, K. Kinetic Study of Laterite Leaching with Sulfuric Acid at Atmospheric Pressure. Bachelor's Thesis, National Technical University of Athens, Athens, Greece, 1983. (In Greek)

28. Komnitsas, K. High Pressure acid Leaching of Laterites. Ph.D. Thesis, National Technical University of Athens, Athens, Greece, 1988. (In Greek)

29. Kontopoulos, A.; Komnitsas, K. Sulphuric acid Pressure Leaching of Low-Grade Greek Laterites. In Proceedings of the 1st International Symposium on Hydrometallurgy, Beijing, China, 12–15 October 1988; Zheng, Y.L., Xu, J.Z., Eds.; Pergamon Press: Oxford, UK, 1988; pp. 140–144.

30. Panagiotopoulos, N.; Agatzini, S.; Kontopoulos, A. Extraction of nickel and cobalt from serpentinic type laterites by atmospheric pressure sulphuric acid leaching. In Proceedings of the Technical Sessions at the 115th TMS-AIME Annual Meeting, New Orleans, LA, USA, 2–6 March 1986; p. A86-30.

31. Agatzini-Leonardou, S.; Dimaki, D. Heap leaching of poor nickel laterites by sulphuric acid at ambient temperature. In *Hydrometallurgy '94*; Springer: Dordrecht, The Netherlands, 1994; pp. 193–208.

32. Agatzini-Leonardou, S.; Zafiratos, I.G. Beneficiation of a Greek serpentinic nickeliferous ore part II. Sulphuric acid heap and agitation leaching. *Hydrometallurgy* **2004**, *74*, 267–275. [CrossRef]

33. Agatzini-Leonardou, S.; Zafiratos, J.G.; Spathis, D. Beneficiation of a Greek serpentinic nickeliferous ore—Part I: Mineral processing. *Hydrometallurgy* **2004**, *74*, 259–265. [CrossRef]

34. Komnitsas, K.; Bartzas, G.; Paspaliaris, I. Inorganic contaminant fate assessment in zero-valent iron treatment walls. *Environ. Forensics* **2006**, *7*, 207–217. [CrossRef]

35. Komnitsas, K.; Bartzas, G.; Paspaliaris, I. Modeling of reaction front progress in fly ash permeable reactive barriers. *Environ. Forensics* **2006**, *7*, 219–231. [CrossRef]

36. Alevizos, G. Mineralogy, Geochemistry and Genesis of the Sedimentary Nickeliferous Iron-Ores of Locris (Central Greece). Ph.D. Thesis, Technical University of Crete, Chania, Greece, 1997.

37. Li, G.; Rao, M.; Jiang, T.; Huang, Q.; Peng, Z. Leaching of limonitic laterite ore by acidic thiosulfate solution. *Miner. Eng.* **2011**, *24*, 859–863. [CrossRef]

38. Mystrioti, C.; Papassiopi, N.; Xenidis, A.; Komnitsas, K. Comparative Evaluation of Sulfuric and Hydrochloric Acid Atmospheric Leaching for the Treatment of Greek Low Grade Nickel Laterites. In *Extraction 2018, Proceedings of the First Global Conference on Extractive Metallurgy, The Minerals, Metals & Materials Series, Ottawa, QC, Canada, 26–29 August 2018*; Davis, B.R., Moats, M.S., Wang, S., Gregurek, D., Kapusta, J., Battle, T.P., Schlesinger, M.E., Flores, G.R.A., Jak, E., Goodall, G., et al., Eds.; Springer: Cham, Switzerland, 2018; pp. 1753–1764.

39. Farrah, H.E.; Lawrance, G.A.; Wanless, E.J. Solubility of calcium sulfate salts in acidic manganese sulfate solutions from 30 to 105 °C. *Hydrometallurgy* **2007**, *86*, 13–21. [CrossRef]

40. Rinaudo, C.; Gastaldi, D.; Belluso, E. Characterization of chrysotile, antigorite, and lizardite by FT-Raman spectroscopy. *Can. Mineral.* **2003**, *41*, 883–890. [CrossRef]

41. Madejov, J.; Janek, M.; Komadel, P.; Herbert, H.J.; Moog, H.C. FTIR analyses of water in MX-80 bentonite compacted from high salinary salt solution systems. *Appl. Clay Sci.* **2002**, *20*, 255–271. [CrossRef]

42. Petit, S.; Decarreau, A. Hydrothermal (200 °C) synthesis and crystal chemistry of iron rich kaolinites. *Clay Miner.* **1990**, *25*, 181–196. [CrossRef]

43. Delineau, T.; Allard, T.; Muller, J.P.; Barres, O.; Yvon, J.; Cases, J.M. FTIR reflectance vs. EPR studies of structural iron in kaolinites. *Clays Clay Miner.* **1994**, *42*, 308–320. [CrossRef]

44. Stopić, S.; Friedrich, B.; Fuchs, R. Sulphuric acid leaching of the Serbian nickel lateritic ore. *Erzmetall* **2003**, *56*, 198–203.

45. Ma, B.; Yang, W.; Pei, Y.; Wang, C.; Jin, B. Effect of activation pretreatment of limonitic laterite ores using sodium fluoride and sulfuric acid on water leaching of nickel and cobalt. *Hydrometallurgy* **2017**, *169*, 411–417. [CrossRef]

46. Luo, W.; Feng, Q.; Ou, L.; Zhang, G.; Chen, Y. Kinetics of saprolitic laterite leaching by sulphuric acid at atmospheric pressure. *Miner. Eng.* **2010**, *23*, 458–462. [CrossRef]

47. Georgiou, D.; Papangelakis, V. Sulphuric acid pressure leaching of a limonitic laterite: Chemistry and kinetics. *Hydrometallurgy* **1998**, *49*, 23–46. [CrossRef]

48. Agatzini-Leonardou, S.; Tsakiridis, P.E.; Oustadakis, P.; Karidakis, T.; Katsiapi, A. Hydrometallurgical process for the separation and recovery of nickel from sulphate heap leach 271 liquor of nickeliferous laterite ores. *Miner. Eng.* **2009**, *22*, 1181–1192. [CrossRef]

49. Mihaylov, I.; Krause, E.; Okita, Y.; Perraud, J.J. The Development of a novel hydrometallurgical process for nickel and cobalt recovery from Goro laterite ore. *CIM Bull.* **2000**, *93*, 124–130.

50. Ritcey, G.M.; Hayward, N.L.; Salinovich, T. The Recovery of Nickel and Cobalt from Lateritic Ores. Australian Patent AU-B-40890, 18 July 1996.

51. Xu, D.; Liu, L.X.; Quast, K.; Addai-Mensah, J.; Robinson, D.J. Effect of nickel laterite agglomerate properties on their leaching performance. *Adv. Powder Technol.* **2013**, *24*, 750–756. [CrossRef]

52. McDonald, R.G.; Whittington, B.I. Atmospheric acid leaching of nickel laterites review. Part I. Sulphuric acid technologies. *Hydrometallurgy* **2008**, *91*, 35–55. [CrossRef]

53. MacCarthy, J.; Nosrati, A.; Skinner, W.; Addai-Mensah, J. Temperature Influence of Atmospheric Acid Leaching Behaviour of Saprolitic Nickel Laterite Ore. In Proceedings of the Chemeca 2013: Challenging Tomorrow, Brisbane, Australia, 29 September–2 October 2013; Wang, L., Ed.; Engineers Australia: Brisbane, Australia, 2013; pp. 439–443.

54. Hunter, H.M.A.; Herrington, R.J.; Oxley, E.A. Examining Ni-laterite leach mineralogy & chemistry—A holistic multi-scale approach. *Miner. Eng.* **2013**, *54*, 100–119.

55. Komnitsas, K.; Zaharaki, D. Geopolymerisation: A review and prospects for the minerals industry. *Miner. Eng.* **2007**, *20*, 1261–1277. [CrossRef]

56. Zaharaki, D.; Komnitsas, K. Valorization of construction and demolition (C&D) and industrial wastes through alkali activation. *Constr. Build. Mater.* **2016**, *121*, 686–693.

minerals

MDPI

Article

Iron Control in Atmospheric Acid Laterite Leaching

Ville Miettinen [1,*], Jarno Mäkinen [1], Eero Kolehmainen [2], Tero Kravtsov [2] and Lotta Rintala [1]

[1]   VTT Technical Research Centre of Finland Ltd., P.O. Box 1000, FI-02044 VTT Espoo, Finland
[2]   Outotec (Finland) Oy, P.O. Box 69, FI-28101 Pori, Finland
*   Correspondence: ville.miettinen@vtt.fi; Tel.: +358-40-833-8096

Received: 7 May 2019; Accepted: 28 June 2019; Published: 30 June 2019

Abstract: Iron control in the atmospheric acid leaching (AL) of nickel laterite was evaluated in this study. The aim was to decrease acid consumption and iron dissolution by iron precipitation during nickel leaching. The combined acid leaching and iron precipitation process involves direct acid leaching of the limonite type of laterite followed by a simultaneous iron precipitation and nickel leaching step. Iron precipitation as jarosite is carried out by using nickel containing silicate laterite for neutralization. Acid is generated in the jarosite precipitation reaction, and it dissolves nickel and other metals like magnesium from the silicate laterite. Leaching tests were carried out using three laterite samples from the Agios Ioannis, Evia Island, and Kastoria mines in Greece. Relatively low acid consumption was achieved during the combined precipitation and acid leaching tests. The acid consumption was approximately 0.4 kg acid per kg laterite, whereas the acid consumption in direct acid leaching of the same laterite samples was approximately 0.6–0.8 kg acid per kg laterite. Iron dissolution was only 1.5–3% during the combined precipitation and acid leaching tests, whereas in direct acid leaching it was 15–30% with the Agios Ioannis and Evia Island samples and 80% with the Kastoria sample.

Keywords: laterite; nickel; leaching; jarosite; precipitation

1. Introduction

Nickel is an essential element for modern industry with uses in stainless steel, nickel-based alloys, casting and alloy steels, electroplating, and rechargeable batteries [1,2]. The global plant production and demand for nickel in 2015 were 1.93 Mt and 1.88 Mt, respectively [3]. Nickel mine production has increased steadily during the past 50 years; however, from 2008 onwards it has been more rapid. A further increase in demand may be encountered due to the higher production numbers of electric vehicles (EVs) as nickel is an essential battery chemical in typical lithium ion batteries [4]. The global reserves of nickel are estimated to be 74–80 Mt including sulphide ores, laterites, and deep-sea nodules [4,5]. Even though laterites represent the vast majority of land-based reserves (approximately 72%), until 2009, less than half of the global nickel production came from nickel laterite ores. Today, more than 60% of the world's nickel is derived from laterite resources [6–9]. To meet the rising nickel demand in the future, it is essential to develop further the methods that allow the economic utilization of nickel laterite ores.

Laterite deposits often contain several laterite types. Limonite laterites are found near the surface, while saprolite ores exist deeper and underneath the limonite ores [10]. The iron content of limonites is usually high and a significant amount of nickel is associated with it, whereas the iron content of saprolites is lower and the nickel is more often associated with silicates. Nickel is also found to be more readily leached from clay-like saprolite ores than limonites [6]. It is typical that Greek laterites contain some of the iron as hematite rather than goethite, which allows the leaching of nickel under conditions mild enough to avoid the high acid consumption associated with the substantial dissolution of iron [6].

At the moment, high-pressure acid leaching (HPAL) and pyrometallurgical treatment are the two main technologies for nickel laterite processing [11,12]. However, they have some disadvantages, such as high autoclave investment costs [6,13], technical problems [14], and high energy consumption [11]. In addition, these processes require rather high-grade nickel laterite ore for feasible operation. Ashok [11] has stated that the nickel content of ore should be over 1.4% for feasible HPAL processing and over 1.7% for pyrometallurgical processing.

The third option for nickel laterite processing is atmospheric acid leaching (AL). Investment costs of the process are lower than for pressure leaching, while the need for energy is significantly lower than for the pyrometallurgical operation [6]. Another advantage of AL is the potential to valorize the leach residues. Komnitsas et al. [15] found that column leaching residues can be alkali activated with the use of NaOH and Na_2SiO_3 as activators. Inorganic polymers with high strength, almost 40 MPa, can be produced, when 10 wt % metakaolin is added to the residue. These low toxicity products can be used as binders or building materials in several applications in the construction sector.

The main disadvantages of the AL process are high acid and neutralization agent consumptions. Usually, a significant fraction of total nickel content of laterite ore is associated with iron. Nickel may substitute iron in iron oxide minerals like goethite and thus selective leaching of nickel is not possible, as the chemical decomposition of the goethite is needed for nickel dissolution [10,11]. In the HPAL process, most of the dissolved iron will eventually precipitate as hematite, but in the case of the AL process, it will end up in the solution and must be removed before nickel recovery [16]. During laterite leaching, iron removal from the pregnant solution is usually carried out by using limestone [17] and, depending on the iron concentration, this can be a significant cost item. Therefore, effective iron control is necessary in order to minimize acid consumption and neutralization costs in atmospheric acid laterite leaching.

The main objective of this study was to combine nickel leaching and iron precipitation as sodium jarosite. The idea of combined leaching of limonite and using saprolite for neutralization has been patented by Arroy and Neudorf [18]. In this patent, limonite is dissolved with mineral acid and the saprolite type of laterite is used as neutralizing agent for jarosite precipitation. This invention decreases the use of additional neutralization agent for iron removal and thus improves the feasibility of the process. A quite similar idea, mostly for the sulfide type of nickel ores, has been patented by Leppinen et al. [19]. The patent's idea is to use poor nickel ore or waste for pH adjustment after the atmospheric acid leaching of nickel, cobalt, copper, or zinc. Iron dissolved in the atmospheric leaching precipitates and acid generated during the iron precipitation reaction is used for nickel dissolution from poor nickel ore. Jarosite precipitation is widely applied in hydrometallurgy to remove iron from the pregnant leach solution (PLS). In 2013, White and Gillaspie [20] reviewed comprehensively the acid leaching of nickel laterites with jarosite precipitation, including a discussion on the laterite profile as well as the disposition of other elements. In sulfuric acid solutions, sodium jarosite ($NaFe_3(SO_4)_2(OH)_6$) forms in the presence of dissolved ferric (Fe^{3+}) and sodium (Na^+) ions according to the following reaction:

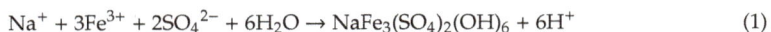

$$Na^+ + 3Fe^{3+} + 2SO_4{}^{2-} + 6H_2O \rightarrow NaFe_3(SO_4)_2(OH)_6 + 6H^+ \qquad (1)$$

Reaction (1) produces acid and has a positive effect on the acid consumption of the process. Favorable conditions for the reaction are pH < 2.5 and temperature 80–100 °C [21,22].

This study's specific aim was to investigate simultaneous nickel leaching and jarosite precipitation using Greek laterites. Investigating the characteristics of laterite ores and their suitability for simultaneous leaching and iron precipitation is the first step for the development of a feasible process. If iron control is successful in batch tests, pilot tests will be designed based on these results in the next phase of the research.

2. Materials and Methods

2.1. Materials

Three nickel laterite ore samples were obtained from the GMM LARCO SA mines. The origin of the ores are the mines of Agios Ioannis (LAI) and Evia Island (LEV), in central Greece, and Kastoria (LK) in northwestern Greece. The comminution of the samples to −300 μm was carried out by Technical University of Crete, Chania, Greece. Each sample (i.e., LAI, LEV, and LK) was divided into sub-samples at VTT by using a sample divider (FRITSCH Rotary Sample Divider LABORETTE 27 and FRITSCH Vibratory Feeder LABORETTE 24, Pittsboro, NC, USA). The dry matter content (% DM) of the samples was determined at VTT from the mass difference after drying the samples at 105 °C. Dry matter contents were LAI 98.6% DM, LEV 99.2% DM, and LK 98.0% DM.

The chemical and mineralogical composition of the samples was studied at Outotec, Espoo, Finland. The composition of the samples was analyzed by inductively coupled plasma optical emission spectrometry (ICP-OES) after total dissolution. Sulfur and carbon analysis was measured using an Eltra CS-2000 automatic analyzer. The quantity of silica was analyzed colorimetrically using a Hach DR 5000 UV-Vis spectrophotometer. The chemical compositions of the laterite samples are presented in Table 1. X-ray diffraction (XRD) analysis for the ores was carried out at Outotec, Espoo, Finland and the analysis for the leach residues was performed at Eurofins Expert Services Oy, Espoo, Finland.

In addition to XRD analysis used to determine the main minerals of each ore sample, polished sections were prepared from each sample, and they were examined by a JEOL JSM-7000F field emission scanning electron microscope (FE-SEM) (Tokyo, Japan) equipped with an Oxford Instruments energy dispersive spectrometer (EDS) and wavelength dispersive spectrometer (WDS). The imaging and both EDS and WDS analyses were performed under routine conditions using 20 kV acceleration voltage and 1 nA beam current for EDS analyses and 20 nA beam current for WDS analyses. The minerals were identified from the EDS analyses and their nickel content was analyzed by WDS using a pure nickel metal standard. Mineral quantification was performed using HSC Chemistry® (Module version 9.3.0, Outotec, Espoo, Finland) using mineralogical analyses gathered from all of the aforementioned methods. The mineralogy of the laterite samples and the distribution of nickel within the minerals are presented in Table 2.

Table 1. Chemical composition of the laterite samples.

Laterite	LAI	LEV	LK
Element	%	%	%
Ni *	1.00	0.843	0.939
Fe *	31.6	26.3	13.2
Al *	9.58	2.62	0.388
Mg *	1.1	2.16	12.8
Na *	<0.1	<0.1	<0.1
P *	0.025	<0.02	<0.02
K *	0.16	0.539	<0.1
Ca *	2.2	2.08	3.37
Ti *	0.55	0.142	0.025
Cr *	1.27	1.72	0.554
Mn *	0.179	0.167	0.179
Co *	0.062	0.042	0.029
Cu *	<0.005	<0.01	<0.01
Zn *	0.015	0.015	0.009
As *	0.02	<0.02	<0.02
C **	0.69	0.76	1.1
S **	0.02	0.09	<0.02
SiO_2 ***	16.3	40.7	37.1

Sample methods: * ICP TOT, ** Eltra, *** Colorimetry.

Table 2. Mineralogy of the samples and distribution of nickel in each main phase.

Mineralogy	LAI	Ni	LEV	Ni	LK	Ni
	wt %	dist. %	wt %	dist. %	wt %	dist. %
Hematite	45.4	16	44.6	47		
Goethite					19.5	24
Al-clays and Al-hydroxides *	21.1	60				
Smectite group minerals	5.5	22	2.1	13		
Primary serpentinite silicates **	12.8	1	8.5	39	67.6	75
Quartz	6		34.4		3.2	
Carbonates	5.8		3.3		7.6	
Chromite	2.6	<1	3.5	<1	1.7	<1
Magnetite	0.8	<1	0.7	<1	0.5	<1

* The content of aluminum clays and aluminum (oxy)hydroxides cannot be assessed with the characterization methods used; ** LAI, chlorite and anthophyllite; LEV, chlorite; LK, lizardite and anthophyllite.

Based on mineralogical analyses, LAI is composed mainly of hematite, clays, and primary serpentinite silicates. Nickel is distributed into mainly (60%) Al-clays and Al-hydroxides and the rest is distributed into smectites and hematite. Based on the mineralogical analyses of the samples, LAI was found to be a mixture of different laterization profiles. It could be defined as an aluminous laterite or a bauxite, based on its high Al content. However, the occurrence of smectites, primary serpentinite silicates, and the high contents of Fe and hematite are typical features of saprolitic and limonitic laterization profiles as well as weathered serpentinite.

LEV is composed mainly of hematite and quartz. Minor chlorite, smectite, chromite, and carbonates comprise the rest of the sample. With respect to the nickel, 47% of it is distributed into hematite, 39% into chlorite, and 13% into the smectites. Based on the mineralogy, LEV was mined from a transitional zone, between limonitic and saprolitic profiles.

In the case of LK, 68% of it is composed of primary serpentinite silicates (i.e., lizardite and anthophyllite), 20% of goethite, 7.6% of calcite and dolomite, 3.2% of quartz, and 1.7% of chromite. LK can be categorized as weathered serpentinite laterite. Its nickel is mainly (75%) distributed into the primary serpentinite silicates and secondarily (24%) into goethite.

Theoretical maximum acid consumptions of the laterite samples are presented in Table 3. The acid consumption is calculated for each metal assuming that the metal dissolution is 100%. It can be seen from the table that acid-consuming metals are quite different in each sample. The main acid-consuming metals in the LAI sample are iron and aluminum, while the aluminum content of the LEV sample is very low and iron is the only significant acid-consuming metal. Acid consumption in both LEV and LAI samples is mainly caused by iron, but in case of the LK sample, the main acid-consuming metal is magnesium.

Table 3. Theoretical maximum acid consumption of Agios Ioannis (LAI), Evia Island (LEV) and Kastoria (LK) laterites as grams sulfuric acid per kg laterite. Metal dissolution to solution is assumed to be 100%.

	LAI	LEV	LK
	H_2SO_4 g/laterite kg	H_2SO_4 g/laterite kg	H_2SO_4 g/laterite kg
Ni	16.7	14.0	15.7
Fe	832	693	348
Al	522	143	21.3
Mg	44.4	87.2	517
Ca	53.8	50.9	82.5
Mn	3.21	3.03	3.21
Total	1470	991	987

2.2. Leaching of Laterite Ores

The aim of the study was to determine whether a suitable feed consisting of LAI, LEV, and LK ores would result in iron control during atmospheric acid leaching. In the first phase of the tests, the effect of pH on nickel and iron dissolution was determined. The leaching test setup is presented in Figure 1. In the experiments, a Mettler Toledo T70 titrator (Columbus, OH, USA) was used for H_2SO_4 feed and a Consort C3040 was used to monitor pH, oxidation/reduction potential (ORP), and temperature on-line. Manufacturer and model of the electrodes were Van London Phoenix PH7V110-10B-300 (pH electrodes) and Van London Phoenix RP75130-10B-300 (oxidation/reduction electrodes) (Houston, TX, USA). The reactor mass was monitored by a balance, Shimadzu BW22KH (Kyoto, Japan) or Hottinger Baldwin Messtechnik GmbH (Darmstadt, Germany), connected to a monitor, Gottl. Kern & Sohn GmbH (Balingen, Germany). In the experiments where LK was added, the amount of LK addition was examined on-line by a balance, Precisa ES 8200C-DR (Aldingen, Germany). Chemicals used in the experiments were H_2SO_4 (Merck EMSURE® ISO sulfuric acid 95–97% for analysis) and Na_2SO_4 (Sigma-Aldrich sodium sulfate ACS reagent ≥99.0% anhydrous granular, St. Louis, MO, USA).

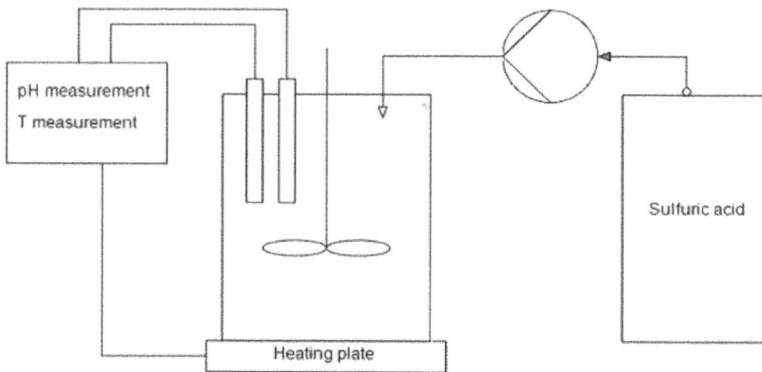

Figure 1. Experimental setup used in the acid consumption test. A 5 L reactor (Ti, Gr. 2) on the heating plate is placed on top of the balance. Titrator is on the right.

The characteristic leaching behavior of each laterite was studied under the following leaching conditions: pH 0.25–1.5, T = 90 °C, S/L = 30%, particle size −300 μm, 800 rpm. Experiments were carried out in 5 L reactors for 8 h. An experiment started (t = 0 h) when the target pH was reached.

Based on the results of the 8 h leaching tests, two-phase leaching tests were designed. These experiments were carried out in pH 0.75–1.8, T = 90 °C, S/L = 30%, particle size −300 μm, 800 rpm, 48 h. In the first part of these tests, LAI or LEV were first leached separately at pH 0.75. LAI was leached for 24 h and LEV for 6 h. In the second part of the tests, leaching was continued for the following 24 h in pH 1.8 with LK addition. In the tests, pH was controlled by LK additions and H_2SO_4. Iron precipitation was ensured by Na_2SO_4 additions before the LK feed. The amounts were 235 g of Na_2SO_4 for LAI and 150 g for LEV.

After leaching, the slurry was filtered (pore size of 0.45 μm) using a Büchner funnel and a filter flask. The leach residue was washed with acidic water (pH 1–1.5), followed by a washing with deionized water. The washed residue was dried and weighed.

2.3. Analytical Methods

Solution samples were analyzed with ICP-OES and ICP-MS. Microwave-assisted digestion was performed with a mixture of HF, HNO_3, and HCl according to EN 13656 and the subsequent determination of elements was applied to solid samples. These analyses were carried out at Labtium Oy, Espoo, Finland.

3. Results and Discussion

3.1. Acid Leaching of Greek Laterites

Sulfuric acid leaching tests were carried out for each LAI, LEV and LK laterite sample. The idea of the leaching tests was to find suitable leaching pH for each laterite and determine acid consumption of direct acid leaching. Nickel dissolution during the 8 h leaching tests is presented in Figure 2 and iron dissolution in Figure 3. The nickel and iron recoveries are shown in Table 4.

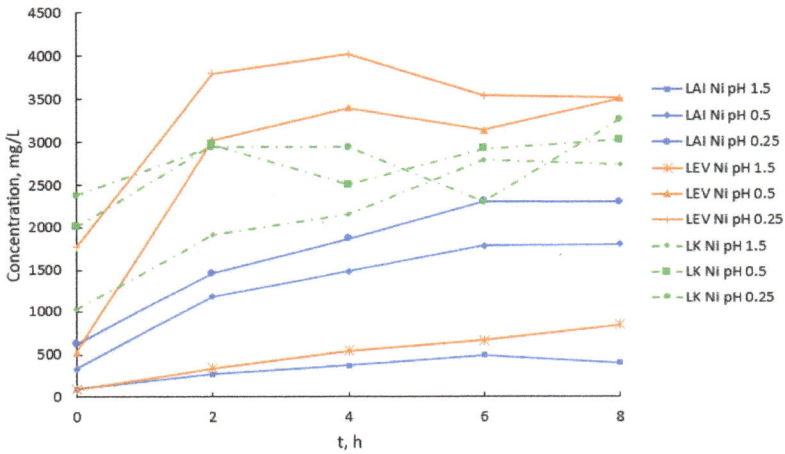

Figure 2. Evolution of Ni concentration vs. time during leaching of LAI, LEV, and LK laterites in pH 0.25, 0.5, and 1.5. Parameters used in the experiments: 90 °C, S/L = 30%, 800 rpm. pH was maintained by H_2SO_4 addition.

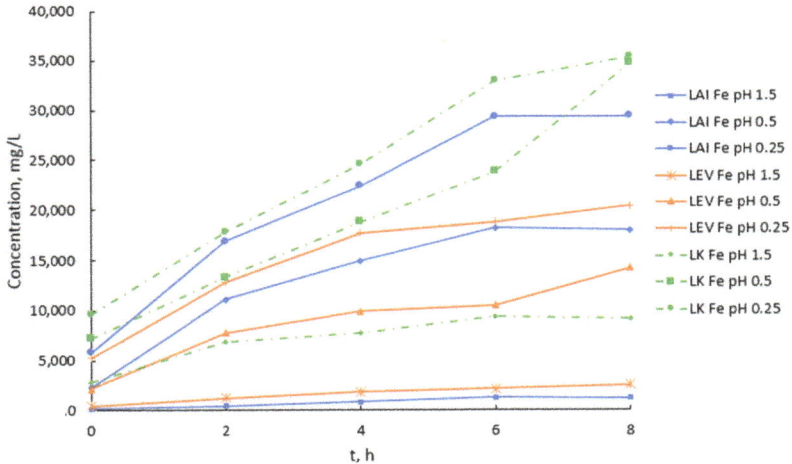

Figure 3. Evolution of Fe concentration vs. time during leaching of LAI, LEV, and LK laterites in pH 0.25, 0.5 and 1.5. Parameters used in the experiments: 90 °C, S/L = 30%, 800 rpm. pH was maintained by H_2SO_4 addition.

Table 4. Nickel and iron yields to solution in acid leaching test of LAI, LEV, and LK laterite in pH 0.25, 0.5 and 1.5. Parameters used in the experiments: 90 °C, S/L = 30%, 800 rpm. pH was maintained by H_2SO_4 addition.

	LAI, pH 1.5	LAI, pH 0.5	LAI, pH 0.25	LEV, pH 1.5	LEV, pH 0.5	LEV, pH 0.25	LK, pH 1.5	LK, pH 0.5	LK, pH 0.25
Ni, %	12.0	54.3	69.0	30.1	>100.0	>100.0	87.5	96.8	>100.0
Fe, %	1.2	17.1	27.9	2.9	16.2	23.3	20.9	78.9	80.5

The mineralogical study shows that a significant portion of nickel in LAI and LEV laterites is associated with iron oxides and iron-containing silicates and it seems, based on the leaching results, that a partial decomposition of these minerals is needed for nickel liberation and dissolution. Nickel dissolution kinetics in LEV leaching tests was fast and maximum nickel dissolution was achieved only four hours from the beginning of the test. Iron dissolution was relatively low in the LEV leaching tests. Based on these results, it can be concluded that, in LEV laterite, nickel is not substituted into goethite like in many limonite laterites [6,7]. In the LAI leaching tests, the nickel was harder to leach and was associated with more refractory host minerals. Nickel dissolution from LK laterite was found to differ compared to nickel dissolution from LAI and LEV samples. Most of the nickel was dissolved already at pH 1.5. Most of the LK (67.6 wt %) consists of primary serpentinite silicates, such as lizardite ($Mg_3Si_2O_5(OH)_4$) and anthophyllite ($Mg_7(Si_8O_{22})(OH)_2$), and 75% of the nickel is associated with these silicates. Only 19.5 wt % of LK consists of hematite and other iron oxides, and 24% of the nickel is associated with these oxides. Therefore, it is clear that iron dissolution is not associated with nickel dissolution.

It can be seen from Figures 2 and 3 that nickel dissolution of LAI and LEV laterites at pH 1.5 is much lower than at pH 0.5 and 0.25. Higher acidity was needed for efficient nickel dissolution for LAI and LEV laterites. In contrast, nickel from LK dissolved considerably already at pH 1.5 and the dissolution was quite similar in all LK leaching tests. Some oscillation in the nickel concentrations was observed; this is most likely related to the addition of acid followed by the subsequent dissolution of metal. Nickel yields at pH 1.5 were 12% for LAI, 30.1% for LEV, and 87.5% for LK (see Table 4). However, the iron dissolution of LK was significantly higher when the pH was 0.25–0.5 compared to the test carried out at 1.5.

When comparing the nickel and iron dissolution at pH 0.5 and 0.25 (see Figures 2 and 3), it is observed that nickel cannot be selectively dissolved because a significant fraction of iron will also dissolve together with the nickel. Theoretical acid consumptions for the laterite samples are presented in Table 3 and actual acid additions after 8 h of leaching are presented in Table 5. The actual amounts of acid consumed in the leaching reactions are presented in Figure 4. Acid addition was calculated based on the amount of sulfuric acid (100%) added to the leaching reactor divided by the amount of dry laterite added to the reactor. Acid consumption was calculated from the metal dissolution rate, assuming that Mg, Mn, and Ni consume 1 mol acid per 1 mol metal and that Fe and Al consume 1.5 mol acid per 1 mol metal. It was found out that acid addition in both LAI and LEV tests was quite low, being only approximately 120–140 kg sulfuric acid per t laterite when the pH was 1.5. Acid additions and calculated acid consumption values are quite similar, indicating that the amount of unreacted acid is very low. Acid addition in the LK leaching test was significantly higher than in LAI and LEV leaching at the same pH, being 650 kg sulfuric acid per t laterite. The metal concentrations in the final leach solutions are shown in Table 6. It can be concluded that the high acid addition of LK laterite at pH 1.5 is mainly due to high magnesium dissolution. Based on the acid consumption calculation, nearly 80% of the acid was consumed in the magnesium dissolution reactions. It seems that dissolution kinetics of magnesium from silicates is relatively fast since most of acid was consumed during the two hours from the beginning of the test. Acid addition in the LK leaching tests was close to the theoretical maximum acid consumption when pH was 0.5 or 0.25. Acid addition in the LAI and LEV leaching tests was low compared to the theoretical maximum consumption. This is mainly explained by the relatively low iron dissolutions. Iron, however, was still the main acid-consuming metal in both the LAI and LEV leaching tests. Approximately 30% of the total acid consumption was caused by iron dissolution in the LEV leaching test and approximately 40% in the LAI leaching test. There are significant differences between acid addition and calculated acid consumption values at pH 0.5 and 0.25. There was some oscillation in the acid addition and calculated acid consumption values, but it seems that the amount of unreacted acid at pH 0.5 is approximately 50 kg per t ore, whereas at pH 0.5 it is somewhat higher at over 100 kg per t ore.

Table 5. Cumulative acid additions (H_2SO_4 kg/t dry ore) during leaching of the LAI, LEV, and LK laterites in pH 0.25, 0.5, and 1.5. Parameters used in the experiments: 90 °C, S/L = 30%, 800 rpm. pH was maintained by H_2SO_4 addition.

pH	Time	LAI	LEV	LK
	h	H_2SO_4 kg/t dry ore	H_2SO_4 kg/t dry ore	H_2SO_4 kg/t dry ore
1.5	2	72.9	89.3	532
	4	92.5	115	597
	6	109	126	647
	8	123	142	647
0.5	2	305	343	787
	4	372	403	832
	6	395	455	1120
	8	442	455	1120
0.25	2	463	4940	880
	4	532	529	911
	6	605	576	945
	8	649	593	951

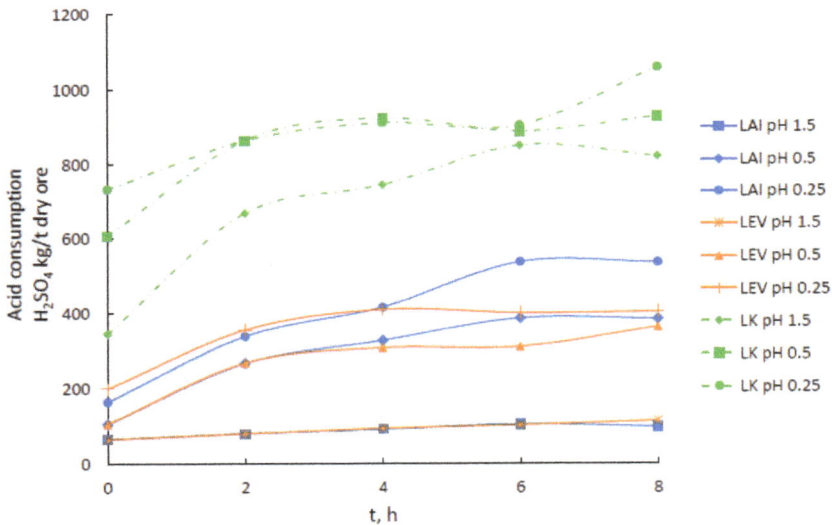

Figure 4. Calculated acid consumptions (H_2SO_4 kg/t dry ore) during leaching of the LAI, LEV, and LK laterites in pH 0.25, 0.5, and 1.5. Parameters used in the experiments: 90 °C, S/L = 30%, 800 rpm. pH was maintained by H_2SO_4 addition. Acid consumptions were calculated based on solution analyses.

Table 6. Dissolutions of Ni, Co, Fe, Ca, Mg, and Al and efficiency of H_2SO_4 leaching of the LAI, LEV, and LK laterites under the following conditions: pH 0.25, 0.5, and 0.25, 90 °C, S/L = 30%, 8 h. Selectivity of Ni is compared over Fe, Mg, and Al using the respective ratios and concentrations in the pregnant leach solution (PLS).

Laterite Ore	pH	Ni (mg/L)	Co (mg/L)	Fe (mg/L)	Ca (mg/L)	Mg (mg/L)	Al (mg/L)	Ni/Fe	Ni/Mg	Ni/Al
LAI	1.5	401	60	1220	447	591	1560	0.33	0.68	0.26
LAI	0.5	1810	156	18,000	515	2630	9020	0.10	0.69	0.20
LAI	0.25	2300	174	29,400	405	2990	12,400	0.08	0.77	0.19
LEV	1.5	845	27	2550	721	2350	1440	0.33	0.36	0.59
LEV [1]	0.5	3510	136	14,200	518	8040	5210	0.25	0.44	0.67
LEV	0.25	3520	77	20,400	417	7540	5070	0.17	0.47	0.69
LK	1.5	2740	63	9210	373	53,400	437	0.30	0.05	6.27
LK	0.5	3030	60	34,700	159	44,900	754	0.09	0.07	4.02
LK	0.25	3270	85	35,400	177	55,300	785	0.09	0.06	4.17

[1] Sample treatment from t = 8 h was unsuccessful. Analysis was done after 26.5 h. In this experiment, 8 h of leaching was performed at 90 °C, and then the slurry was agitated without heating overnight.

The 8 h leaching experiments showed that nickel from LK laterite can be dissolved at pH 1.5, which is also a suitable pH for jarosite precipitation. It was found that LAI and LEV need more acidic conditions, while simultaneous nickel leaching and jarosite precipitation in these cases is not possible. The combined leaching and precipitation tests were designed based on these facts.

3.2. Combined Leaching and Iron Precipitation

The aim of the combined leaching and iron precipitation tests was to dissolve nickel from LEV or LAI laterite first in a lower pH and then to increase the pH by LK addition. This would allow for the simultaneous dissolution of nickel from LK and precipitation of iron from the solution.

Combined tests began by leaching LAI or LEV laterites at pH 0.75, followed by an increase of pH to 1.8 by LK laterite addition. Leaching was continued until a total leaching time of 48 h was reached. The duration of the first leaching phase was decided for LAI and LEV based on the results of the 8 h experiments. As metallurgical process design involves balancing costs and yields, the question of whether leaching at pH 0.75 would be sufficient for nickel dissolution form LAI and LEV was studied. In Figure 2, it was seen that at pH 0.5 and 0.25, a significant dissolution of nickel from LEV occurred already after 2 h leaching. Therefore, the leaching time for LEV was determined to be 6 h. Based on the 8 h test, nickel dissolution from LAI was slower than from LEV. As seen from Table 4, the nickel yield from LAI after 8 h increased from 54.3% to 69.0% when pH was decreased from 0.5 to 0.25. Thus, the leaching time selected for LAI at pH 0.75 was 24 h.

Ni, Co, Fe, Ca, Mg, and Al dissolution, acid addition, and LK consumption during the combined leaching, and the iron precipitation experiments are shown in Figures 5 and 6 as well as in Table 7. A clear decrease in iron concentration is noticed after LK addition. In both cases, iron concentration after leaching of LAI or LEV was over 10 g/L, and after the LK addition the iron concentration decreased to approximately 2 g/L. Table 7 shows the efficiency of H_2SO_4 leaching of laterites in combined nickel leaching and iron precipitation tests, by comparing the selectivity of Ni over Fe, Mg, and Al using the respective ratios and concentrations in the PLS. As can been seen in Figures 5 and 6, iron precipitation was clearly observed after LK addition. In Table 7, this is shown also by the high Ni/Fe selectivity values.

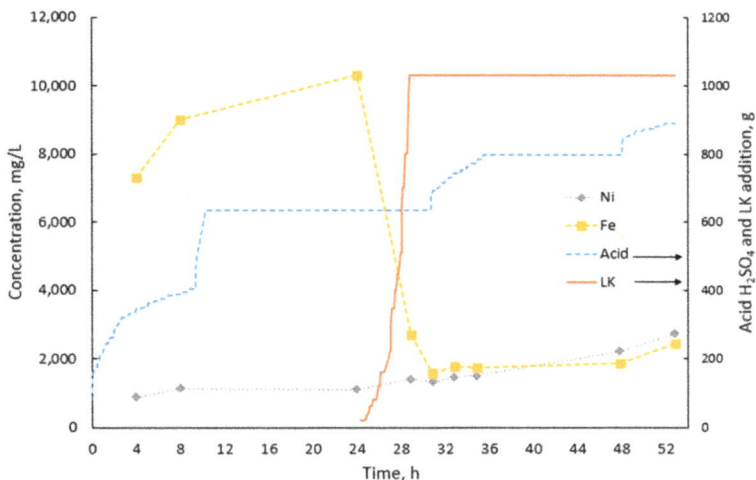

Figure 5. Leaching of LAI for 24 h at pH 0.75, followed by LK laterite addition increasing the pH to 1.8. Concentrations of Ni, Co, Fe, acid, and LK additions. Parameters used in the experiments: 90 °C, S/L = 30%, 800 rpm.

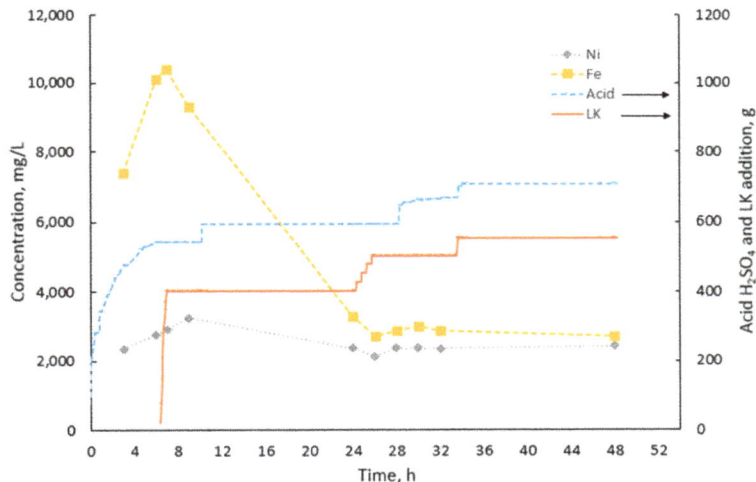

Figure 6. Leaching of LEV for 6 h at pH 0.75, followed by LK laterite addition increasing the pH to 1.8. Concentrations of Ni, Co, Fe, acid, and LK additions. Parameters used in the experiments: 90 °C, S/L = 30%, 800 rpm.

Table 7. Efficiency of H_2SO_4 leaching of LAI (24 h) and LEV (6 h) laterites under the following conditions: pH 0.75, 90 °C, S/L = 30%, followed by LK addition to increase pH to 1.8 (until 48 h), by comparing the selectivity of Ni over Fe, Mg, and Al using the respective ratios and concentrations in the PLS.

Laterite Ore	pH	Ni (mg/L)	Co (mg/L)	Fe (mg/L)	Ca (mg/L)	Mg (mg/L)	Al (mg/L)	Ni/Fe	Ni/Mg	Ni/Al
LAI	0.75	1100	82	10300	595	1630	6040	0.11	0.68	0.18
LAI + LK	1.8	2220	124	1850	456	24100	2000	1.2	0.09	1.11
LEV	0.75	2770	119	10100	-	6510	4560	0.27	0.43	0.61
LEV + LK	1.8	2430	94	2710	-	19300	2520	0.90	0.13	0.96

The Ni/Fe ratio that defines selectivity, when LAI was separately leached at pH 0.75, was 0.11. Komnitsas et al. [13] also studied leaching of ore from a similar origin. Their column leaching test for a lower grade Agios Ioannis ore (0.58% Ni) resulted in Ni/Fe ratios varying between 0.15 and 0.24. Mystrioti et al. [23] applied a counter-current mode of operation for HCl leaching of a similar ore which suppressed iron dissolution to 0.6%. However, they found out that the operation had a negative effect on Ni and Co extraction, which was limited to 55% and 63%, respectively. At the moment, the sulfuric acid leaching approach applied in this study appears to be more favorable in terms of Ni leaching.

Nickel concentration after leaching at pH 0.75 was 1100 mg/L for LAI and 2770 mg/L for LEV. This is in line with the dissolution results after 8 h leaching. Ni concentrations for LAI were 401 mg/L at pH 1.5 and 1810 mg/L at pH 0.5. Ni concentrations for LEV were 845 mg/L at pH 1.5 and 3510 mg/L at pH 0.5. After 48 h leaching, Ni concentrations were 2220 mg/L for LAI + LK and 2430 mg/L for LEV + LK.

Acid additions of the combined tests are presented in Table 8. Acid additions were 370 kg sulfuric acid per 1 t ore for the LAI + LK test and 420 kg sulfuric acid per 1 t ore for the LAI + LK test, whereas in direct acid leaching acid additions were approximately 500–900 kg sulfuric acid per 1 t laterite ore. Acid consumptions calculated using solution analyses were quite close to the actual acid consumptions. The calculated acid consumption in the LAI + LK test was 415 kg sulfuric acid per t ore, and in the LEV + LK test the acid consumption was 322 sulfuric acid per t ore. Calculated acid consumptions in direct leaching was 400–500 kg sulfuric acid per t ore for LAI and LEV laterites and 900 kg sulfuric acid per t ore for LK laterite. Reduction of acid consumption and iron concentration is significant. Furthermore, the obtained acid consumptions are competitive to those in Komnitsas et al.'s [13] column leaching tests for Agios Ioannis ore, which is lower in Ni grade. Their acid consumptions were 0.69–0.76 kg sulfuric acid per 1 kg ore, when nickel extractions were 60.2–73.5 %.

Table 8. Cumulative acid consumptions (H_2SO_4 kg/t dry ore) from H_2SO_4 leaching of LAI (24 h) and LEV (6 h) laterites under the following conditions: pH 0.75, 90 °C, S/L = 30%, followed by LK addition to increase pH to 1.8 (until 48 h).

Laterite	pH	Time, h	H_2SO_4 kg/t Dry Ore
LAI	0.75	24	565.25
LAI + LK	1.8	48	369.69
LEV	0.75	6	481.71
LEV + LK	1.8	48	421.94

At the end of the precipitation phase, iron yield to solution in the LAI + LK test was 1.4%, and in the LEV + LK test it was 2.8%. However, nickel yield to solution was not as high in the combined leaching and iron precipitation tests as in direct acid leaching of the laterites. The iron precipitation as jarosite in the LEV + LK and LAI + LK tests was verified by XRD analysis from the leach residues. The XRD analyses showed that iron was precipitated as sodium jarosite ($NaFe_3(SO_4)_2(OH)_6$) in both cases. Based on the calculations in Table 3, it is assumed that the major acid-consuming element in the two-stage leaching approach is aluminum. This is consistent with the fact that iron precipitated as sodium jarosite and not sodium alunite.

Leach residues contained also iron oxides, gypsum ($CaSO_4 \cdot 2H_2O$), and quartz (SiO_2). Hematite was the main form of iron oxide, but some goethite was found from the leach residue of the LAI + LK leaching test.

This study's aim was to investigate the simultaneous nickel leaching and jarosite precipitation as a phenomenon. The overall aim was to improve the feasibility of atmospheric acid leaching by combining different types of laterites and specifically using a low-grade silicate laterite for neutralization. Using a low-grade laterite for neutralization decreases the need for limestone to enable iron precipitation, while also dissolving some additional nickel from the low-grade laterite. This study is the first step in the laterite process development project and the results presented here will be used for designing a pilot test run. In future research, the feasibility of the process will be further investigated. The aim is to find process conditions, which allow maximizing nickel yield while minimizing acid consumption and final iron concentration in the pregnant leach solution.

4. Conclusions

A process involving atmospheric acid leaching of nickel laterites and simultaneous iron precipitation was investigated for the treatment of Greek LAI, LEV, and LK laterites. It was observed that iron control is possible when nickel leaching is combined with jarosite precipitation These tests involved two steps, namely, (i) acidic leaching of LAI or LEV laterite at pH 0.75 and (ii) leaching of nickel from LK laterite in higher pH with simultaneous iron precipitation as jarosite. The aim of the jarosite precipitation is to decrease iron concentration in the leaching solution and at the same time to use the acid generated from the iron precipitation to obtain a higher nickel dissolution.

Atmospheric acid leaching tests of LAI, LEV, and LK laterite samples in 8 h experiments showed that pH lower than 1.5 was needed to effectively dissolve Ni from LAI and LEV, while Ni dissolved from LK already at pH 1.5. The potential for combined nickel leaching and iron precipitation was observed after 8 h of leaching. Based on the results, combined leaching and precipitation tests were planned. Tests began by leaching LAI (24 h) or LEV (6 h) laterites at pH 0.75, followed by increasing pH to 1.8 by LK laterite addition.

Acid consumption of atmospheric laterite leaching was high and the 8 h leaching tests showed that approximately 600–800 g acid per 1 kg laterite was needed for the Greek laterite samples LAI, LEV, and LK for an efficient nickel dissolution. The combined leaching and precipitation process has a potential to reduce acid consumption and decreased iron removal costs. Consumptions were 0.37 kg sulfuric acid per 1 kg for the LAI + LK test and 0.42 kg sulfuric acid per 1 kg for the LAI + LK test. Iron concentration decreased from 10 g/L to approximately 2–3 g/L during the jarosite precipitation stage. That indicates significant savings in iron removal costs since, in the conventional process, all dissolved iron has to be precipitated out from the leach solution using lime or limestone.

Author Contributions: V.M. and L.R. designed the experiments and critically analyzed the results. E.K. and V.M. provided industrial insight to the study. J.M. performed the literature review for the paper. L.R. supervised the carrying out of the experiments. T.K. studied the mineralogy of the samples. V.M. and L.R. wrote the paper in cooperation with the co-authors.

Funding: The authors would like to acknowledge the financial support of the European Commission in the frame of the Horizon 2020 project intitled "Metal recovery from low-grade ores and wastes", www.metgrowplus.eu, Grant Agreement No. 690088.

Acknowledgments: The authors would like to acknowledge the GMM LARCO SA Mines for providing laterite samples used in the leaching tests.

Conflicts of Interest: The authors declare no conflict of interest.

References

1. Mudd, G.M. Global trends and environmental issues in nickel mining: Sulfides versus laterites. *Ore Geol. Rev.* **2010**, *38*, 9–26. [CrossRef]
2. Barnett, S. Nickel—A key material for innovation in a sustainable future. In Proceedings of the 2nd Euro Nickel Conference, London, UK, 18–19 March 2010; Informa Pty Ltd.: London, UK.
3. McRae, M.E. *Minerals Yearbook, Nickel (Advance Release)*; U.S. Geological Survey: Reston, VA, USA, June 2018.
4. European Commission. Commission Staff Working Document. Report on Raw Materials for Battery Applications. Available online: https://ec.europa.eu/transport/sites/transport/files/3rd-mobility-pack/swd20180245.pdf (accessed on 24 March 2019).
5. McRae, M.E. Nickel: U.S. Geological Survey, Mineral Commodity Summaries 2018. U.S. Department of Interior, U.S. Geological Survey; pp. 112–113. Available online: https://min-erals.usgs.gov/minerals/pubs/mcs/2018/mcs2018.pdf (accessed on 24 March 2019).
6. McDonald, R.G.; Whittington, B.I. Atmospheric acid leaching of nickel laterites review Part, I. Sulfuric acid technologies. *Hydrometallurgy* **2008**, *91*, 35–55. [CrossRef]
7. Sudol, S. The thunder from down under: everything you wanted to know about laterites but were afraid to ask. *Can. Min. J.* **2005**, *126*, 8–12.
8. Oxley, A.; Smith, M.; Caceres, O. Why heap leach nickel laterites? *Miner. Eng.* **2016**, *88*, 53–60. [CrossRef]
9. Li, G.; Zhou, Q.; Zhu, Z.; Luo, J.; Rao, M.; Peng, Z.; Jiang, T. Selective leaching of nickel and cobalt from limonitic laterite using phosphoric acid: An alternative for value-added processing of laterite. *J. Clean. Prod.* **2018**, *189*, 620–626. [CrossRef]
10. Whittington, B.I.; Muir, D. Pressure Acid Leaching of nickel Laterites: A Review. *Miner. Process. Extr. Metall. Rev.* **2000**, *21*, 527–599. [CrossRef]
11. Dalvi, A.D.; Bacon, W.G.; Osborne, R.C. The Past and Future of Nickel Laterites, PDAC 2004 Convention, March 7–10, 2004. Available online: http://citeseerx.ist.psu.edu/viewdoc/download?doi=10.1.1.732.7854&rep=rep1&type=pdf (accessed on 24 March 2019).
12. Rodrigues, F.; Pickles, C.A. Factors Affecting the Upgrading of a Nickeliferous Limonitic Laterite Ore by Reduction Roasting, Thermal Growth and Magnetic Separation. *Minerals* **2017**, *7*, 176. [CrossRef]
13. Komnitsas, K.; Petrakis, E. Column Leaching of Greek Low-Grade Limonitic Laterites. *Minerals* **2018**, *8*, 377. [CrossRef]
14. Agatzini-Leonardou, S. Hydrometallurgical Process for the separation and recovery of nickel from sulfate heap leach liquor of nickelferrous latetrite ores. *Miner. Eng.* **2009**, *22*, 1181–1192. [CrossRef]
15. Komnitsas, K.; Petrakis, E.; Bartzas, G.; Karmali, V. Column leaching of low-grade saprolitic laterites and valorization of leaching residues. *Sci. Total Environ.* **2019**, *665*, 347–357. [CrossRef] [PubMed]
16. Wang, K. Iron, Aluminium and chromium co-removal from atmospheric nickel laterite leach solutions. *Miner. Eng.* **2018**, *116*, 35–45. [CrossRef]
17. Cheng, C.Y. Purification of synthetic laterite leach solution by solvent extraction using D2EHPA. *Hydrometallurgy* **2000**, *56*, 369–386. [CrossRef]
18. Arroyo, J.C.; Neudorf, D.A. Atmospheric Leaching Process for the Recovery of Nickel and Cobalt from Limonite and Saprolite Laterite Ores. Patent no: 6,261,527, 2001.
19. Leppinen, J.; Riihimäki, T.; Ruonala, M. Method for Separating Nickel from Material with Low Nickel Content. Patent no: CA2819224A1, 2011.
20. White, D.T.; Gillaspie, J.D. Acid leaching of nickel laterites with jarosite precipitation. In *Ni-Co 2013*; Springer: Cham, Switzerland, 2013; pp. 75–95.
21. Dutrizac, J.E. The effectiveness of jarosite species for precipitating sodium jarosite. *Jom* **1999**, *51*, 30–32. [CrossRef]
22. Das, G.K.; Acharya, S. Jarosites: A Review. *Miner. Process. Extr. Metall. Rev.* **1996**, *16*, 185–210. [CrossRef]
23. Mystrioti, C.; Papassiopi, N.; Xenidis, A.; Komnitsas, K. Counter-Current Leaching of Low-Grade Laterites with Hydrochloric Acid and Proposed Purification Options of Pregnant Solution. *Minerals* **2018**, *8*, 599. [CrossRef]

minerals

MDPI

Article

Leaching of Primary Copper Sulfide Ore in Chloride-Ferrous Media

Karina E. Salinas [1], Osvaldo Herreros [2] and Cynthia M. Torres [3,*]

1 Departamento de Ingeniería Química y Procesos de Minerales, Universidad de Antofagasta, Antofagasta 1240000, Chile; karinasalinasc@gmail.com
2 Departamento Ingeniería de Minas, Universidad de Antofagasta, Antofagasta 1240000, Chile; osvaldo.herreros@uantof.cl
3 Departamento de Ingeniería en Metalurgia y Minas, Universidad Católica del Norte, Antofagasta 1240000, Chile
* Correspondence: cynthia.torres@ucn.cl; Tel.: +56-552-651-022

Received: 28 June 2018; Accepted: 23 July 2018; Published: 25 July 2018

Abstract: Copper extraction from primary copper sulfide ore from a typical porphyry copper deposit from Antofagasta, Chile, was investigated after leaching with a chloride-ferrous media at two temperatures. The study focused on whether this chemical leaching system could be applied at an industrial scale. Leaching tests were conducted in columns loaded with approximately 50 kg of agglomerated ore; the ore was first cured for 14 days and then leached for 90 days. The highest copper extraction, 50.23%, was achieved at 32.9 °C with the addition of 0.6 kg of H_2SO_4 per ton of ore, 0.525 kg of NaCl per ton of ore, and 0.5 kg of $FeSO_4$ per ton of ore. In respect to copper extraction, the most effective variables were temperature and the addition of NaCl.

Keywords: leaching; primary sulfide copper; chalcopyrite; chloride

---

1. Introduction

Copper is considered the third most important metal for industry after iron and aluminum, with a yearly production of 10 Mt [1]. In the north of Chile there are many mining industries that produce copper in cathodes from oxidized minerals with the traditional process of leaching, followed by solvent extraction (SX), and electrowinning (LIX-SX-EW). Due to the deepening of pits [2], depletion of oxidized ores and the presence of sulfide ore is noted, a situation that puts at risk the continuity of hydrometallurgical processes due to the lack of leachable mineral resources. Hence, different leaching techniques for copper sulfide ores, especially chalcopyrite, need to be investigated in the future.

Presently, the mining industry faces the challenge of extracting copper from low-grade copper ores. In 2002 the average copper ore grade was 1.13%, while in 2011 it was 0.84% [3].

Many chalcopyrite leaching studies have been performed with leachants such as ferric sulfate [4–7], ferric chloride [7,8], sodium nitrate-sulfuric acid [9], hydrochloric acid [10], and sodium chloride-sulfuric acid [11], among others [10].

The slow copper from chalcopyrite dissolution has been attributed to the formation of a passivating layer on the mineral surface. This phenomenon has been widely studied, with no consensus regarding its cause or how to avoid its formation [12,13]. So far, three possible causes have been identified: (i) the formation of a layer of precipitated ferric iron that prevents contact between the chalcopyrite and the leaching solution [14,15], (ii) the formation of a layer of sulfur, which reduces the flow of electrons and the diffusion of the leaching agent to the mineral surface [4,16,17], and/or (iii) the formation of an intermediate layer during the dissolution of chalcopyrite, such as a sulfide, disulfide, or polysulfide compounds, which have a slow kinetic dissolution [12,18,19]. Recent studies showed that passivation of chalcopyrite is dependent on redox potential (Eh) since high redox potentials (high concentration of ferric ions) promote passivation of chalcopyrite [5]. On the other hand, ferrous ions may increase the leaching kinetics significantly compared to ferric ions [6,20,21]. This is in contrast through to other leaching processes, such as the bacterial leaching of gold sulfide concentrates, where ferric iron promotes the dissolution of pyrite and arsenopyrite [22].

In addition, numerous studies showed that leaching of chalcopyrite in chloride media results in higher copper extractions than in sulfate media [16,23,24]. The main reasons are the formation of chloride complexes, which change the electrochemical behavior of chalcopyrite passivation and/or the higher porosity of the mineral surface [8,16,25]. For example, Dutrizac [7] associated the high percentages of copper extraction in a chloride medium to the lower activation energy of the medium, approximately 42 kJ/mol, compared with that in a sulfate medium (75 kJ/mol).

Li et al. [26] and Kaplun et al. [27] demonstrated that the redox potential (Eh) significantly affects the leaching rate of chalcopyrite. To understand the effects of passivation during leaching processes, the passivation potential (Epp) has been used to determine the adequate potential to effectively leach the chalcopyrite. Chalcopyrite leached in a medium with sulfuric acid maintains its reactivity under 685 mV (SHE) regardless of impurities, acidity, or temperature. Between 685 and 755 mV (SHE), chalcopyrite may be in between the passive and active state, depending on the form at which that potential was reached. At 755 mV (SHE) the leaching of chalcopyrite becomes slow due to the effect of passivation [28]. Sandström et al. [29] suggested that chalcopyrite leaching should be carried out in sulfuric acid medium at 620 mV SHE, rather than at 800 mV SHE.

The leaching of chalcopyrite at high potentials (high ferric concentration) has been questioned because some studies have shown that ferric ions inhibit the leaching of chalcopyrite [5] while ferrous ions increase its leaching kinetics [6]. Kametani and Aoki [21] established a critical reduction potential of 420 mV versus Ag/AgCl at which chalcopyrite dissolution reaches its maximum, demonstrating that there is an optimal concentration of ferrous ions that increase the leaching of chalcopyrite [6,30]. In addition, temperature is one of the most important parameters in the kinetics of chalcopyrite leaching [31], with the dissolution rate increasing at higher temperatures [23,32].

Nicol and coworkers [33–35] performed chalcopyrite leaching experiments in chloride solutions containing cupric ions and dissolved oxygen. As part of that overall program on the fundamental aspects of the heap leaching of copper sulfide minerals, Nicol [36] compared the anodic behavior of chalcopyrite at high potentials (transpassive region) in both sulfate and chloride solutions. Nicol et al. [37] studied the voltammetric characteristics of chalcopyrite under various conditions appropriate to low temperature leaching with acidic chloride solutions while Nicol and Basson [38] described the anodic behavior of covellite over a wide range of chloride concentrations at potentials relevant to its dissolution under ambient conditions similar to heap leaching. The other important aspect in column or heap leaching is the solution flowrate [39]. According to Ilankoon and Neethling [40], the way the solution spreads from drippers and the subsequent development of flow paths and any associated channeling needs to be properly investigated.

Actually, the current heap leaching process at mining companies involves a curing stage with the addition of 12 kg of sulfuric acid per ton of ore and a determined amount of Intermediate Pregnant Leach Solution (IPLS). The agglomerated ore is transported by conveyor belts to form 107 leach pads weighing a total of approximately 200,000 tons. Once the heap is formed, irrigation begins with a continuous leaching cycle of approximately 300 days. The modification proposed in this study includes the addition of reagents such as sodium chloride and ferrous sulfate in the agglomeration stage, together with the acid, and carrying out the leaching process at an average temperature of approximately 30 °C to increase the percentage of copper extraction from primary sulfide ore.

The main objective of this work is to determine the effects of the modifications of the agglomeration-curing stages and the temperature on the copper extractions of the sulfidic ore.

2. Materials and Experimental Procedure

2.1. Materials

2.1.1. Ore Preparation and Sampling

500 kg of ore containing primary copper sulfide was obtained from a mine stockpile and crushed using a jaw crusher with a 152.4 mm aperture. Subsequently, sieve analysis was performed to obtain an ore sample with $P_{(80)}$ of 11.5 mm.

The ore retained in the meshes was classified by the cone and quartering technique to obtain the quantity needed to load the nine columns.

2.1.2. Mineralogical Analysis

The composition of sample of the ore was determined by mineralogical analysis using scanning electron microscopy and elemental analysis (SEM-EDX). To determine the composition of the gangue minerals and the ore, the sample particle size was reduced to 2 mm and was characterized by optical microscopy using point-counting method.

Most of the gangue is composed of potassium feldspar (23.7%), plagioclase (21.05%), biotite (17.37%), quartz (14.21%), sericite (9.47%), and chlorite (7.89%) (Table 1), which are characteristics of porphyry copper deposits.

Table 1. Composition of the gangue minerals.

Species	Chemical Formula	% *w/w*
Kaolinite		1.05
Smectite		1.05
Albite	$NaAlSi_3O_8$	0.53
Biotite	$K(Mg,Fe^{2+})_3(AlSi_3O_{10}(OH,F)_2$	17.37
Chlorite	$(Fe,Mg_5Al)(Si,Al)_4O_{10}(OH)_8$	7.89
Quartz	SiO_2	14.21
Epidote	$Ca_2Al_3Si_3O_{12}(OH)$	1.05
Feld. k	$KAlSi_3O_8$	23.7
Opaque minerals		2.63
Plagioclase	$(Ca,Na)(Al,Si)_3O_8$	21.05
Sericite	$KAl_2(Si_3Al)O_{10}(OH,F)_2$	9.47
Total		100

The mineralogical composition of the feed ore, Table 2, indicates that the ore consists of chalcopyrite (0.55%), pyrite (0.28%), covellite (0.15%), and magnetite (0.63%).

Table 2. The mineralogical composition of the feed ore determined by optical microscopy.

Feed Ore	% *w/w*
Chalcocite	0.04
Covelite	0.15
Chalcopyrite	0.55
Bornite	0.01
Pyrite	0.28
Hematite	0.02
Magnetite	0.63
Ilmenite	0.03
Rutile	0.01
Gangue	98.29
Total	100.00

The mineralogical composition indicates 74.47% primary sulfides (i.e., chalcopyrite) and 25.52% secondary sulfides such as covellite and chalcocite.

2.1.3. SEM-EDX Analysis of the Feed Ore

Analysis of the feed ore sample was performed using a scanning electron microscope (JEOL model JSM 6360 LV, JEOL Ltd., Tokyo, Japan) equipped with a microscope-coupled X-ray dispersive energy analyzer (SEM-EDX).

The elemental composition of the feed ore indicates that it contains mostly oxygen, aluminum, and silica, so it can be concluded that it contains a large amount of aluminum silicates such as albite and biotite according to Table 1.

Figures 1 and 2 show the SEM-EDX analysis of chalcopyrite (CPY) in the feed ore. In the sample it was possible to detect the presence of chalcopyrite, as shown in Figure 2.

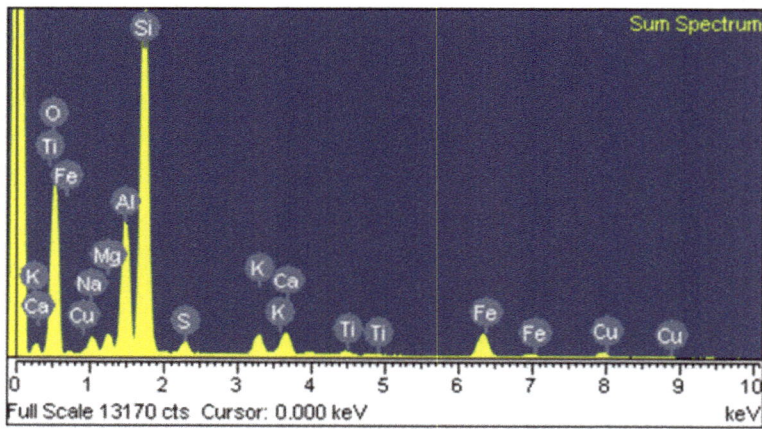

Figure 1. Elemental composition obtained by SEM-EDX analysis of Chalcopyrite (CPY) in the feed ore sample.

Figure 2. SEM image of chalcopyrite (CPY) in the feed ore sample showing chalcopyrite inclusion among gangue minerals.

2.1.4. Chemical Analysis of the Feed Ore

A sample of ore was taken prior to loading each column to carry out chemical analysis by atomic absorption spectrometry (AA), using a Varian Model SpectrAA220 instrument (LabX Media Group, Midland, ON, Canada), and used to determine the ore grade through the sequential leaching method of copper. This analysis allows the estimation of the degree of solubility of copper-rich minerals in sulfuric acid and cyanide. The sulfuric acid leaching enables the dissolution of soluble copper oxides. The copper contained in the secondary sulfides and bornite was dissolved using a cyanide solution (CNsCu). Residual copper (Cu_r) corresponds to the copper present in the chalcopyrite.

The chemical analysis of the feed ore used in each column is shown in Table 3, from which it can be deduced that most of the copper is present as chalcopyrite.

Table 3. Chemical analysis of the feed ore used in each column *, showing the percentage of soluble copper (Cu S); Cyanidable copper (CNsCu), and residual copper (Cu_r).

Column Number	1	2	3	4	5	6	7	8	9
Total Copper	0.35	0.38	0.43	0.36	0.47	0.38	0.37	0.41	0.45
Soluble Copper	0.04	0.04	0.04	0.04	0.04	0.06	0.04	0.04	0.05
Cyanidable Copper	0.13	0.14	0.16	0.14	0.19	0.14	0.13	0.15	0.17
Residual Copper	0.18	0.20	0.23	0.18	0.24	0.18	0,20	0.22	0.23
Total Iron	4.88	4.95	4.90	4.82	4.97	4.94	4.64	4.87	4.67
Soluble Iron	0.45	0.37	0.38	0.39	0.40	0.39	0.32	0.46	0.45

* See Table 4 for conditions of agglomeration in each column.

Table 4. Ore agglomeration conditions in columns prior to leaching.

Column	H_2SO_4 (kg/50 kg of Ore)	NaCl (kg/50 kg of Ore)	$FeSO_4$ (kg/50 kg of Ore)
1	0.6	0	0
2	0.6	0	0.5
3	0.6	0	1
4	0.6	0.525	0
5	0.6	0.525	0.5
6	0.6	0.525	1
7	0.6	0	0
8	0.6	0.525	0
9	0.6	0.525	0.5

2.2. Experimental Procedure

2.2.1. Agglomeration-Curing

The agglomeration-curing stages were carried out in a drum, with a diameter of 30 cm and length of 40 cm, during 5 min by adding sulfuric acid equivalent to 12 kg per ton of ore and the reagents (Table 4).

Once the ore was agglomerated, it was transferred to the columns and cured for 14 days under two different conditions: for columns 1–6, the curing was temperature-controlled (32.9 °C) in a container equipped with an electric hot air generator, while columns 7–9 were kept at 14.5 °C.

2.2.2. Column Loading and Pumping System Installation

PVC columns 1.5 m in height and 18 cm of diameter were used. Each column was loaded with 50 kg of ore ($P_{(80)}$ 11.5 mm). An irrigation system with peristaltic pumps (Master Flex, 7557-14/1-100 RPM Model) and hoses of 13 mm diameter was installed to continuously irrigate the columns with IPLS or raffinate (Table 5). The temperature sensors were configured to measure the temperature every two hours.

Table 5. Chemical composition of feed solutions.

Element	Pregnant Leach Solution (PLS)	Raffinate
Sulfuric Acid (g/L)	8.21–8.64	5.33–7.25
Total iron (g/L)	4.14–4.58	4.24–4.62
pH	1.2	1.38–1.50

2.2.3. Leaching

Leaching was conducted in nine columns using IPLS (Intermediate Pregnant Leach Solution) at an irrigation rate of 8 L/m^{2}h up to 1.6 m^3/ton, and later, with raffinate at an irrigation rate of 5 L/m^{2}h. In order to monitor the leaching process, daily samples of the feed solution and the discharge solution of each column were collected for analysis (Cu, total Fe, and acidity). Table 5 shows the chemical composition of the IPLS and raffinate feed solutions. The concentrations of copper and iron were determined by atomic absorption spectrometry (AA) and the acidity was determined by titration with NaOH.

The irrigation time for columns 1–6 was 90 days and for columns 7–9 it was 45 days. At the end of the experiment, the columns were irrigated with 3 liters of water, simulating a washing step for the remaining copper solution. The residual ore was then unloaded onto an HDPE folder and representative samples were taken by incremental division. The samples were dried at 90 °C for 12 h and then crushed and pulverized. Sub-samples were subjected to sequential copper, total, and soluble iron analyses. The chemical analyses of the leachate and solid residues were used to calculate the amount of copper in extractions.

Leached residues from column 1 (agglomerated only with H$_2$SO$_4$) and column 5 (agglomerated with H$_2$SO$_4$, FeSO$_4$, and NaCl) were sampled for scanning electron microscopy analysis (SEM-EDX).

3. Results and Discussion

3.1. Leaching Temperature, Oxidation-Reduction Potential and Acid Consumption

An average temperature of 32.9 °C was recorded in the heated columns (columns 1–6), and 14.5 °C in columns 7–9. Acid consumption increased with temperature [6]. The maximum consumption calculated for the tests was approximately 36.5 kg acid/ton ore for the heated columns at 32.9 °C and 20 kg acid/ton ore for the columns operating at 14.5 °C.

The leaching solutions remained at a potential of about 600 mV (vs. SHE) in all columns with no major variations in order to prevent the formation of a passivation layer that hinders the dissolution of copper.

3.2. Copper Extraction

3.2.1. Chemical and SEM-EDX Analysis of the Residue

The results of the chemical analyses of the residual ore in each column are presented in detail in Table 6.

Table 6. Chemical analysis of the residual ore in each column.

Column Number	1	2	3	4	5	6	7	8	9
Total copper	0.23	0.24	0.29	0.20	0.23	0.19	0.34	0.36	0.40
Soluble copper	0.02	0.02	0.02	0.02	0.02	0.02	0.02	0.02	0.03
Cyanidable copper	0.06	0.07	0.08	0.05	0.04	0.04	0.12	0.13	0.16
Residual copper	0.15	0.16	0.19	0.14	0.17	0.13	0.20	0.21	0.21
Total iron	4.6	4.77	4.65	4.79	4.91	4.97	4.74	4.71	4.68

The analysis of the residues in column 1 (ore agglomerated with sulfuric acid and leached at 32.9 °C) using SEM shows that the ore still contains chalcopyrite but with a less crystalline appearance as a result of leaching (Figure 3 and Table 7).

Figure 3. Chalcopyrite in the leach residues of column 1.

Table 7. Chalcopyrite in the leached residues of column 1.

Element	Weight (%)
S	66.58
Fe	19.09
Cu	14.33
Total	100

The results in Table 7 indicate that the grains comprise partly reacted chalcopyrite, coated with a layer of sulfur. This is deduced from the presence of 66.58% sulfur, compared with the theoretical sulfur content in chalcopyrite of 34.8%.

3.2.2. Extraction of Copper

Table 8 shows the percentage of total copper extraction, with the respective percentages of extraction from copper oxides and sulfides in each column based on the analysis of feed ore and residue of each column.

Table 8. Percentage of copper extraction from oxides and sulfides.

Column Number	1	2	3	4	5	6	7	8	9
Extraction of total copper (%)	35.4	35.8	33.7	44.2	51.9	50.3	9.9	13.0	12.4
Extraction of oxide copper (%)	57.5	57.5	62.5	62.5	60.0	71.7	45.0	50.0	50.0
Extraction of sulfide copper (%)	32.6	33.2	30.8	41.9	51.2	46.3	5.7	9.1	7.7

From Table 8, it is noted that the use of NaCl and $FeSO_4$ during leaching carried out at 32.9 °C resulted in an increase in the extraction of copper from sulfides and a slight increase in the respective extraction from oxides. However, the use of these reagents during leaching at 14.5 °C results in the low extraction of copper from sulfides, thus indicating the effect of temperature on leaching.

In Figure 4, the results of copper extraction during the stages of agglomeration and curing (from 0 to 14 days), and leaching (from day 14 onwards) are compared. The results indicate that the extraction of copper during the leaching stage at 14.5 °C is not significant, and the curves tend to be asymptotic. However, when the temperature was increased to 32.9 °C, copper extractions during the leaching stage improved.

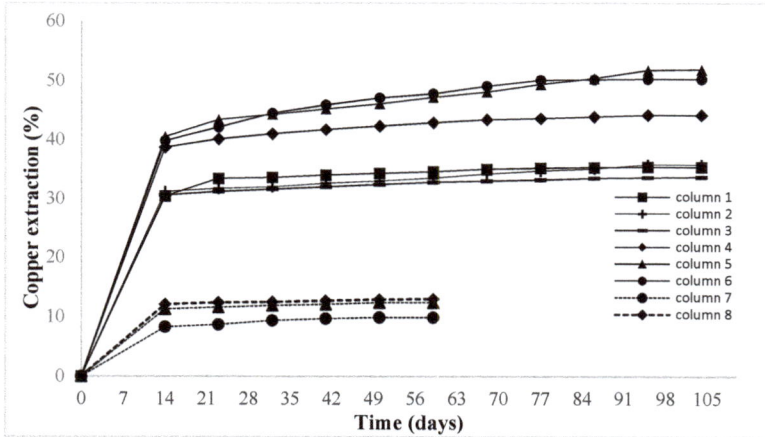

Figure 4. Extractions of copper during agglomeration-curing and leaching stages. Black (-) and dotted (—) lines correspond to the experiments performed at 32.9 and 14.5 °C in columns, respectively. Discontinuous vertical lines show the division between the agglomeration-curing and leaching stages.

3.2.3. Effect of Temperature, and Sulfuric Acid Used during Agglomeration

Table 9 shows the extraction of copper for columns 1 and 7 as calculated by chemical analysis of the feed ore and residue. Columns 1 and 7 were leached at 32.9 °C and 14.5 °C, respectively.

Table 9. Copper extractions in column 1 (32.9 °C, 0.6 kg H_2SO_4/50 kg of ore, 0 NaCl, 0 $FeSO_4$) and column 7 (14.5 °C, 0.6 kg H_2SO_4/50 kg of ore, 0 NaCl, 0 $FeSO_4$).

	Column 1			Column 7		
	Initial	Residue	% Ext	Initial	Residue	% Ext
Total copper	0.35	0.23	35.4	0.37	0.34	9.9
Soluble copper	0.04	0.02	50.0	0.04	0.02	50.0
Cyanidable copper	0.13	0.06	53.8	0.13	0.12	7.7
Copper in residue	0.18	0.15	16.7	0.20	0.20	0.0
Total iron	4.88	4.60		4.64	4.74	

The effect of temperature is significant in the leaching of cyanidable copper, which corresponds to the dissolution of chalcocite and covellite present in the ore. On the other hand, the copper remaining in the residues, which corresponds to the chalcopyrite, shows that no dissolution takes place in column 7.

3.2.4. Effect of Temperature and Sulfuric Acid/Sodium Chloride Used during Agglomeration

Table 10 shows the comparison of the columns 4 and 8, which were agglomerated with both H_2SO_4 and NaCl. Column 4 was leached at 32.9 °C, column 8 at 14.5 °C.

Column 4 shows 64.3% extraction from chalcocite-covellite, while column 8 shows only 13.3% extraction. With regard to the leaching of chalcopyrite (based on the copper present in the residue) column 4 shows an extraction of 22.2%, while column 8 only 4.5%. These results are according with those reported in the literature (Al-Harahsheh et al. [20], Kimball et al. [28], Lu et al. [29]).

Table 10. Copper extractions in column 4 (32.9 °C, 0.6 kg H_2SO_4/50 kg of ore, 0.525 kg NaCl/50 kg of ore, 0 $FeSO_4$) and column 8 (14.5 °C, 0.6 kg H_2SO_4/50 kg of ore, 0.525 kg NaCl/50 kg of ore, 0 $FeSO_4$).

	Column 4			Column 8		
	Initial	Residue	% Ext	Initial	Residue	% Ext
Total copper	0.36	0.20	44.2	0.41	0.36	13.0
Soluble copper	0.04	0.02	50.0	0.04	0.02	50.0
Cyanidable copper	0.14	0.05	64.3	0.15	0.13	13.3
Copper in residue	0.18	0.14	22.2	0.22	0.21	4.5
Total iron	4.82	4.79		4.87	4.71	

3.2.5. Effect of Temperature and Sulfuric Acid/Sodium Chloride/Ferrous Sulfate Used during Agglomeration

By comparing the results obtained after leaching (see Table 11) when H_2SO_4, NaCl, and $FeSO_4$ added during agglomeration, in columns 5 (32.9 °C) and 9 (14.5 °C), it can be deduced that temperature is the predominant factor that improves dissolution of copper sulfides. These results are consistent for column leaching systems with those reported by Kimball et al. [31].

Table 11. Copper extractions in column 5 (32.9 °C, 0.6 kg H_2SO_4/50 kg of ore, 0.525 kg NaCl/50 kg of ore, 0.5 kg $FeSO_4$/50 kg of ore) and column 9 (14.5 °C, 0.6 kg H_2SO_4/50 kg of ore, 0.525 kg NaCl/50 kg of ore, 0.5 kg $FeSO_4$/50 kg of ore).

	Column 5			Column 9		
	Initial	Residue	% Ext	Initial	Residue	% Ext
Total copper	0.47	0.23	51.9	0.45	0.40	12.4
Soluble copper	0.04	0.02	50.0	0.05	0.03	40.0
Cyanidable copper	0.19	0.04	78.9	0.17	0.16	5.9
Copper in residue	0.24	0.17	29.2	0.23	0.21	8.7
Total iron	4.97	4.91		4.67	4.68	

Column 5 shows 78.9% chalcocite-covellite, while column 9 shows only 5.9% extraction. Regarding the leaching of chalcopyrite (based on the copper present in the residue), column 5 shows an extraction of 29.2%, while column 9 shows only 8.7%.

3.2.6. Effect of Ferrous Ion

Similar copper extractions for the two columns agglomerated with H_2SO_4, NaCl, and $FeSO_4$, specifically column 5 (0.5 kg ferrous sulfate added; 51.9% Cu extracted) compared with column 6 (1 kg ferrous sulfate added; 50.3% Cu extracted), indicate that a greater amount of $FeSO_4$ in the agglomeration stage does not improve leaching kinetics. Similarly, copper extractions for columns 2 and 3 indicate that $FeSO_4$ alone has no notable effect on copper leaching, as there is little difference between them and column 1 (only agglomerated with acid).

Column 4 shows a greater extraction compared to columns 1, 2, and 3 as a consequence of NaCl addition during agglomeration (Figure 4). However, a comparison between columns 4, 5, and 6, indicates that the addition of $FeSO_4$ results in higher extraction efficiency in the presence of NaCl.

Columns operated at 14.5 °C show less than 15% copper extractions, although they contain NaCl and $FeSO_4$ added during agglomeration (Figure 4). This clearly indicates that the copper extraction in these tests is strongly temperature dependent.

3.2.7. Proposed Reactions in the Agglomeration-Curing Stage

During the agglomeration-curing stage the following reactions may occur:

(a) Agglomeration only with the use of acid (columns 1 and 7): as the acid is not able to leach sulfide copper ore, ferric ion was supplied by the raffinate solution. The action of the ferric ion has been proposed by Nicol and Basson [38] and Dutrizac [7] as follows:

Reactions	ΔG^0 14.5 °C	ΔG^0 32.9 °C
$CuS + 2Fe^{3+} = Cu^{2+} + 2Fe^{2+} + S$	−5.936	−6.858
$Cu_2S + 4Fe^{3+} = 2Cu^{2+} + 4Fe^{2+} + S$	−18.411	−20.136
$CuFeS_2 + 4Fe^{3+} = Cu^{2+} + 5Fe^{2+} + 2S$	−30.819	−32.671

(b) Agglomeration with acid and chloride (columns 4 and 8): The addition of chloride enhances the leaching action of the ferric ion [41]. The following reactions could occur:

Reactions	ΔG^0 14.5 °C	ΔG^0 32.9 °C
$CuS + 2Fe^{3+} + Cl^- = CuCl^- + 2Fe^{2+} + S$	−6.438	−7.416
$Cu_2S + 4Fe^{3+} + 2Cl^- = 2CuCl^- + 4Fe^{2+} + S$	−19.377	−21.251
$CuFeS_2 + 4Fe^{3+} + Cl^- = CuCl^- + 5Fe^{2+} + 2S$	−31.321	−33.229

Agglomeration with acid, chloride, and ferrous (columns 5, 6, and 7): in this case, the addition of ferrous ions improves the total copper extractions compared to the leaching of agglomerates produced without the use of this reagent. Possibly its presence prevents or minimizes the formation of the passivation layer on the surface of the sulfides.

4. Conclusions

The effect of the modification of the agglomeration-curing stages and temperature on copper leaching kinetics of a copper sulfide ore was reached using H_2SO_4, NaCl, and $FeSO_4$.

In the column tests, the largest copper extractions were obtained when leaching was carried out at a temperature of 32.9 °C. As temperature increases, for the same agglomeration conditions, extraction is up to five times higher compared to leaching at 14.5 °C.

The highest copper extraction was obtained in column 5 ore agglomerated with H_2SO_4, NaCl, and 1% w/w $FeSO_4$, in which an extraction of 50.23% of copper was achieved, followed by column 6 (agglomerated with H_2SO_4, NaCl, and 2% w/w $FeSO_4$), where an extraction of 48.5% was obtained.

The next highest extraction, 41.84%, was obtained in column 4 (agglomerated with H_2SO_4 and NaCl). Copper extractions in the columns leached at 14.5 °C did not exceed 15%, despite the addition of NaCl and $FeSO_4$.

This medium, using ferrous-chloride, could be applied in the agglomerate stage, prior to heap leaching of copper sulfide minerals at an industrial scale.

Author Contributions: The authors contributions are the following: K.E.S. conceived, designed, and performed the experiments; O.H. and C.M.T. analyzed the data; K.E.S. also contributed with reagents/materials/analysis tools; O.H., K.E.S. and C.M.T. wrote the paper.

Funding: Research office from Universidad Católica del Norte.

Acknowledgments: The authors are grateful to Laboratorio de Investigación de Procesos (LIP) from Universidad de Antofagasta for supporting the present study.

Conflicts of Interest: The authors declare no conflicts of interest.

References

1. Ekman Nilsson, A.; Macias Aragonés, M.; Arroyo Torralvo, F.; Dunon, V.; Angel, H.; Komnitsas, K.; Willquist, K. A Review of the Carbon Footprint of Cu and Zn Production from Primary and Secondary Sources. *Minerals* **2017**, *7*, 168. [CrossRef]
2. Watling, H. Chalcopyrite hydrometallurgy at atmospheric pressure: 1. Review of acidic sulfate, sulfate–chloride and sulfate—Nitrate process options. *Hydrometallurgy* **2013**, *140*, 163–180. [CrossRef]
3. Villarino, J. *Reporte Anual del Consejo Minero 2011–2012*; Consejo Minero: Santiago, Chile, 2012.
4. Munoz, P.; Miller, J.D.; Wadsworth, M.E. Reaction mechanism for the acid ferric sulfate leaching of chalcopyrite. *Met. Trans. B* **1979**, *10*, 149–158. [CrossRef]
5. Hiroyoshi, N.; Hirota, M.; Hirajima, T.; Tsunekawa, M. A case of ferrous sulfate addition enhancing chalcopyrite leaching. *Hydrometallurgy* **1997**, *47*, 37–45. [CrossRef]
6. Hiroyoshi, N.; Miki, H.; Hirajima, T.; Tsunekawa, M. Enhancement of chalcopyrite leaching by ferrous ions in acidic ferric sulfate solutions. *Hydrometallurgy* **2001**, *60*, 185–197. [CrossRef]
7. Dutrizac, J. The dissolution of chalcopyrite in ferric sulfate and ferric chloride media. *Met. Trans. B* **1981**, *12*, 371–378. [CrossRef]
8. Dutrizac, J. Elemental sulphur formation during the ferric chloride leaching of chalcopyrite. *Hydrometallurgy* **1990**, *23*, 153–176. [CrossRef]
9. Sokić, M.D.; Marković, B.; Živković, D. Kinetics of chalcopyrite leaching by sodium nitrate in sulphuric acid. *Hydrometallurgy* **2009**, *95*, 273–279. [CrossRef]
10. Senanayake, G. A review of chloride assisted copper sulfide leaching by oxygenated sulfuric acid and mechanistic considerations. *Hydrometallurgy* **2009**, *98*, 21–32. [CrossRef]
11. Flett, D. Chloride hydrometallurgy for complex sulphides: A review. *CIM Bull.* **2002**, *95*, 95–103.
12. Klauber, C. A critical review of the surface chemistry of acidic ferric sulphate dissolution of chalcopyrite with regards to hindered dissolution. *Int. J. Miner. Process.* **2008**, *86*, 1–17. [CrossRef]
13. Debernardi, G.; Carlesi, C. Chemical-electrochemical approaches to the study passivation of chalcopyrite. *Miner. Process. Extr. Met. Rev.* **2013**, *34*, 10–41. [CrossRef]
14. Stott, M.; Watling, H.; Franzmann, P.; Sutton, D. The role of iron-hydroxy precipitates in the passivation of chalcopyrite during bioleaching. *Miner. Eng.* **2000**, *13*, 1117–1127. [CrossRef]
15. Córdoba, E.; Muñoz, J.; Blázquez, M.; González, F.; Ballester, A. Leaching of chalcopyrite with ferric ion. Part II: Effect of redox potential. *Hydrometallurgy* **2008**, *93*, 88–96. [CrossRef]
16. Carneiro, M.F.C.; Leão, V.A. The role of sodium chloride on surface properties of chalcopyrite leached with ferric sulphate. *Hydrometallurgy* **2007**, *87*, 73–82. [CrossRef]
17. Vilcáez, J.; Inoue, C. Mathematical modeling of thermophilic bioleaching of chalcopyrite. *Miner. Eng.* **2009**, *22*, 951–960. [CrossRef]
18. Harmer, S.L.; Thomas, J.E.; Fornasiero, D.; Gerson, A.R. The evolution of surface layers formed during chalcopyrite leaching. *Geochim. Cosmochim. Acta* **2006**, *70*, 4392–4402. [CrossRef]
19. Ammou-Chokroum, M.; Cambazoglu, M.; Steinmez, D. Oxydation menagée de la chalcopyrite en solution acide: Analyses cinétique de réactions. II. Modéles diffusionales. *Bull. Soc. Fr. Miner. Cristal.* **1977**, *100*, 161–177.
20. Hiroyoshi, N.; Miki, H.; Hirajima, T.; Tsunekawa, M. A model for ferrous-promoted chalcopyrite leaching. *Hydrometallurgy* **2000**, *57*, 31–38. [CrossRef]
21. Kametani, H.; Aoki, A. Effect of suspension potential on the oxidation rate of copper concentrate in a sulfuric acid solution. *Met. Trans. B* **1985**, *16*, 695–705. [CrossRef]
22. Komnitsas, C.; Pooley, F. Optimization of the bacterial oxidation of an arsenical gold sulphide concentrate from Olympias, Greece. *Miner. Eng.* **1991**, *4*, 1297–1303. [CrossRef]
23. Al-Harahsheh, M.; Kingman, S.; Al-Harahsheh, A. Ferric chloride leaching of chalcopyrite: Synergetic effect of $CuCl_2$. *Hydrometallurgy* **2008**, *91*, 89–97. [CrossRef]
24. Sato, H.; Nakazawa, H.; Kudo, Y. Effect of silver chloride on the bioleaching of chalcopyrite concentrate. *Int. J. Miner. Process.* **2000**, *59*, 17–24. [CrossRef]
25. Lu, Z.; Jeffrey, M.; Lawson, F. An electrochemical study of the effect of chloride ions on the dissolution of chalcopyrite in acidic solutions. *Hydrometallurgy* **2000**, *56*, 145–155. [CrossRef]
26. Li, J.; Kawashima, N.; Kaplun, K.; Absolon, V.J.; Gerson, A.R. Chalcopyrite leaching: The rate controlling factors. *Geochim. Cosmochim. Acta* **2010**, *74*, 2881–2893. [CrossRef]

27. Kaplun, K.; Li, J.; Kawashima, N.; Gerson, A. Cu and Fe chalcopyrite leach activation energies and the effect of added Fe^{3+}. *Geochim. Cosmochim. Acta* **2011**, *75*, 5865–5878. [CrossRef]

28. Viramontes-Gamboa, G.; Peña-Gomar, M.M.; Dixon, D.G. Electrochemical hysteresis and bistability in chalcopyrite passivation. *Hydrometallurgy* **2010**, *105*, 140–147. [CrossRef]

29. Sandström, Å.; Shchukarev, A.; Paul, J. XPS characterisation of chalcopyrite chemically and bio-leached at high and low redox potential. *Miner. Eng.* **2005**, *18*, 505–515. [CrossRef]

30. Hiroyoshi, N.; Kuroiwa, S.; Miki, H.; Tsunekawa, M.; Hirajima, T. Synergistic effect of cupric and ferrous ions on active-passive behavior in anodic dissolution of chalcopyrite in sulfuric acid solutions. *Hydrometallurgy* **2004**, *74*, 103–116. [CrossRef]

31. Kimball, B.E.; Rimstidt, J.D.; Brantley, S.L. Chalcopyrite dissolution rate laws. *Appl. Geochem.* **2010**, *25*, 972–983. [CrossRef]

32. Lu, Z.; Jeffrey, M.; Lawson, F. The effect of chloride ions on the dissolution of chalcopyrite in acidic solutions. *Hydrometallurgy* **2000**, *56*, 189–202. [CrossRef]

33. Nicol, M.; Miki, H.; Velásquez-Yévenes, L. The dissolution of chalcopyrite in chloride solutions: Part 3. Mechanisms. *Hydrometallurgy* **2010**, *103*, 86–95. [CrossRef]

34. Yévenes, L.V.; Miki, H.; Nicol, M. The dissolution of chalcopyrite in chloride solutions: Part 2: Effect of various parameters on the rate. *Hydrometallurgy* **2010**, *103*, 80–85. [CrossRef]

35. Velásquez-Yévenes, L.; Nicol, M.; Miki, H. The dissolution of chalcopyrite in chloride solutions: Part 1. The effect of solution potential. *Hydrometallurgy* **2010**, *103*, 108–113. [CrossRef]

36. Nicol, M.J. The anodic behaviour of chalcopyrite in chloride solutions: Overall features and comparison with sulfate solutions. *Hydrometallurgy* **2017**, *169*, 321–329. [CrossRef]

37. Nicol, M.; Miki, H.; Zhang, S. The anodic behaviour of chalcopyrite in chloride solutions: Voltammetry. *Hydrometallurgy* **2017**, *171*, 198–205. [CrossRef]

38. Nicol, M.; Basson, P. The anodic behaviour of covellite in chloride solutions. *Hydrometallurgy* **2017**, *172*, 60–68. [CrossRef]

39. McBride, D.; Croft, T.; Cross, M.; Bennett, C.; Gebhardt, J. Optimization of a CFD–Heap leach model and sensitivity analysis of process operation. *Miner. Eng.* **2014**, *63*, 57–64. [CrossRef]

40. Ilankoon, I.; Neethling, S. Liquid spread mechanisms in packed beds and heaps. The separation of length and time scales due to particle porosity. *Miner. Eng.* **2016**, *86*, 130–139. [CrossRef]

41. Herreros, O. *Lixiviación de Especies Sulfuradas de Cobre en Medios Clorurados: Aspectos Cinéticos y Termodinámicos*; Editorial Académica Española: Beau Bassin, Mauritius, 2017.

minerals

Article

Leaching of Chalcopyrite in Acidified Nitrate Using Seawater-Based Media

Pía C. Hernández [1,*], María E. Taboada [1,2], Osvaldo O. Herreros [3], Teófilo A. Graber [1,2] and Yousef Ghorbani [4]

[1]   Departamento de Ingeniería Química y Procesos de Minerales, Universidad de Antofagasta, Avda. Angamos 601, Antofagasta 1270300, Chile; mariaelisa.taboada@uantof.cl (M.E.T.); teofilo.graber@uantof.cl (T.A.G.)

[2]   Centro de Investigación Científico Tecnológico Para la Minería (CICITEM), Sucre 220, Of. 602, Antofagasta 1270300, Chile

[3]   Departamento de Ingeniería en Minas, Universidad de Antofagasta, Avda. Angamos 601, Antofagasta 1270300, Chile; osvaldo.herreros@uantof.cl

[4]   School of Natural and Built Environment, Faculty of Science, Engineering and Computing, Kingston University, London KT1 2EE, UK; y.ghorbani@kingston.ac.uk

*   Correspondence: pia.hernandez@uantof.cl; Tel.: +56-55-2-637-313

Received: 4 May 2018; Accepted: 30 May 2018; Published: 1 June 2018

Abstract: The leaching of copper from industrial copper ore with 4.8 wt % chalcopyrite by acidified nitrate with seawater based media was investigated. Water quality (pure water and seawater), temperature (25–70 °C), reagent concentration, and nitrate type (sodium and potassium) were studied variables. Leaching conditions were: 100 g ore/1 L solution; P_{80} of 62.5 μm; 400 rpm and leaching time, varying between 3 and 7 days. Nitrates in sulfuric acid are known to be good oxidants for sulfide ores. This study showed that up to 80 wt % copper could be extracted at 45 °C in 7 days. In the absence of nitrate, under the same leaching conditions, only a 28 wt % copper extraction was achievable. The extraction rate increased to 97.2 wt % when leach temperature was increased to 70 °C in nitrate-chloride-acid media. The presence of chloride ions also increased the copper extraction rate. The copper extraction achieved in seawater systems were higher than in water systems under the same leaching conditions (increased by an average of 18 wt %). This effect can be attributable to the contribution of chloride that increases proton (H^+) activity.

Keywords: nitrate; chalcopyrite; chloride; leaching; seawater

1. Introduction

Given that copper oxide minerals will probably be depleted in the near future and hydrometallurgical plants could then be left unused, hydrometallurgical methods to process copper sulfide represent an alternative use for solvent extraction and electro-winning plants.

Many researchers have argued that chloride ions are beneficial to leaching in the hydrometallurgical processing of chalcopyrite [1–7], owing to the formation of soluble copper chloro complexes [8,9]. Cupric and ferric ions have been tested as oxidants in chloride media [10–13] with beneficial results. Mining companies in desert areas (e.g., Las Luces, Taltal, Chile) use seawater in their operations because of limited access to freshwater [14–16]. The use of seawater in sulfuric acid leaching is expected to increase the dissolution rate of the copper sulfide ores as seawater contains approximately 19 g·L^{-1} chloride [17,18].

There are large reserves of nitrate (caliche) in Northern Chile [19–21]. Sodium nitrate is obtained by caliche leaching with water and then crystallizing the leaching solution by cooling or evaporation. In the case of evaporation, discarded salts (tailings) from the solar pond still contains significant amount (4.6 wt % $NaNO_3$) of nitrate salts [22]. Such tailings can be used as oxidants in chalcopyrite leaching. Habashi [23] proposed the use of nitric acid (HNO_3) and indicated that in sulfide ore leaching nitrate ions act as an oxidant by forming NO or NO_2 gases and oxygen, from the decomposition of HNO_3. Other researchers [24,25] related the better performance of HNO_3 as an oxidant due to its high redox potential. In a system of $CuFeS_2$, HNO_3 and H_2SO_4, the reaction products are Cu^{2+}, Fe^{3+}, S^0 and NO gas. This gas (NO) can be captured and oxidized with air and water to form HNO_3. At higher temperatures, sodium nitrate converts to nitric acid in presence of sulfuric acid (Equations (1)–(2); Kazakov et al. [26]):

$$NaNO_3 + H_2SO_4 = HNO_3 + NaHSO_4 \tag{1}$$

$$2NaNO_3 + H_2SO_4 = 2HNO_3 + Na_2SO_4 \tag{2}$$

There are several leaching studies [27–31] that use nitrogen species (e.g., NO_3^-, nitrate, NO_2^-, nitrite, HNO_3, nitric acid, or HNO_2, nitrous acid, and NO_2, nitrogen dioxide) as they increase the oxidization capacity, and enhance kinetics and/or dissolution of ores. Nitrogen species catalyzed (NSC) process uses nitric and sulfuric acid to leach sulfide ores at high temperature and pressure. This process has been successfully used in the mining industry for pressure leaching of copper sulfides [32]. The main advantage of this process is faster reaction rates. Tsogtkhankhai et al. [33] studied leaching copper concentrates using nitric acid, and determined the effects of nitric acid strength, liquid solid ratio, time and temperature on the copper extraction rate. Depending on the activation energy, temperature has a slight effect on the rate of copper extraction from chalcopyrite. The outset of leaching is controlled by an external diffusion regime. As time passes, the dissolution reaction is controlled by an internal diffusion regime due to the formation of a film around the particles. The same authors [34] determined optimal sulfide dissolution conditions for the Cu-Fe-S-N-O-H system in nitric acid at different temperatures (20 to 100 °C) using thermodynamic diagrams. They found that the best conditions were: high initial oxidizing potentials (high concentration of nitric acid), pH lower than 1 and high temperature. Arias [35] patented a hydrometallurgical process of copper sulfide heap leaching using H_2SO_4 and $NaNO_3$ at pH ≤ 2, with a sulfide ore/nitrate ratio (by weight) of 1:1. Sodium nitrate was mixed with sulfuric acid and water. Prater and Queneau [36] patented a process for dissolving sulfide ores (copper, iron, cobalt, nickel, silver) using nitric acid which is continuously added during leaching. Iron (in the form of hydrogen jarosite) and sulfur compounds were precipitated; these products are highly filterable and are subsequently removed. It is mentioned that the formation of jarosites, which is followed by drop of pH, may cause losses of valuable ions from solution as a result of co-precipitation [37].

Leaching of copper ores in sulfuric acid with sodium nitrate/nitrite solutions resulted in copper extraction rates of 80–99 wt % [28,29]. Vračar et al. [30] leached Cu_2S at different temperatures, $NaNO_3$ and H_2SO_4 concentrations, stirring speeds and solid-liquid ratios. The authors determined that Cu_2S was leached in two stages, forming CuS, followed by elemental S, according to X-ray diffraction (XRD) analysis of leaching residues. The activation energy for the process was 60 $kJ \cdot mol^{-1}$. Ore leaching was a first order reaction and a second order reaction with respect to $NaNO_3$ and H_2SO_4 concentrations, respectively. Sokić et al. [28] leached a chalcopyrite concentrate and studied the effect of different temperatures, particle sizes, concentrations of $NaNO_3$ and H_2SO_4 and stirring speeds. The authors proposed the following reactions to explain the dissolution of chalcopyrite in acid-nitrate media (Equations (3)–(5)). These reactions are thermodynamically viable at 25 °C and atmospheric pressure due to the negative value of the Gibbs energy:

$$CuFeS_2 + 5NaNO_3 + 5H_2SO_4 = CuSO_4 + \tfrac{1}{2}Fe_2(SO_4)_3 + \tfrac{5}{2}Na_2SO_4 + 2S + 5NO_2 + 5H_2O \quad \Delta G_{25°C} = -422.7 \ kJ \cdot mol^{-1} \tag{3}$$

$$CuFeS_2 + 4NaNO_3 + 4H_2SO_4 = CuSO_4 + FeSO_4 + 2Na_2SO_4 + 2S + 4NO_2 + 4H_2O \quad \Delta G_{25°C} = -352.3 \ kJ \cdot mol^{-1} \tag{4}$$

$$FeSO_4 + \tfrac{1}{3}NaNO_3 + \tfrac{2}{3}H_2SO_4 = \tfrac{1}{2}Fe_2(SO_4)_3 + \tfrac{1}{6}Na_2SO_4 + \tfrac{1}{3}NO + \tfrac{2}{3}H_2O \quad \Delta G_{25°C} = -65.1 \ kJ \cdot mol^{-1} \tag{5}$$

The authors found that the modeled kinetics of the reaction is dominated by a chemical surface reaction, followed by diffusion in the product layer with an activation energy of 83 kJ·mol^{-1}. The residues produced were unreacted chalcopyrite and elemental sulfur. Gok and Anderson [29] leached chalcopyrite using an acidic solution containing a nitrite salt (Equation (6)). They obtained a 5 wt % increase in copper extraction rate by using nitrite instead of nitrate. The overall copper extraction rate was positively affected by the system temperature and the newly formed elemental sulfur in the residue.

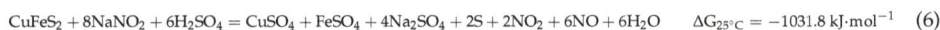

$$CuFeS_2 + 8NaNO_2 + 6H_2SO_4 = CuSO_4 + FeSO_4 + 4Na_2SO_4 + 2S + 2NO_2 + 6NO + 6H_2O \quad \Delta G_{25°C} = -1031.8 \text{ kJ·mol}^{-1} \quad (6)$$

This study considers the use of nitrate instead of nitrite because Chile has an industry based in nitrate production. Moreover, nitrite is more expensive than nitrate.

Shiers et al. [38] carried out leaching tests using nitrate as an oxidant in chalcopyrite concentrate and ore at 50 °C. The authors determined the effect of nitrate, ferric nitrate, ferric chloride and ferric sulfate on copper extraction. A mixed chloride-nitrate system was favorable for extracting copper from chalcopyrite while the presence of ferric chloride was also beneficial.

However, the use of nitrate as an oxidant in an industrial process could be a problem in the solvent extraction stage (SX) due to the degradation of oxime-type extractants by nitration (nitration is the degradation of organic compounds caused by the presence of high nitrate concentration in pregnant leaching solution, PLS) [39–41]. Nowadays, chemical industries have developed nitration resistant reagents to solve this problem [42–44].

In this study, seawater and nitrate salts (NaNO$_3$, KNO$_3$) were used to investigate the acid dissolution of commercial copper ore that contains 4.8 wt % chalcopyrite (1.6 wt % Cu). The effect of different physical and chemical conditions and variables were studied, including temperature, sulfuric acid and nitrate concentrations, nitrate sources, leaching time and water quality.

2. Materials and Methods

2.1. Ore Sample

In this study, an ore from the Atacama Region, Chile was used. Table 1 provides the mineralogical and chemical composition of the ore sample.

Table 1. Mineralogy and chemical composition of the ore sample.

Mineralogy		Chemical Analysis	
Minerals	(wt %)	Element	(wt %)
Magnetite (Fe$_3$O$_4$)	36.7	Iron (Fe)	33.6
Quartz (SiO$_2$)	17.1	Silicon (Si)	10.9
Plagioclase ((Ca,Na)(Al,Si)AlSi$_2$O$_8$)	9.1	Calcium (Ca)	4.1
Pyrite (FeS$_2$)	8.1	Sulphur (S)	3.8
Chalcopyrite (CuFeS$_2$)	4.8	Aluminium (Al)	2.8
Calcite (CaCO$_3$)	4.6	Sodium (Na)	1.8
Kaolinite (Al$_4$(Si$_4$O$_{10}$)(OH)$_8$)	4.3	Copper (Cu) *	1.6
Garnet (Ca$_3$Fe$_2$Si$_3$O$_{12}$)	3.7	Magnesium (Mg)	0.9
Actinolite (Ca$_2$(Mg,Fe)$_{2.5}$Si$_8$O$_{22}$(OH)$_2$)	3.5	Manganese (Mn)	0.6
Amphibole (NaCa$_2$(Mg,Fe,Al)$_5$(Si,Al)$_8$O$_{22}$(OH)$_2$)	3.5	Potassium (K)	0.6
Epidote (Ca$_2$Al$_2$FeSi$_3$O$_{12}$(OH))	2.6	Cobalt (Co)	0.4
Sericite (KAl$_2$(AlSi$_3$O$_{10}$)(OH)$_2$)	≤1.0	Chromium (Cr)	0.1
Chlorite ((Mg,Fe)$_3$(AlSi)$_4$O$_{10}$(OH)$_2$(Mg,Fe)$_3$(OH)$_6$)	≤1.0	Nickel (Ni)	0.1
		Zinc (Zn)	<0.1
		Molybdenum (Mo)	<0.1
		Silver (Ag)	<0.1

* Chemical analysis by decomposing the ore in a nitric-perchloric acid solution and atomic absorption spectrometry (AAS).

The mineral characterization was carried out by semi-quantitative X-ray diffraction (Siemens/Bruker, Semi-QXRD, model D5000, Germany). The mineralogical data show that chalcopyrite is the only copper mineral present in the ore. Chemical analysis was performed by inductively coupled plasma atomic emission spectroscopy (ICP-AES, ICPE-9000, Shimadzu, Tokyo, Japan). The chemical analysis by atomic absorption spectrometry (AAS, Perkin-Elmer 2380, Perkin Elmer, Wellesley, MA, USA) measured the copper grade as 1.6 wt %. The particle size of the sample (P_{80}) in all tests was 62.5 μm, which was determined using a Microtrac Particle Analyzer (Microtrac S3500, Microtrac, Montgomeryville, PA, USA).

2.2. Reagents

Analytical grade H_2SO_4, $NaNO_3$, KNO_3 and $NaCl$ were used in all leaching tests. Distilled water (referred to as "water" in the text) and seawater, obtained from San Jorge Bay, Antofagasta, Chile, were used to prepare dilute sulfuric acid. The seawater was filtered using quartz sand (50 μm) and a polyethylene membrane (1 μm). The seawater composition is shown in Table 2. One test used a synthetic saline solution composed of distilled water and 35 $g \cdot L^{-1}$ $NaCl$. That value was chosen because the salinity of seawater is 3.5%, so all salinity of seawater was considered as $NaCl$.

Table 2. Composition of seawater obtained from San Jorge Bay, Antofagasta, Chile.

Chemical Method	ICP-AES					AAS		Volumetric Analysis		Gravimetric Analysis
Ionic Species	Na^+	Mg^{2+}	Ca^{2+}	K^+	B^{3+}	Cu^{2+}	NO_3^-	Cl^-	HCO_3^-	SO_4^{2-}
$mg \cdot L^{-1}$	9480	1190	386	374	4.6	0.07	2.05	18,765	142	2771

2.3. Experimental Procedure

The leaching tests used 2 L jacketed glass reactors equipped with lids to prevent evaporative loss (the reactors are not hermetically sealed). Temperature was controlled by a thermostatic bath. Stirring was mechanical at 400 rpm, which provided stable suspension. The solid/liquid ratio used for all tests was 100 g ore in 1 L of solution. 1 L of leach solution was prepared by adding sodium nitrate (solid) and sulfuric acid to seawater or water. The tests were performed separately using both seawater and water. The stirring was periodically interrupted to collect 10 mL samples of supernatant solution from the reactors to analyze the copper content using AAS method. The copper extraction over time was calculated by dividing the copper concentration in the solution at time t and the initial copper concentration (1.6 wt %). The evaporation loss and the volume of samples removed during the tests are considered in calculating copper extraction (loss of mass). The redox potential (ORP) and pH were monitored during the leaching time. ORP was measured with a redox electrode (Ag/AgCl reference electrode with 3.5 M KCl as electrolyte) and pH was measured with a glass membrane electrode calibrated to buffers at pH 1 and pH 4. At the end of the leaching time, the pulp was filtered and washed with distilled water and dried at 60 °C. The leaching residues were characterized (Section 2.4). The total copper extracted from solid residues were compared with the leached copper extractions with an average standard deviation of ±2%.

The performed leaching tests were summarized in Table 3. The variables studied were: sulfuric acid concentration (0.25, 0.5 and 1 M), sodium nitrate concentration (0, 0.25, 0.5 and 1 M), temperature (25, 30, 35, 40, 45 and 70 °C), nitrate sources (KNO_3 and $NaNO_3$), water type (water, seawater and saline solution) and leaching time (3, 5 and 7 days).

In test 14, a synthetic saline solution was used in the preparation of a H_2SO_4 (1 M) + $NaNO_3$ (1 M) leach liquor. The objective of the test was to determine whether $NaCl$ is the only component of seawater that aids higher copper extraction.

Table 3. Experimental conditions used in chalcopyrite leaching tests

No.	H_2SO_4 (M)	$NaNO_3$ (M)	Water Type	T (°C)	t (Days)	Final Cu Extraction (wt %)
1	1.0	1.0	seawater	45	7	80.2
	1.0	1.0	water	45	7	63.9
2	1.0	0.0	seawater	45	7	27.9
	1.0	0.0	water	45	7	14.8
3	1.0	1.0	seawater	45	3	78.8
	1.0	1.0	water	45	3	61.0
4	1.0	1.0	seawater	40	3	65.3
	1.0	1.0	water	40	3	45.6
5	1.0	1.0	seawater	35	3	62.4
	1.0	1.0	water	35	3	37.2
6	1.0	1.0	seawater	30	3	57.1
	1.0	1.0	water	30	3	35.9
7	0.5	1.0	seawater	45	3	66.5
	0.5	1.0	water	45	3	49.4
8	0.25	1.0	seawater	45	3	52.6
	0.25	1.0	water	45	3	42.1
9	1.0	0.5	seawater	45	3	77.3
	1.0	0.5	water	45	3	47.4
10	1.0	0.25	seawater	45	3	68.6
	1.0	0.25	water	45	3	43.1
11	1.0	1.0	seawater	45	5	79.3
	1.0	1.0	water	45	5	62.9
12	0.5	0.5	seawater	45	5	61.2
	0.5	0.5	water	45	5	47.7
13	0.5	0.5 *	seawater	45	5	58.3
	0.5	0.5 *	water	45	5	49.1
14	1.0	1.0	saline solution	45	5	75.7
15	1.0	1.0	seawater	70	3	97.2
	1.0	1.0	water	70	3	83.2
16	1.0	1.0	seawater	25	3	31.8
	1.0	1.0	water	25	3	11.0

* 0.5 M as KNO_3.

2.4. Characterization of Ore Residues

The solid residues were characterized by AAS, optical microscopy with reflected light, scanning electron microscopy (SEM-EDX, JEOL 6260 LV, Tokyo, Japan) and semi-quantitative XRD (QXRD) methods.

3. Results and Discussion

3.1. Effect of Sodium Nitrate and Concentrations

Figure 1 shows copper extraction from the leach liquors using H_2SO_4 (1 M) with and without $NaNO_3$ in seawater and water-based media (tests 1 and 2 in Table 3).

It can be seen that the addition of $NaNO_3$ (1 M) to systems with H_2SO_4 (1 M), increased the copper extraction from 27.9 to 80.2% in seawater-based media and from 14.8 to 63.9% in water- based media. This improvement was related to the oxidizing potential of nitrate ions. The use of seawater clearly had a positive effect on these two sets of tests, yielding higher copper extraction rates than with water. Copper extraction in seawater reached its maximum at about 96 h (4 days).

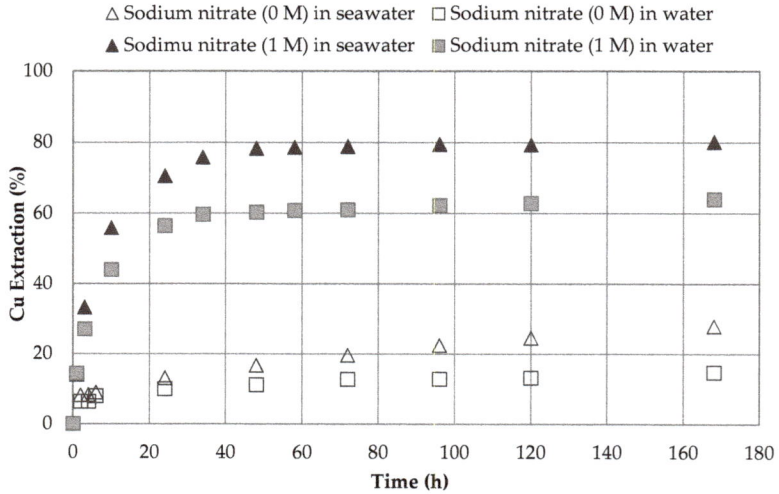

Figure 1. Copper extraction in the leach liquors using H_2SO_4 (1 M) + $NaNO_3$ (0 or 1 M) in seawater and water based media (at 45 °C and 7 days).

Figure 2a shows that a concentration of 0.5 M of sodium nitrate in seawater was sufficient to achieve a reasonably good level of copper extraction (77.3%) in 3 days. For the water system, a much higher concentration of nitrate (1 M) was needed to extract only 60.9% of copper in the same period (Figure 2b). This indicates that nitrate performs better with chloride ions during chalcopyrite leaching.

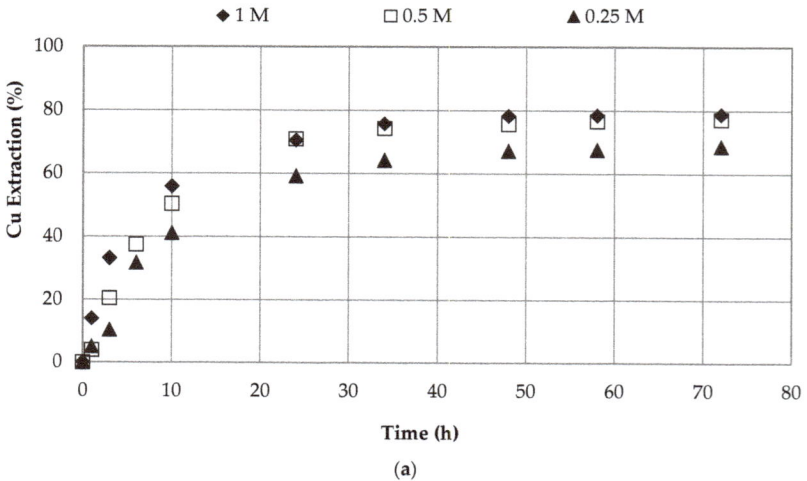

(a)

Figure 2. *Cont.*

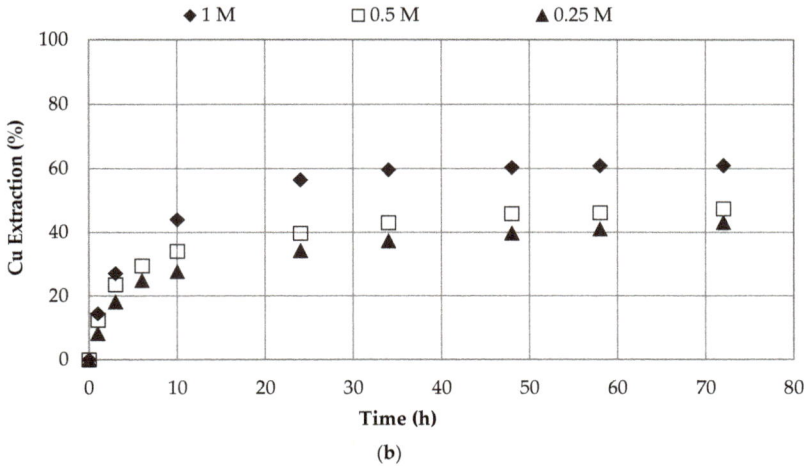

Figure 2. Effect of sodium nitrate concentration on copper extraction using H_2SO_4 (1 M), 3 days of leaching and 45 °C: (**a**) in seawater and (**b**) in water.

3.2. Effect of the Nitrate Source

Figure 3 shows the effect of the type of nitrate. Similar copper extraction rates were obtained when $NaNO_3$ or KNO_3 was used. Copper extraction reached 60% in seawater systems and 48% in water systems (0.5 M H_2SO_4) in 5 days. Thus, the type of nitrate ($NaNO_3$ or KNO_3) does not affect copper extraction, which is consistent with Shiers et al. [38].

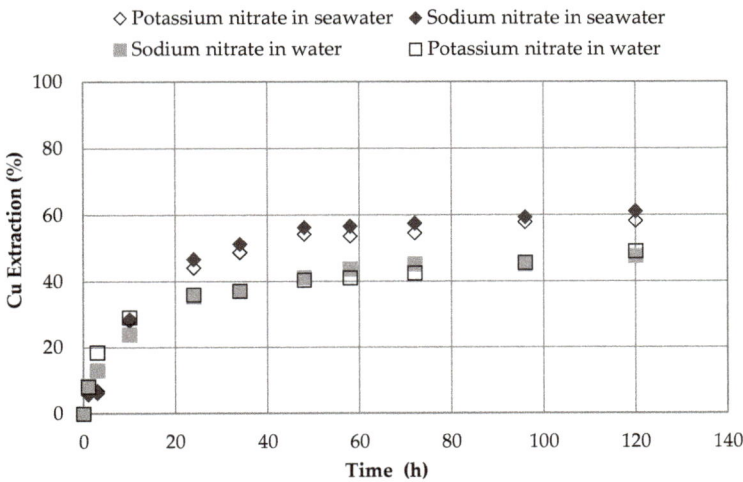

Figure 3. Effect of nitrate type on copper extraction. Experimental conditions: H_2SO_4 (0.5 M), $NaNO_3$ (0.5 M), KNO_3 (0.5 M), 5 days of leaching and 45 °C.

3.3. Effect of Sulfuric Acid Concentration

Figure 4a,b show the influence of sulfuric acid concentration. The highest copper extraction rate was obtained at higher concentrations of acid in seawater and water systems. The results indicate that the oxidation power of nitrate ions increases with increased in sulfuric acid concentrations. This observation is in good agreement with the findings of Sokić et al. [28]. In this study, copper extraction increased by 50% and 45% when the sulfuric acid concentration increased from 0.25 to 1.0 M in seawater and water systems, respectively.

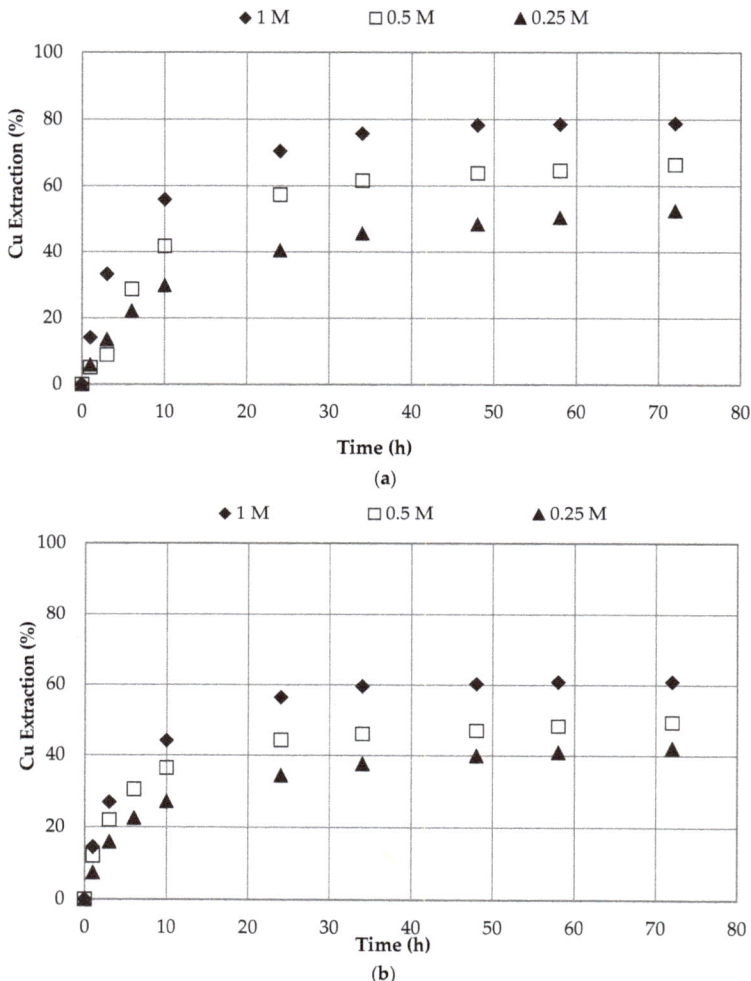

Figure 4. Effect of sulfuric acid concentration on copper extraction using $NaNO_3$ (1 M), 3 days of leaching and 45 °C: (**a**) seawater and (**b**) water.

As Figure 5 shows, copper extraction using 0.5 M of H_2SO_4 in a seawater-based media is similar to that with 1 M of H_2SO_4 in a water-based media. The same behavior is observed in the kinetic curves of copper extraction using 0.25 M of H_2SO_4 in seawater and 0.5 M of H_2SO_4 in water. This confirms that chloride ions from seawater increase the acid activity of the systems. Puvvada et al. [45] and Senanayake [46] obtained similar results. These authors indicated that the presence of salts such as NaCl, $CaCl_2$ or $MgCl_2$ increase acid activity. This shows that a lower acid concentration can be used when seawater is present in the system.

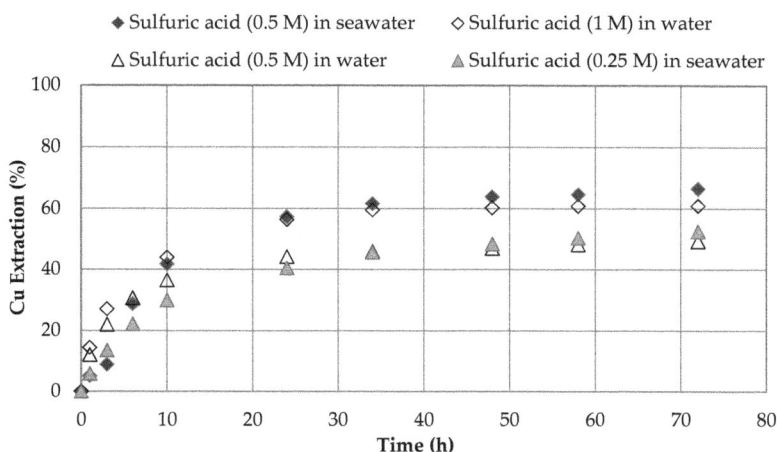

Figure 5. Copper extraction in the leach liquors with experimental conditions of: $NaNO_3$ (1 M), 3 days of leaching, 45 °C.

3.4. Redox Potential and pH

Figure 6 shows the copper extraction rates and redox potential values obtained during the tests at different concentrations of reagents (1 and 0.5 M of H_2SO_4 and $NaNO_3$). Because of the high acid concentrations (1 and 0.5 M) used in the tests, all the pH values were < 1. The redox potential (ORP) ranged from 742 mV to 793 mV in seawater and from 739 mV to 789 mV in water when 1 M of sulfuric acid and sodium nitrate were used. When 0.5 M was used, the range of redox potential was 701 mV to 738 mV in seawater and 696 mV to 729 mV in water. In both cases, the redox potential was higher in the seawater system than in water system. This indicates that copper leaching is more effective at higher redox potentials and in a strongly acid medium with seawater.

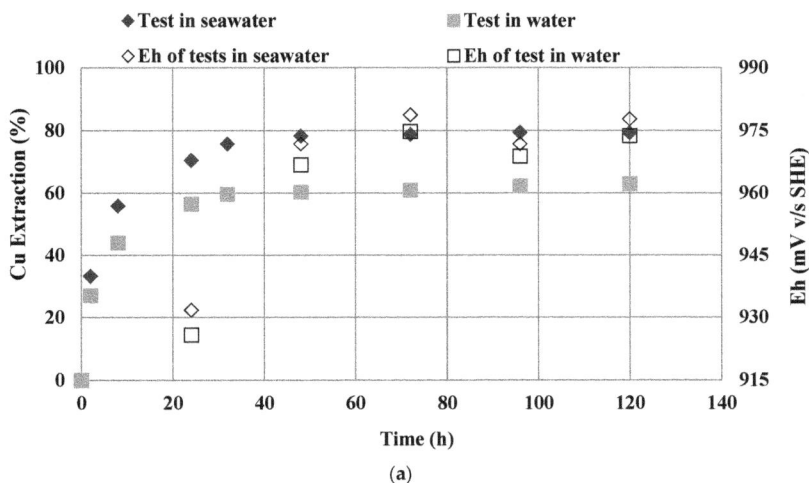

(a)

Figure 6. *Cont.*

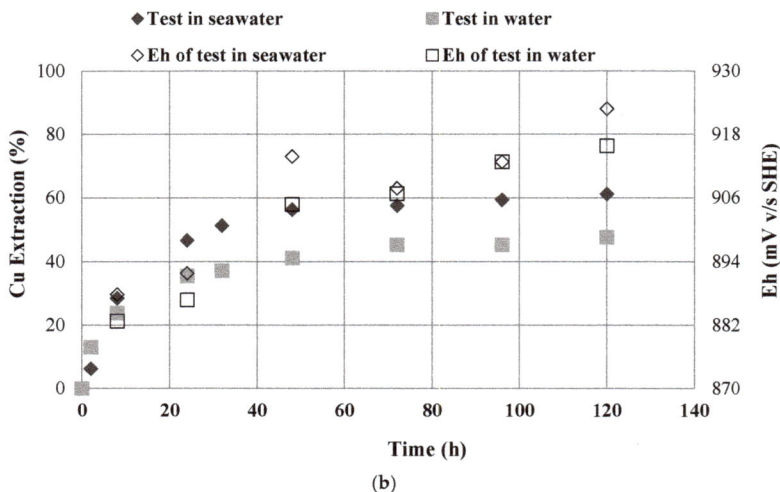

(b)

Figure 6. Values of redox potential (mV) against time (h) for the systems: (**a**) H_2SO_4 (1 M), $NaNO_3$ (1 M) in seawater and water and (**b**) H_2SO_4 (0.5 M), $NaNO_3$ (0.5 M) in seawater and water. Experimental conditions: 5 days of leaching and 45 °C.

3.5. Effect of Temperature

Figure 7a,b show the effect of temperature. The highest copper extraction rate of 97.2% was obtained at 70 °C after 3 days of leaching in a seawater media. This high rate of copper extraction was due to the presence of chloride in solution, higher acid and nitrate content, and higher temperature. The results obtained of the effect of the temperature are in agreement with the results reported in the literature [28,30].

Copper extraction was higher in the seawater-based media than in the water-based media independent of temperature. The increase in temperature from 25 to 70 °C resulted in an increase in the copper extraction rate from ≈31% to ≈97% when seawater was used.

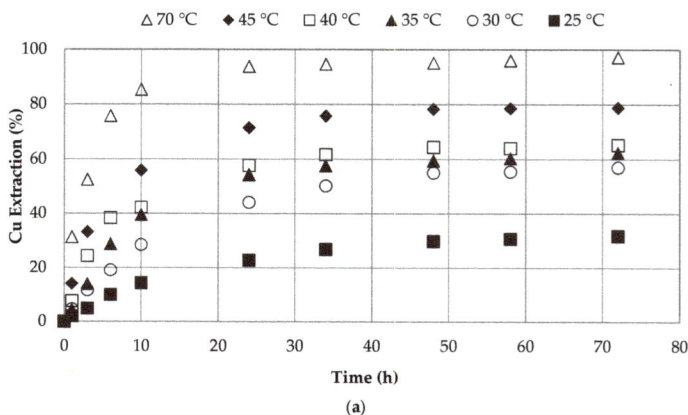

(a)

Figure 7. *Cont.*

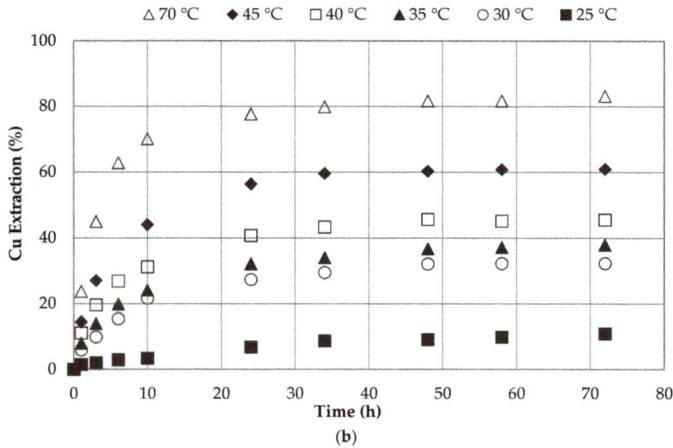

Figure 7. Effect of temperature on copper extraction using H_2SO_4 (1 M), $NaNO_3$ (1 M), 3 days of leaching: (**a**) seawater and (**b**) water.

3.6. Effect of Synthetic Saline Solution

Figure 8 compares the kinetic curves of copper extraction in seawater and a synthetic saline solution. The kinetics curves in Figure 8 show that the two extraction rates are similar. The small variations may be related to the presence of SO_4^{2-} ions in the seawater. Other ions in seawater, such as calcium, potassium, magnesium, do not contribute significantly to leaching under the conditions in this work. Figure 8 shows that chloride ions, present in both solvents, helped in the leaching process.

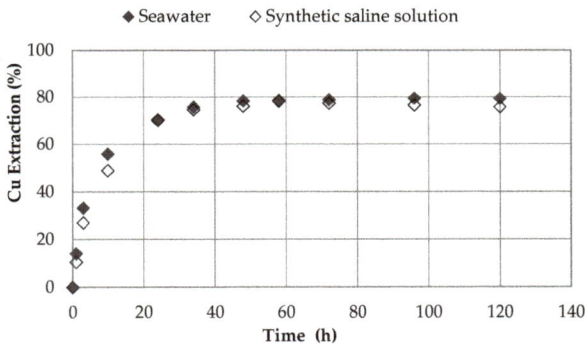

Figure 8. Comparison of copper extraction rates when leach solution was prepared in synthetic saline water and seawater. Experimental conditions: NaCl (35 g·L^{-1}), H_2SO_4 (1 M), $NaNO_3$ (1 M), 5 days of leaching and 45 °C.

3.7. Characterization of Ore and Residue Samples

The residue characterization (test 11) confirms that chalcopyrite was significantly leached in the seawater media. Covellite was found in small quantities, 1% and 3%, in the solid residues of the leaching test in the seawater and water media, respectively (Figure 9). It is supposed that covellite in the solid residue is formed as an intermediate product during chalcopyrite dissolution.

The formation of elemental sulfur as a product of chalcopyrite leaching in both seawater and water-based media was confirmed by SEM-EDS analysis (red region, Figure 10).

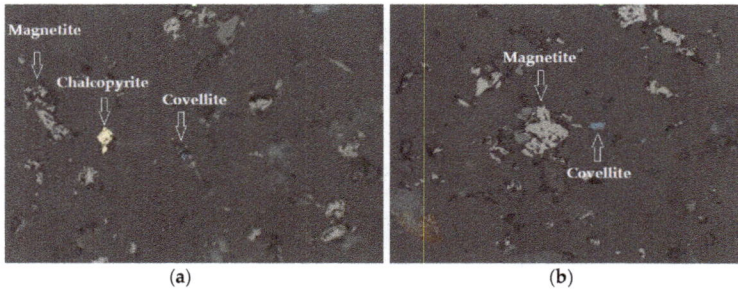

Figure 9. Optical microscope images of solid residue after leaching from (**a**) seawater, and (**b**) water system.

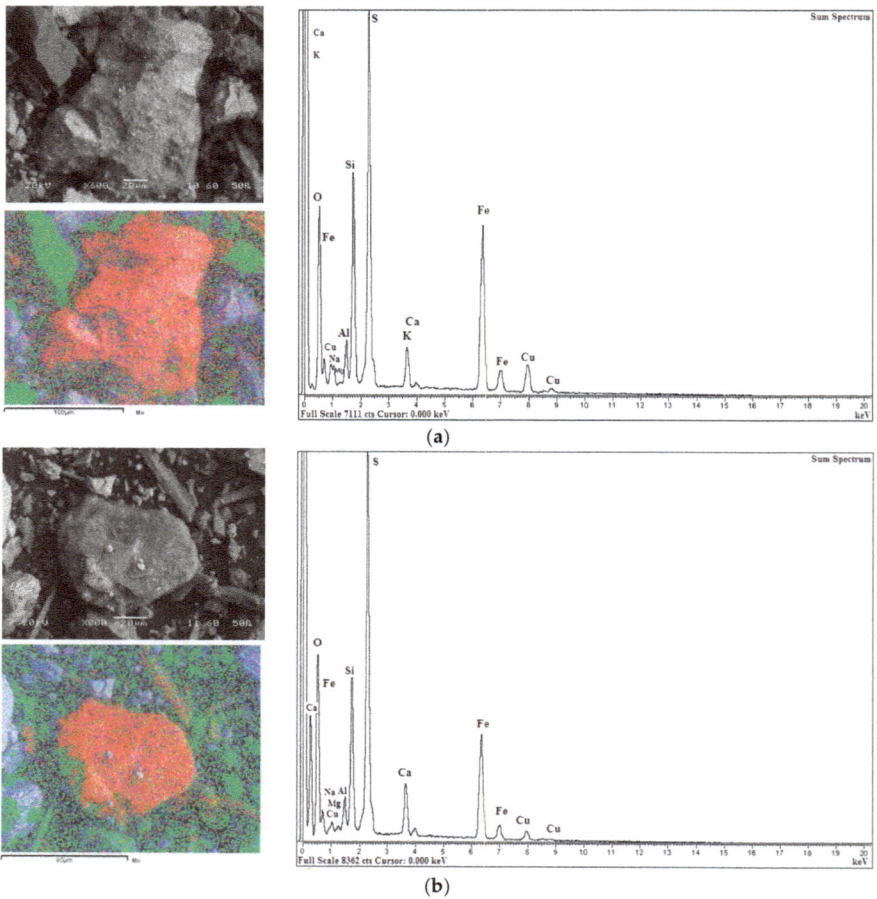

Figure 10. Results of scanning electron microscopy (SEM) analysis carried out using the solid residue. The ore sample was leached in: (**a**) seawater medium and (**b**) water medium. The false red color shows the sulfur presence.

3.8. Thermodynamic and Chemical Reactions

Equations (7)–(10) are chemical reactions proposed for the system $CuFeS_2$-NO_3^--H^+-Cl^-. In the literature, the system $CuFeS_2$-NO_3^--H^+ was analyzed and respective chemical reactions were provided [28,47], but these studies did not include the presence of chloride ions.

$$6CuFeS_2 + 10NO_3^- + 40H^+ + 12Cl^- = 6CuCl_2 + 6Fe^{3+} + 12S + 10NO + 20H_2O \tag{7}$$

$$6CuFeS_2 + 10NO_3^- + 40H^+ + 6Cl^- = 6CuCl^+ + 6Fe^{3+} + 12S + 10NO + 20H_2O \tag{8}$$

$$3CuFeS_2 + 15NO_3^- + 30H^+ + 6Cl^- = 3CuCl_2 + 3Fe^{3+} + 6S + 15NO_2 + 15H_2O \tag{9}$$

$$3CuFeS_2 + 15NO_3^- + 30H^+ + 3Cl^- = 3CuCl^+ + 3Fe^{3+} + 6S + 15NO_2 + 15H_2O \tag{10}$$

The presence of covellite in the leached residue can be explained by Equations (11) and (12):

$$2CuFeS_2 + 8NO_3^- + 16H^+ = Cu^{2+} + CuS + 2Fe^{3+} + 3S + 8NO_2 + 8H_2O \tag{11}$$

$$CuS + 2NO_3^- + 4H^+ = Cu^{2+} + S + 2NO_2 + 2H_2O \tag{12}$$

These equations have negative Gibbs energy (ΔG) indicating their thermodynamic viability at the temperatures used in this work (See Table 4).

Table 4. Gibbs energy (ΔG) value at different temperature

No.	$\Delta G_{25 \,°C}$ (kJ·mol^{-1})	$\Delta G_{70 \,°C}$ (kJ·mol^{-1})
(7)	−1195.8	−1205.8
(8)	−1351.5	−1338.8
(9)	−340.0	−413.6
(10)	−417.9	−480.2
(11)	−245.7	−274.8
(12)	−28.5	−38.2

4. Conclusions

In this study, leaching of chalcopyrite ore with 1.6 wt % Cu was investigated using a nitrate-acid-seawater system. A higher rate of copper extraction was obtained in an acid nitrate media when chloride ions were present in the leaching tests. The highest copper extraction rate of 97.2% was obtained at 70 °C with high sulfuric acid (1 M) and sodium nitrate (1 M) concentrations.

The addition of both nitrate types ($NaNO_3$ or KNO_3) to sulfuric acid in seawater or water-based media provided similar levels of copper extraction. The changes in the cation types did not affect copper extraction under the studied conditions.

Without chloride, in the form of seawater or synthetic brine, the nitrate system requires a higher acid concentration to extract the same amount of copper.

The residue characterization showed that sulfur formed during dissolution.

This study presents an alternative method of leaching chalcopyrite ores using seawater and/or brine with nitrate ions at moderate temperatures (45 °C). The main disadvantage of this method is the production of NOx gases that should be controlled in industrial leaching systems.

Proposals for future work include the use of discarded brines (liquid waste with high concentration of chloride ions) from reverse osmosis plants and discarded salts from caliche industry, as sources of oxidants and chloride media during chalcopyrite leaching.

Author Contributions: M.E.T. and P.C.H. designed the experiments; P.C.H. performed the experiments; P.C.H., M.E.T., T.A.G., O.O.H. and Y.G. analyzed the data; P.C.H. prepared and wrote the paper.

Acknowledgments: This work was financially supported by PAI/Concurso Nacional Insercion en la Academia, Convocatoria 2015 folio No. 79150003 and Fondecyt iniciación No. 11170179, CONICYT.

Conflicts of Interest: The authors declare no conflict of interest.

References

1. Liddicoat, J.; Dreisinger, D. Chloride leaching of chalcopyrite. *Hydrometallurgy* **2007**, *89*, 323–331. [CrossRef]
2. Lu, Z.Y.; Jeffrey, M.I.; Lawson, F. The effect of chloride ions on the dissolution of chalcopyrite in acidic solutions. *Hydrometallurgy* **2000**, *56*, 189–202. [CrossRef]
3. Velásquez-Yévenes, L.; Nicol, M.; Miki, H. The dissolution of chalcopyrite in chloride solutions: Part 1. The effect of solution potential. *Hydrometallurgy* **2010**, *103*, 108–113. [CrossRef]
4. Herreros, O.; Viñals, J. Leaching of sulfide copper ore in a NaCl-H_2SO_4-O_2 media with acid pre-treatment. *Hydrometallurgy* **2007**, *89*, 260–268. [CrossRef]
5. Ibáñez, T.; Velásquez, L. Lixiviación de la calcopirita en medios clorurados. *Rev. Metal.* **2013**, *49*, 131–144. [CrossRef]
6. Yoo, K.; Kim, S.-K.; Lee, J.-C.; Ito, M. Tsunekawa, and N. Hiroyoshi. Effect of chloride ions on leaching rate of chalcopyrite. *Miner. Eng.* **2010**, *23*, 471–477. [CrossRef]
7. Watling, H. Chalcopyrite hydrometallurgy at atmospheric pressure: 2. Review of acidic chloride process options. *Hydrometallurgy* **2014**, *146*, 96–110. [CrossRef]
8. Winand, R. Chloride hydrometallurgy. *Hydrometallurgy* **1991**, *27*, 285–316. [CrossRef]
9. Senanayake, G. A review of chloride assisted copper sulfide leaching by oxygenated sulfuric acid and mechanistic considerations. *Hydrometallurgy* **2009**, *98*, 21–32. [CrossRef]
10. Al-Harahsheh, M.; Kingman, S.; Al-Harahsheh, A. Ferric chloride leaching of chalcopyrite: Synergetic effect of $CuCl_2$. *Hydrometallurgy* **2008**, *91*, 89–97. [CrossRef]
11. Lundström, M.; Aromaa, J.; Forsén, O.; Hyvärinen, O.; Barker, M.H. Leaching of chalcopyrite in cupric chloride solution. *Hydrometallurgy* **2005**, *77*, 89–95. [CrossRef]
12. Lundström, M.; Liipo, J.; Aromaa, J. Dissolution of copper and iron from sulfide concentrates in cupric chloride solution. *Int. J. Miner. Process.* **2012**, *102–103*, 13–18. [CrossRef]
13. O'Malley, M.; Liddell, K. Leaching of $CuFeS_2$ by aqueous $FeCl_3$, HCl, and NaCl: Effects of solution composition and limited oxidant. *Metall. Mater. Trans. B.* **1987**, *18*, 505–510. [CrossRef]
14. Moreno, P.; Aral, H.; Cuevas, J.; Monardes, A.; Adaro, M.; Norgate, T.; Bruckard, W. The use of seawater as process water at Las Luces copper–molybdenum beneficiation plant in Taltal (Chile). *Miner. Eng.* **2011**, *24*, 852–858. [CrossRef]
15. Aroca, F.; Backit, A.; Jacob, J. CuproChlor®, a hydrometallurgical technology for mineral sulphides leaching. In *Hydroprocess 2012, Proceedings of the 4th International Seminar on Process Hydrometallurgy*; Casas, S.J.C., Ciminelli, V., Montes-Atenas, G., Stubina, N., Eds.; GECAMIN LTDA: Santiago, Chile, 2012; pp. 96–180.
16. COCHILCO. *Consumo de Agua en La minería del Cobre al 2016*; Gobierno de Chile, Ministerio de Minería, Ed.; Registro Propiedad Intelectual: Santiago, Chile, 2017.
17. Torres, C.; Taboada, M.; Graber, T.; Herreros, O.; Ghorbani, Y.; Watling, H. The effect of seawater based media on copper dissolution from low-grade copper ore. *Miner. Eng.* **2015**, *71*, 139–145. [CrossRef]
18. Hernández, P.; Taboada, M.; Herreros, O.; Torres, C.; Ghorbani, Y. Chalcopyrite dissolution using seawater-based acidic media in the presence of oxidants. *Hydrometallurgy* **2015**, *157*, 325–332. [CrossRef]
19. Taboada, M.; Hernández, P.; Galleguillos, H.; Flores, E.; Graber, T. Behavior of sodium nitrate and caliche mineral in seawater: Solubility and physicochemical properties at different temperatures and concentrations. *Hydrometallurgy* **2012**, *113*, 160–166. [CrossRef]
20. Valencia, J.A.; Méndez, D.A.; Cueto, J.Y.; Cisternas, L.A. Saltpeter extraction and modelling of caliche mineral heap leaching. *Hydrometallurgy* **2008**, *90*, 103–114. [CrossRef]
21. Ordóñez, J.I.; Moreno, L.; Gálvez, E.D.; Cisternas, L.A. Seawater leaching of caliche mineral in column experiments. *Hydrometallurgy* **2013**, *139*, 79–87. [CrossRef]
22. Torres, M.A.; Meruane, G.E.; Graber, T.A.; Gutiérrez, P.C.; Taboada, M.E. Recovery of nitrates from leaching solutions using seawater. *Hydrometallurgy* **2013**, *133*, 100–105. [CrossRef]

23. Habashi, F. Nitric Acid in the Hydrometallurgy of Sulfides. In *EPD Congress 1999*; TMS: Baltimore, MD, USA, 1999.
24. Peters, E. Hydrometallurgical process innovation. *Hydrometallurgy* **1992**, *29*, 431–459. [CrossRef]
25. Prasad, S.; Pandey, B. Alternative processes for treatment of chalcopyrite—A review. *Miner. Eng.* **1998**, *11*, 763–781. [CrossRef]
26. Kazakov, A.; Rubtsov, Y.I.; Andrienko, L.; Manelis, G. Kinetics and mechanism of thermal decomposition of nitric acid in sulfuric acid solutions. *Bull. Acad. Sci. USSR Div. Chem. Sci.* **1987**, *36*, 1999–2002. [CrossRef]
27. Baldwin, S.A.; van Weert, G. On the catalysis of ferrous sulphate oxidation in autoclaves by nitrates and nitrites. *Hydrometallurgy* **1996**, *42*, 209–219. [CrossRef]
28. Sokić, M.D.; Marković, B.; Živković, D. Kinetics of chalcopyrite leaching by sodium nitrate in sulphuric acid. *Hydrometallurgy* **2009**, *95*, 273–279. [CrossRef]
29. Gok, O.; Anderson, C.G. Dissolution of low-grade chalcopyrite concentrate in acidified nitrite electrolyte. *Hydrometallurgy* **2013**, *134–135*, 40–46. [CrossRef]
30. Vračar, R.Ž.; Vučković, N.; Kamberović, Ž. Leaching of copper(I) sulphide by sulphuric acid solution with addition of sodium nitrate. *Hydrometallurgy* **2003**, *70*, 143–151. [CrossRef]
31. Watling, H. Chalcopyrite hydrometallurgy at atmospheric pressure: 1. Review of acidic sulfate, sulfate-chloride and sulfate-nitrate process options. *Hydrometallurgy* **2013**, *140*, 163–180. [CrossRef]
32. Anderson, C.G. Treatment of copper ores and concentrates with industrial nitrogen species catalyzed pressure leaching and non-cyanide precious metals recovery. *JOM* **2003**, *55*, 32–36. [CrossRef]
33. Tsogtkhangai, D.; Mamyachenkov, S.V.; Anisimova, O.S.; Naboichenko, S.S. Kinetics of leaching of copper concentrates by nitric acid. *Russ. J. Non Ferr. Met.* **2011**, *52*, 469–472. [CrossRef]
34. Tsogtkhangai, D.; Mamyachenkov, S.V.; Anisimova, O.S.; Naboichenko, S.S. Thermodynamics of reactions during nitric acid leaching of minerals of a copper concentrate. *Russ. J. Non Ferr. Met.* **2011**, *52*, 135–139. [CrossRef]
35. Arias, J.A. Heap leaching copper ore using sodium nitrate. U.S. Patent US6,569,391B1, 27 May 2003.
36. Queneau, P.; Prater, J. Nitric acid process for recovering metal values from sulfide ore materials containing iron sulfides. U.S. Patent US3,793,429, 19 Februar 1974.
37. Antivachis, D.N.; Chatzitheodoridis, E.; Skarpelis, N.; Komnitsas, K. Secondary sulphate minerals in a cyprus-type ore deposit, Apliki, Cyprus: mineralogy and its implications regarding the chemistry of Pit Lake waters. *Mine Water Environ.* **2017**, *36*, 226–238. [CrossRef]
38. Shiers, D.; Collinson, D.; Kelly, N.; Watling, H. Copper extraction from chalcopyrite: Comparison of three non-sulfate oxidants, hypochlorous acid, sodium chlorate and potassium nitrate, with ferric sulfate. *Miner. Eng.* **2016**, *85*, 55–65. [CrossRef]
39. Alguacil, F.J.; Cobo, A.; Alonso, M. Copper separation from nitrate/nitric acid media using Acorga M5640 extractant: Part I: solvent extraction study. *Chem. Eng. J.* **2002**, *85*, 259–263. [CrossRef]
40. Eyzaguirre, D. Efecto del nitrato en la extracción por solventes de Cía. Minera Lomas Bayas. In *Proceedings of the VI Exposición Mundial Para la Minería Latinoamericana EXPOMIN, Seminario Innovación Tecnológica en Minería*; Instituto de Ingenieros de Minas de Chile: Santiago, Chile, 2000.
41. Yáñez, H.; Soto, A.; Soderstrom, M.; Bednarski, T. Nitration in copper SX? Cytec Acorga provides a new reagent. In *Hydrocopper 2009*; Domic, E., Casas, J., Eds.; GECAMIN LTDA: Antofagasta, Chile, 2009; pp. 332–341.
42. Zambra, R.; Quilodrán, A.; Rivera, G.; Castro, O. Use of NR®reagents in presence of nitrate ion in SX: A revision of the present moment. In *Hydroprocess 2013*; GECAMIN LTDA: Santiago, Chile, 2013.
43. Hamzah, B.; Jalaluddin, N.; Wahab, A.W.; Upe, A. Copper (II) Extraction from Nitric Acid Solution with 1-Phenyl-3-methyl-4-benzoyl-5-pyrazolone as a Cation Carrier by Liquid Membrane Emulsion. *J. Chem.* **2010**, *7*, 239–245. [CrossRef]
44. Virnig, M.J.; Mattison, P.L.; Hein, H.C. Processes for the recovery of copper from aqueous solutions containing nitrate ions. U.S. Patent 6,702,872, 9 March 2004.
45. Puvvada, G.V.K.; Sridhar, R.; Lakshmanan, V. Chloride metallurgy: PGM recovery and titanium dioxide production. *J. Miner. Met. Mater. Soc.* **2003**, *55*, 38–41. [CrossRef]

46. Senanayake, G. Review of theory and practice of measuring proton activity and pH in concentrated chloride solutions and application to oxide leaching. *Miner. Eng.* **2007**, *20*, 634–645. [CrossRef]

47. Narangarav, T.; Nyamdelger, S.; Ariunaa, G.; Azzaya, T.; Burmaa, G. Dissolution behavior of copper concentrate in acidic media using nitrate ions. *Mong. J. Chem.* **2014**, *15*, 79–84. [CrossRef]

![minerals logo] *minerals*

MDPI

Article

Accelerating Copper Leaching from Sulfide Ores in Acid-Nitrate-Chloride Media Using Agglomeration and Curing as Pretreatment

Pía C. Hernández [1,*], Junior Dupont [1], Osvaldo O. Herreros [2], Yecid P. Jimenez [1] and Cynthia M. Torres [3]

[1] Departamento de Ingeniería Química y Procesos de Minerales, Universidad de Antofagasta, Avda. Angamos 601, 1240000 Antofagasta, Chile; junycool30@hotmail.com (J.D.); yecid.jimenez@uantof.cl (Y.P.J.)
[2] Departamento de Ingeniería en Minas, Universidad de Antofagasta, Avda. Angamos 601, 1240000 Antofagasta, Chile; osvaldo.herreros@uantof.cl
[3] Departamento de Ingeniería Metalúrgica y Minas, Universidad Católica del Norte, Avda. Angamos 0610, 1240000 Antofagasta, Chile; cynthia.torres@ucn.cl
* Correspondence: pia.hernandez@uantof.cl; Tel.: +56-55-2637313

Received: 29 March 2019; Accepted: 24 April 2019; Published: 25 April 2019

Abstract: This work investigates the effect of an agglomeration and curing pretreatment on leaching of a copper sulfide ore, mainly chalcopyrite, using mini-columns in acid-nitrate-chloride media. Ten pretreatment tests were conducted to evaluate different variables, namely the addition of nitrate as $NaNO_3$ (11.7 and 23.3 kg/ton), chloride as NaCl (2.1 and 19.8 kg/ton), curing time (20 and 30 days) and repose temperature (25 and 45 °C). The optimum copper extraction of 58.6% was achieved with the addition of 23.3 kg of $NaNO_3$/ton, 19.8 kg of NaCl/ton, and after 30 days of curing at 45 °C. Under these pretreatment conditions, three samples of ore were leached in mini-columns. The studied parameters were temperature (25 and 45 °C) and chloride concentration (20 and 40 g/L). The optimum copper extraction of 63.9% was obtained in the mini-column leaching test at 25 °C, with the use of 20 g/L of chloride. A higher temperature (45 °C) and a higher chloride concentration (40 g/L) negatively affected the extraction. The pretreatment stage had favorable effects, in terms of accelerating copper dissolution and improving leaching of copper sulfide ore in acid-nitrate-chloride media. Waste salts from caliche industry and waste brine from reverse osmosis can be used for providing the nitrate and chloride media.

Keywords: agglomeration; curing; copper sulfide; chalcopyrite; pretreatment; nitrate; chloride

1. Introduction

Industrial scale leaching of copper sulfide minerals like chalcopyrite ($CuFeS_2$) remains a metallurgical challenge. Chalcopyrite is mined worldwide [1], given the depths at which mining operations can now be carried out [2]. Copper sulfide is extracted by flotation, followed by the application of pyrometallurgical methods [3]. Leaching is an alternative metallurgical treatment for chalcopyrite, although chalcopyrite tends to form passive layers around particles during leaching, which results in a slow dissolution rate and low levels of copper extraction [4,5]. Many researchers have reported a higher percentage of copper extraction in acid-chloride media than acid-sulfate media to leach chalcopyrite [6–12]. A chloride-acid media promotes the formation of copper chloride-complexes [13] and iron chloride-complexes [14] that modify the redox potential (ORP), according Equations (1) and (2):

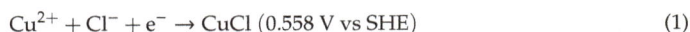

$$Cu^{2+} + Cl^- + e^- \rightarrow CuCl \ (0.558 \ V \ vs \ SHE) \tag{1}$$

$$Fe^{3+} + Cl^- + e^- \rightarrow FeCl^+ \ (0.761 \ V \ vs \ SHE) \tag{2}$$

Cu(I) ions are stabilized in the solution by adding a new redox pair Cu(I)/Cu(II). Porous layer of sulfur is generated around the particle when chloride ions are present. This porous layer facilitates the diffusion of reagents that improves the leaching kinetics [15,16]. A chloride medium can be provided by seawater or processing brines, such as waste solution from reverse osmosis [17–22].

Oxidizing agents like oxygen [23], chlorine [24], ferric [25], cupric [26], nitrate [27], and others are required to dissolve chalcopyrite. Some researchers have tested nitric acid [28,29], and nitrite [30] to leach chalcopyrite, with good results in terms of copper extraction. Nitrogen species tend to be very strong oxidizing agents, and there are reports in the literature of obtaining high copper extraction with the use of nitrogen ions [31].

Mining of copper, lithium, nitrates and other minerals represents an important part of the Chilean economy [32]. Nitrate comes from the caliche industry concentrated in northern Chile. Nitrate salts, waste salts and intermediate processing solutions can be used in chalcopyrite leaching. Sokić et al. [27] studied leaching of chalcopyrite concentrate in acid-nitrate media and determined that increasing temperature and acid and nitrate concentrations increases copper extraction to over 75%. Gok and Anderson [30] investigated the effect of addition of sodium nitrite in an acid medium during leaching of chalcopyrite and found that the extraction increased by 5% by using nitrite instead of nitrate, while other conditions remained the same. Shiers et al. [33] studied leaching chalcopyrite using different oxidants: hypochlorous acid, sodium chlorate and sodium nitrate with the addition of ferric ions (ferric sulfate, ferric nitrate and ferric chloride), and obtained a copper extraction of 92% in chloride-nitrate-ferric mixture, which proved favorable for chalcopyrite leaching at 50 °C. Tsogtkhankhai et al. [34] studied the kinetics of leaching copper with nitric acid, and with the estimation of the activation energy determined that temperature has a strong influence on the system. Castellón et al. [35] leached chalcopyrite concentrate in acid-nitrate media using seawater as a dissolvent and obtained a copper extraction of 97% in 24 h under the studied conditions ([H_2SO_4] = [$NaNO_3$] = 0.5 M) at 45 °C. The nitrogen species catalyzed (NSC) method uses sulfuric and nitric acids to leach sulfide minerals with the use of high pressure and temperature. The main advantage of this process is rapid extraction [31]. Many patents have been filed for leaching sulfide minerals in acid-nitrate media. Queneau and Prater [36] patented a process for leaching copper, iron, cobalt, silver and nickel sulfide in nitric acid. Arias [37] developed a process for leaching copper sulfide ore in heaps with an acid-nitrate media. Hard [38] patented a copper leaching process using sulfuric acid in the presence of nitrate ions. The optimal pH range (0.5 and 1.5), nitrate concentration (1–10%) and ratio between H^+ and NO_3^- ions (ratio of 4 to 1) have been determined. Carnahan and Heinen [39] patented an in situ copper leaching process that uses a diluted acid medium and oxygen with nitrate ions that can come from nitric acid, an alkaline metal or ammonium nitrate. Lueders and Frankiewicz [40] patented a two-step leaching process in which sulfide ores are oxidized with nitrogen dioxide in an acid medium. Anderson et al. [41] patented a process to leach sulfide ores under conditions of temperature and pressure, using an acid medium with sodium nitrite. Hernández et al. [42] studied leaching chalcopyrite by agitation in acid-nitrate media at temperatures between 25 and 70 °C, and compared the use of seawater to that of freshwater. Copper extraction close to 98% were obtained with three days of leaching at 70 °C in a seawater medium, with a high concentration of nitrate and acid (1 M), and an 80% extraction was obtained with seven days of leaching at 45 °C.

Ore agglomeration and curing are carried out prior to leaching in heaps [43–45]. From the physical point of view, smaller particles of ore adhere to larger ones, resulting in relatively homogenous and stable material. This in turn results in a significant number of holes in piles, which is essential for adequate liquid and gas permeability of the agglomerated mineral bed [46,47]. The objective of curing is to attack the mineral chemically with concentrated sulfuric acid to dissolve most of the surface copper on particles, which creates a favorable condition for leaching. In this way, solutions with high levels of copper are obtained. Curing also solubilizes iron, contributing the ferric ions necessary to dissolve sulfides and inhibit the formation of colloidal silica to minimize silica passing to the leaching solutions [45]. Cerda et al. [48] studied the effect of a pretreatment stage at the laboratory level on dissolving a copper sulfide ore in a chloride-acid medium, and obtained a copper extraction of 93% leaching ore pretreated with 90 kg Cl^-/ton of ore and 40 days of repose at 50 °C. A copper extraction of only 55% can be achieved without pretreatment, indicating the significant effect of the pretreatment stage. Bahamonde et al. [49] determined the positive effect of pretreating bornite and chalcopyrite in acid-chloride media to extract copper from these minerals. Taboada et al. [50] evaluated the effect of a pretreatment stage on leaching a mixed copper ore in mini-columns with the addition of ferric and ferrous ions in acid-chloride media and a repose period of 30 days. The process involved a 30-day cycle of pretreatment and 15 additional days of leaching (for a total of 45 days), resulting in 60 and 80% copper extraction, respectively. The addition of ferric ions in the pretreatment enhances copper extraction. Velásquez-Yévenes and Quezada-Reyes [21] studied leaching chalcopyrite in columns with a chloride medium, and a pretreatment with repose periods of 30, 50, 80 and 100 days. For the pretreatment, 5 kg H_2SO_4/ton and 60 kg/ton of seawater or waste brine were used, according to the test, with 0.5 g Cu^{2+}/L. Intermittent irrigation began following the repose period. The authors found that the pretreatment improved extraction, which at room temperature and with 100 days of repose were 43 and 37%, respectively with brine and seawater. Velásquez Yévenes et al. [51] studied the effect of adding chloride in agglomeration and a curing process to improve the copper extraction from sulfide ore in leaching by columns. The copper dissolution increased with the curing stage with chloride and acid addition.

The objective of this study is to determine the effects of copper sulfide pretreatment (agglomeration and curing) on leaching by percolation at a moderate temperature (≤ 45 °C) in acid-nitrate-chloride media in mini-columns. The pretreatment variables studied were the addition of sodium nitrate (11.7 and 23.3 kg/ton), and sodium chloride (2.1 and 19.8 kg/ton), curing time duration (20 and 30 days) and repose temperature (25 and 45 °C). The leaching variables studied were the temperature of the system (25 and 45 °C) and the chloride concentration (20 and 40 g/L).

2. Materials and Methods

2.1. Ore Sample

Copper sulfide from a mine in the Antofagasta Region was used in this work, with particle sizes under 9.53 mm, and a P_{80} of 7.9 mm. Atomic absorption spectrometry (AAS, Perkin-Elmer 2380, Perkin Elmer, Wellesley, MA, USA) showed that the chemical composition of the ore was 0.70% total Cu, 0.04% soluble Cu and 5.65% total Fe. Ore acid consumption was 33.5 kg/ton ore as determined by a leaching aliquot and titration with NaOH. The mineralogy was determined by quantitative X-ray diffraction (QXRD, Siemens/Bruker, Semi-QXRD, model D5000, Germany) and validated by optical microscopy using point-counting method (Table 1).

Table 1. Mineralogy of the ore determined by quantitative X-ray diffraction.

Mineral	Chemical Formula	Amount (wt. %)	Cu (wt. %)
Chalcopyrite	$CuFeS_2$	1.69	0.59
Bornite	Cu_5FeS_4	0.10	0.06
Covellite	CuS	0.02	0.01
Atacamite	$CuCl_2 \cdot 3Cu(OH)_2$	0.07	0.04
Chrysocolla	$Cu_{1.75}Al_{0.25}H_{1.75}(Si_2O_5)(OH)_4 \cdot 0.25(H_2O)$	0.01	<0.01
Pyrite	FeS_2	3.81	
Magnetite	Fe_3O_4	0.40	-
Hematite	Fe_2O_3	0.08	-
Limonite	$FeOOH$	0.07	
Quartz	SiO_2	38.39	
Gangues	-	55.36	-
Total Cu			0.70

2.2. Reagents

Technical grade sulfuric acid (H_2SO_4, 95%), sodium chloride (NaCl, 95%), sodium nitrate ($NaNO_3$, 95%) and seawater were used as reactive agents. Seawater was obtained from San Jorge Bay, Antofagasta, Chile. It was pumped from the coast and then filtered (pore size 0.001 mm) through a polyethylene membrane. The chemical analyses of seawater shown in Table 2 were obtained by volumetry, gravimetry, inductively coupled plasma atomic emission spectroscopy (ICP-AES, ICPE-9000, Shimadzu, Tokyo, Japan) and AAS.

Table 2. Seawater composition.

Method of Analysis	ICP-AES				AAS		Volumetric Analysis	Gravimetric Analysis
Ionic Species	Na^+	Mg^{2+}	Ca^{2+}	K^+	Cu^{2+}	NO_3^-	Cl^-	SO_4^{2-}
Amount (g/L)	10.73	1.29	0.40	0.39	7.2×10^{-5}	3×10^{-3}	20.06	2.81

2.3. Experimental Procedure

2.3.1. Description of Pretreatment Stage

Table 3 shows the variables used to study the effect of pretreatment on copper extraction

Table 3. Variables studied in the pretreatment step.

Sodium Nitrate Addition (kg/ton)	Sodium Chloride Addition (kg/ton)	Curing Time (Days)	Repose Temperature (°C)
11.7	2.1 *	20	25
23.3	19.8	30	45

* chloride contributed by seawater (humidity) without any additives.

Ten copper sulfide ore samples were prepared (500 g), all with the same granulometry. The ore was placed on a plastic sheet where the sodium nitrate and sodium chloride were added, both in solid form. The ore and reactive agents were homogenized by rolling, after which 50% of the sulfuric acid consumption of the ore (16.8 kg/ton) was added and 7% moisture was obtained by adding seawater. The ore was homogenized and the formation of stable agglomerates was observed. The agglomerated ore was stored in sealed plastic bags, the weight of which was registered using an analytical balance (Mettler Toledo C. AX-204, precision of $\pm 7 \times 10^{-5}$ g). To test the effect of a repose temperature of 45 °C, pretreated samples were placed in an oven (Binder model ED115) that remained constant at that temperature throughout the repose period. To test the effect of a repose temperature of 25 °C, the samples were left in the laboratory with temperature control set at that temperature. The moisture level in the samples was checked every two days. Variation in weight indicated moisture loss by evaporation, the samples were sprayed with distilled water until the original weight was obtained. Following the repose period, the pretreated ore was transferred to a beaker, washed with distilled water at a solid/liquid ratio of 1:3 at room temperature, with mechanical agitation (400 rpm) for two hours. The suspension was then filtered using a vacuum pump. The pH and redox potential (ORP) of the resulting solution were then measured (Hanna portable pH/ORP meter, model HI991003, accuracy ±0.02 pH and ±2 mV). The solid residue was dried in an oven at 60 °C until reaching a constant weight. The copper content in the solution and in the solid residue was obtained by AAS. The copper extraction rate was calculated using the copper grades of the head and residual ore, which was corroborated with the results obtained by solutions. A standard deviation of ±2% was obtained by calculating the copper extraction rates from solid and solution for all the tests. These tests were carried out in duplicate.

2.3.2. Description of Leaching Tests

The pretreatment conditions that resulted in the highest copper extraction rate were repeated in three experiments using 3 kg of ore with the same size distribution as in the previous tests. The pretreated ore was placed in acrylic mini-columns 40 cm high and 9.5 cm in diameter, similar dimensions used by Taboada et al. [50] and Velasquez-Yévenez et al. [51], and left in repose for a curing time determined by the previous tests. Following the curing time, the ore was leached using a solution of 6.3 g/L NaNO$_3$ and seawater, with a pH of 1 (using H$_2$SO$_4$), and the experimental variables as shown in Table 4. The mini columns at 45 °C were wrapped in plastic hoses through which water circulated at a desired temperature, controlled by a thermostatic bath (Julabo bath F25-ME Refrigerated/Heating Circulator) to keep the temperature constant (Figure 1).

Figure 1. Mini-column leaching tests.

Table 4. Experimental conditions of mini-column leaching tests.

Conditions	C1	C2	C3
Irrigation temperature (°C)	25	45	25
Chloride (g/L)	20	20	40

The mini-columns were irrigated with a peristaltic pump in an open cycle at a rate of 8 L/h·m^2 (Master Flex, Model N° 7557-14/1-100 RPM, Cole-Parmer International, Vernon Hills, IL, USA). At certain intervals, the output solution was removed and mass, pH, ORP and copper content were determined. Following the test (leaching for 8 days), the ore was washed in the mini-columns with an acid solution (pH 1) and tap water for 24 h and then left in repose for another 24 h to completely drain the solution. The leached ore was then removed from the mini-columns and dried in an oven at 60 °C until reaching a constant weight. The solid residue was analyzed for copper content (AAS). A standard deviation of ±1.9% was obtained by calculating copper extraction using solid and solution for all the tests. The tests were carried out in duplicate. In addition, the solid residue was analyzed by optical microscope for the determination of mineralogical species.

3. Results and Discussion

3.1. Effect of Pretreatment on Copper Extraction

Table 5 shows the pretreatment tests, with all the variables used and the average of copper extraction.

The highest copper extraction was obtained in test 8, which involved the highest values of the studied variables. The final pH levels in all the tests were in the range of 2.2 to 3.3, which indicates complete consumption of the aggregate acid in the pretreatment. Redox potential values ranged between 637 and 792 mV (vs SHE), which were higher than the values reported by Cerda et al. [48]. This is attributed to the presence of nitrate, which has a high redox potential. Temperature and the addition of chloride have significant effects on dissolving copper sulfides. The residue obtained from test 8 was analyzed by optical microscope. Figure 2 shows the percentages of the copper species extracted in test 8 by leaching with pretreatment, which was determined by mass balance.

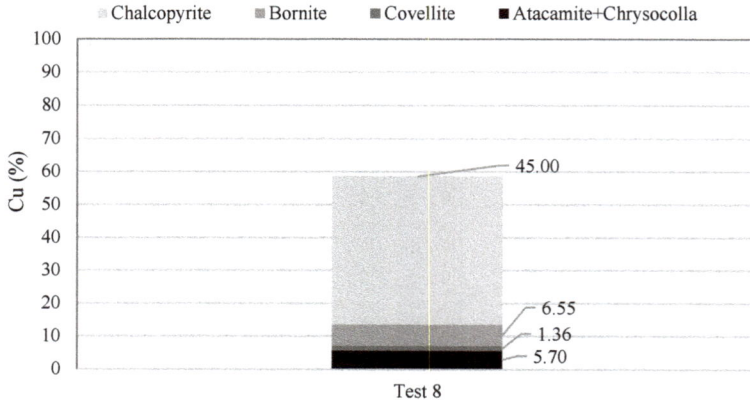

Figure 2. Copper extraction by sequential leaching mineralogical copper species obtained in test 8. Experimental conditions: 23.3 kg $NaNO_3$/ton, 19.8 kg NaCl/ton, 30 days of repose at 45 °C.

Table 5. Pretreatment tests and maximum copper extraction (wt. %).

Test	Sodium Nitrate Addition (kg/ton)	Sodium Chloride Addition (kg/ton)	Curing Time (Days)	Temperature (°C)	Cu Extraction (wt. %)
1	11.7	2.1	20	45	19.8
2	23.3	2.1	20	45	22.9
3	11.7	19.8	20	45	34.9
4	23.3	19.8	20	45	36.6
5	11.7	2.1	30	45	28.7
6	23.3	2.1	30	45	30.7
7	11.7	19.8	30	45	44.9
8	23.3	19.8	30	45	58.6
9	11.7	2.1	30	25	11.8
10	23.3	2.1	30	25	22.8

Table 6 shows the copper extraction per mineralogical species and the percentages of the minerals present at the beginning and end of the pretreatment stage.

According to the results shown in Table 6, part of bornite (75%), covellite (95%) and chalcopyrite (53.3%) were only leached with the pretreatment stage. Copper oxide minerals were completely leached.

Table 6. Copper extraction (%) per mineralogical species and presence (%) in the feed ore and solid residue of pretreatment (test 8).

Mineral	Extraction (%)	Feed Ore (%)	Pretreatment Ore (%)
Chalcopyrite	53.3	1.69	0.79
Bornite	75.0	0.10	0.03
Covellite	95.0	0.02	<0.01
Atacamite	100.0	0.07	0
Chrysocolla	100.0	0.01	0

3.1.1. Effect of Adding Chloride

The addition of chloride increased copper extraction in all the tests, which concurs with what was reported by Senanayake [52] and Watling [6], where chalcopyrite and bornite oxidized more readily in saline water media. The highest extraction of 58.6% was obtained in the test with the addition of 23.3 kg of sodium nitrate and 19.8 kg of sodium chloride per ton at 45 °C, with 30 days of repose. According to Velásquez-Yévenes and Quezada-Reyes [21], the pretreatment with acid and chloride resulted in a more homogenous distribution of reactive agents in the ore, resulting in the dissolution reaction beginning earlier and the formation of soluble species, with better liquid/solid interaction and a higher degree of porosity in the bed, so that moisture is retained in the pores. Velásquez-Yévenes [51] determined that adding sodium chloride in a curing stage improves copper extraction from copper sulfate, based on a test comparing extraction with and without the addition of chloride. They demonstrated that increasing the quantity of chloride from 20 to 70 kg/ton does not significantly increase the copper extraction. This finding concurs with the results of other authors [15,16].

3.1.2. Effect of Adding Nitrate

The addition of sodium nitrate had a positive effect in all the tests, particularly in the test with 30 days of repose at 45 °C and the addition of a high level of chloride. The additional nitrate resulted in more oxidizing ions being available to leach copper sulfides. This concurs with what was reported by Sokić et al. [27], who confirmed that a system with sulfuric acid without the addition of oxidants does not react with chalcopyrite. A mixture with high concentrations of sodium chloride and sodium nitrate in an acid medium was more effective, which concurs with Shiers et al. [33], who reported a 92% extraction for the $NO_3^-/FeCl_3$ (50 °C, 168 h) oxidant. The authors suggested that nitrate is the most cost-effective oxidant.

3.1.3. Effect of Curing Time

Increasing the curing time from 20 to 30 days increased copper extraction in all the tests by providing more time for the dissolution reaction to occur. This is because the mineral is exposed to a high ionic charge provided by sodium nitrate, sodium chloride and sulfuric acid, which, along with a low amount of water, allow dissolution reactions to occur rapidly. This concurs with Cerda et al. [48], Velásquez-Yévenes and Quezada-Reyes [21], and Velásquez-Yévenes et al. [51] who found that a higher extraction was obtained with a longer repose period.

3.1.4. Effect of Repose Temperature

Dissolution reactions are more rapid with higher temperatures because less energy is necessary to activate dissolution and to break molecular bonds and thus dissolve mineralogical species in contact with the leaching agents [48].

3.1.5. Chemical Reactions in Pretreatment

Table 7 shows the main proposed reactions that can occur during agglomeration and curing.

Table 7. Chemical reactions proposed for the pretreatment stage in acid-chloride-nitrate media.

Mineral	Proposed Chemical Reaction	N°
$CuFeS_2$	$2CuFeS_2 + 10H_2SO_4 + 10NaNO_3 + 4NaCl \rightarrow$ $2CuCl_2 + Fe_2(SO_4)_3 + 10NO_2 + 4S + 10H_2O + 7Na_2SO_4$	(3)
CuS	$3CuS + 4H_2SO_4 + 2NaNO_3 + 6NaCl \rightarrow$ $3CuCl_2 + 2NO + 3S + 4H_2O + 4Na_2SO_4$	(4)
Cu_5FeS_4	$Cu_5FeS_4 + 12H_2SO_4 + 12NaNO_3 + 10NaCl \rightarrow$ $5CuCl_2 + FeSO_4 + 12NO_2 + 4S + 12H_2O + 11Na_2SO_4$	(5)
FeS_2	$3FeS_2 + 4H_2SO_4 + 2NaNO_3 + 6NaCl \rightarrow$ $3FeCl_2 + 2NO + 6S + 4H_2O + 4Na_2SO_4$	(6)

Table 8 shows the standard Gibbs energy of proposed reactions 3 to 6 at temperatures of 25 and 45 °C, which were calculated with HSC software [53].

Table 8. Standard Gibbs energy values at 25 and 45 °C of proposed chemical reactions for the pretreatment stage.

Equation N°	$\Delta G°_{25 °C}$ (kcal/mol)	$\Delta G°_{45 °C}$ (kcal/mol)
3	−209.5	−216.8
4	−97.8	−98.8
5	−234.4	−243.0
6	−115.5	−117.4

As can be observed, all the reactions are thermodynamically feasible at atmospheric pressure and in a temperature range of 25 to 45 °C.

3.2. Effect on Mini-Column Leaching

Test C1 obtained a copper extraction of 63.9%, while test C2 obtained one of 55.0%, and test C3 one of 55.4%. Figure 3 shows the copper extraction curves versus the repose period and subsequent leaching for columns 1 to 3. Table 4 shows the leaching conditions.

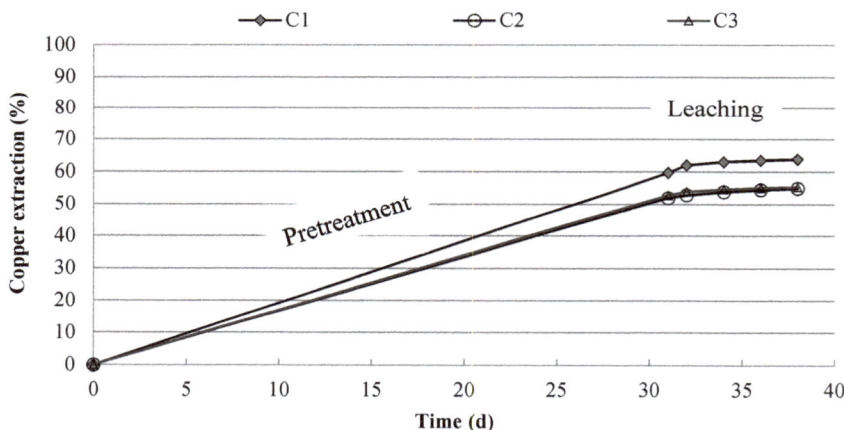

Figure 3. Copper extraction rate (%) vs. time (d) in mini-columns, leaching pretreated ore with 16.8 kg sulfuric acid/ton, 23.3 kg sodium nitrate/ton, 19.8 kg sodium chloride/ton, 30 days of curing time at 45 °C. Irrigation conditions: pH 1, 6.3 g sodium nitrate/L, seawater. C1: 25 °C and 20 g chloride/L, C2: 45 °C and 20 g chloride/L, and C3: 25 °C and 40 g chloride/L.

It can be seen from Figure 3 that the highest copper extraction of 63.9% was obtained with C1 at 38 d (30 d of pretreatment and 8 d of leaching) at 25 °C. The pretreatment resulted in high dissolution rates in all the experiments. Velásquez-Yévenes and Quezada-Reyes [21] conducted chalcopyrite leaching tests in columns at room temperature, with 30 days of repose and 55–60 days of irrigation with sulfuric acid, cupric ions and seawater or brine. The highest copper extraction obtained was less than 25%. Comparing this to the results with C1, it is evident that nitrate is the most favorable oxidizing agent for extracting copper. Figure 3 shows that the C2 and C3 curves have similar kinetics. The two columns show differences in temperature and chloride concentrations. According to the results, temperature can be replaced by chloride concentration. This concurs with what was found in the pretreatment stage, where the two variables are the most significant in the dissolution system.

C1 reached a copper extraction of 63.9% with 8 days of leaching. Comparing this result to the extraction obtained in the pretreatment stage alone (58.6%), it can be observed that the leaching stage is only improved by 9%. C2 and C3 obtained copper extraction close to that obtained in the pretreatment, from which it can be inferred that the leaching only washed the solubilized species during the pretreatment. No contribution from the leaching process was observed.

Based on our results, pretreatment is recommended, followed by washing with an acid solution (pH 1) in the leaching process, without other additives like nitrate or chloride. Other options include using higher concentrations of nitrate (greater oxidation potential), other oxidants, or higher concentrations of acid, among others. The process could also be carried out in two stages, first an acid wash of the pretreated ore, followed by oxidative leaching.

The pH levels of the output solutions of the three columns were ≥ 1.5, indicating acid consumption. The patterns of redox potential values (Figure 4) were similar for C1 and C3, and higher than those obtained for C2. The redox potential values of C2 indicate that there are fewer oxidants in the system at 45 °C than at 25 °C.

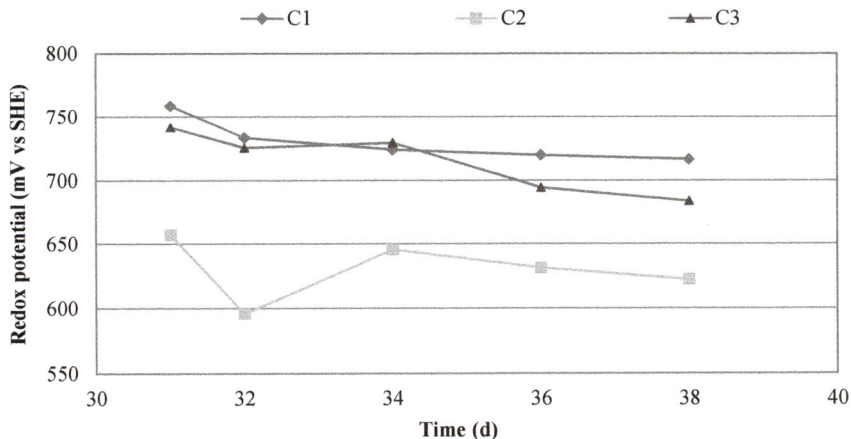

Figure 4. Redox potential (mV vs. SHE) vs. time (d) in mini-columns leaching of pretreated ore with 16.8 kg sulfuric acid/ton, 23.3 kg sodium nitrate/ton, 19.8 kg sodium chloride/ton, 30 days of curing at 45 °C. Irrigation conditions: pH 1, 6.3 g sodium nitrate/L, seawater. C1: 25 °C and 20 g chloride/L, C2: 45 °C and 20 g chloride/L, and C3: 25 °C and 40 g chloride/L.

3.2.1. Effect of Temperature on the Leaching Stage

The copper extraction was lower at 45 °C than at 25 °C, which could be because the mini-columns are open to the air, so that NOx gas formed by the reaction between H_2SO_4 and $NaNO_3$ escape from the leaching system, which results at 45 °C in fewer nitrate ions being available for the copper dissolution reaction, as shown in Equation (7):

$$2NaNO_3 + H_2SO_4 \rightarrow 2HNO_{3(g)} + Na_2SO_4 \tag{7}$$

With $\Delta G_{25\,°C} = 1.3$ kcal/mol and $\Delta G_{45\,°C} = -0.1$ kcal/mol.

Oxidant loss could not occur at 25 °C, therefore the availability of oxidants would be higher than at 45 °C which would affect the copper dissolution.

The redox potential values in the output solutions of C1 and C3 were higher than that of C2 (see Figure 4), which could be due to a smaller quantity of oxidant in the C2 medium than in those of C1 or C3. It could also be due to ferric sulfate being generated in C2 by the pretreatment, which would have reacted with chalcopyrite in the leaching stage, forming ferrous sulfate and resulting in a lower redox potential.

$$CuFeS_2 + 4Fe^{3+} \rightarrow Cu^{2+} + 5Fe^{2+} + 2S \tag{8}$$

$$CuFeS_2 + 4Fe^{3+} + Cl^- \rightarrow CuCl^+ + 5Fe^{2+} + 2S \tag{9}$$

With $\Delta G_{25\,°C}$ = −31.9 kcal/mol equation 8 and $\Delta G_{25\,°C}$ = −32.4 kcal/mol Equation (9).

The results obtained indicate that under the conditions used in this study, copper dissolution is not favored by increasing the temperature from 25 to 45 °C.

3.2.2. Effect of Chloride Concentration on Leaching Stage

Copper extraction rates was higher in the column with 20 g/L of chloride (C1) than in the column with 40 g/L (C3), which could be due to greater saturation of the leaching medium (more ionic strength) in a higher chloride concentration, which is unfavorable for copper dissolution under the studied conditions. Several authors have determined that the presence of chloride ions in leaching systems improves copper extraction [6,8,9]. However, the increase in chloride concentration is not necessarily proportional to the copper extraction rate [15,51].

3.2.3. Analyses of Solid Residues

Table 9 shows the residual mineral species in C1 obtained by leaching test (observed with an optical microscope) in comparison to the original mineralogical state (head ore).

Table 9. Mineralogy of initial ore and solid residue of C1.

Mineral	Chemical Formula	Ore		C1 Residue	
		Amount (wt. %)	Cu (wt. %)	Amount (wt. %)	Cu (wt. %)
Chalcopyrite	$CuFeS_2$	1.69	0.59	0.64	0.22
Bornite	Cu_5FeS_4	0.10	0.06	<0.02	<0.01
Covellite	CuS	0.02	0.01	<0.02	<0.01
Atacamite	$CuCl_2 \cdot 3Cu(OH)_2$	0.07	0.04	0.00	0.00
Chrysocolla	$Cu_{1.75}Al_{0.25}H_{1.75}(Si_2O_5)(OH)_4 \cdot 0.25(H_2O)$	0.01	<0.01	0.00	0.00
Pyrite	FeS_2	3.81	-	3.29	-
Magnetite	Fe_3O_4	0.40	-	0.60	-
Hematite	Fe_2O_3	0.08	-	0.04	-
Limonite	$FeOOH$	0.07	-	0.05	-
Quartz	SiO_2	38.39	-	33.12	-
Gangues	-	55.36	-	62.23	-
	Total Cu		0.70		0.24

Table 9 shows the presence of copper sulfide in the residue, with reduced levels of chalcopyrite, bornite and covellite, while copper oxide was totally leached. It is possible to dissolve 62% of the chalcopyrite in the ore using a pretreatment/leaching process in acid-nitrate-chloride media.

4. Conclusions

We studied the effects of a pretreatment stage, adding nitrate and chloride and varying repose time and temperature on chalcopyrite dissolution in acid-nitrate-chloride media. The main findings are:

- The copper sulfide pretreatment provided satisfactory results, with 58.6% Cu extraction with the addition of 23.3 kg sodium nitrate and 19.8 kg sodium chloride per ton, and 30 days of curing time at 45 °C.
- The acid-nitrate-chloride media at high concentrations was effective in the pretreatment, providing oxidizing ions in the repose period.

- A copper extraction of 63% was obtained by leaching pretreated ore in mini-columns using a leaching solution of 6.3 g/L of sodium nitrate and 20 g/L of chloride at pH 1 and a temperature of 25 °C. Raising the temperature and increasing the chloride concentration did not improve copper extraction rates under the studied conditions.
- Chalcopyrite, bornite and covellite were only partially leached in the mini-columns, while copper oxides were completely leached.
- A favorable copper dissolution from sulfide ore was obtained in acid-nitrate-chloride media with the inclusion of a pretreatment stage in the leaching system. Waste salts from the caliche industry can be used for the nitrate medium, while waste brine from reverse osmosis can provide the chloride medium.
- An emphasis on the pretreatment stage is proposed given that this results in copper extraction rates in the range of 60%, so in the next phase only an acid wash will be required.

Author Contributions: Conceptualization, P.C.H. and Y.P.J.; Formal analysis, O.O.H. and C.M.T.; Investigation, J.D.; Supervision, P.C.H. and Y.P.J.; Validation, O.O.H. and C.M.T.; Writing—original draft, P.C.H.

Funding: The authors are grateful for the financial support provided by CONICYT through Fondecyt Iniciación Project N° 11170179 and PAI/Concurso Nacional Inserción en la Academia, Convocatoria 2015 folio N° 79150003.

Acknowledgments: Special thanks to Departamento de Ingeniería en Minas and Laboratorio de Investigación de procesos of Departamento de Ingeniería Química y Procesos de Minerales, of Universidad de Antofagasta for the assistance in ore preparation and technical support with equipment and materials.

Conflicts of Interest: The authors declare no conflict of interest.

References

1. Baba, A.A.; Ayinla, K.I.; Adekola, F.A.; Ghosh, M.K.; Ayanda, O.S.; Bale, R.B.; Sheik, A.R.; Pradhan, S.R. A review on novel techniques for chalcopyrite ore processing. *Int. J. Min. Eng. Miner. Process.* **2012**, *1*, 1–16. [CrossRef]
2. Watling, H. Chalcopyrite hydrometallurgy at atmospheric pressure: 1. Review of acidic sulfate, sulfate–chloride and sulfate–nitrate process options. *Hydrometallurgy* **2013**, *140*, 163–180. [CrossRef]
3. Wang, S. Copper leaching from chalcopyrite concentrates. *JOM J. Miner. Met. Mater. Soc.* **2005**, *57*, 48–51. [CrossRef]
4. Barriga, F.; Palencia, I.; Carranza, F. The passivation of chalcopyrite subjected to ferric sulfate leaching and its reactivation with metal sulfides. *Hydrometallurgy.* **1987**, *19*, 159–167. [CrossRef]
5. Klauber, C. A critical review of the surface chemistry of acidic ferric sulphate dissolution of chalcopyrite with regards to hindered dissolution. *Int. J. Miner. Process.* **2008**, *86*, 1–17. [CrossRef]
6. Watling, H. Chalcopyrite hydrometallurgy at atmospheric pressure: 2. Review of acidic chloride process options. *Hydrometallurgy* **2014**, *146*, 96–110. [CrossRef]
7. Ruiz, M.; Montes, K.; Padilla, R. Chalcopyrite leaching in sulfate–chloride media at ambient pressure. *Hydrometallurgy* **2011**, *109*, 37–42. [CrossRef]
8. Velásquez-Yévenes, L.; Nicol, M.; Miki, H. The dissolution of chalcopyrite in chloride solutions: Part 1: The effect of solution potential. *Hydrometallurgy* **2010**, *103*, 108–113. [CrossRef]
9. Velásquez, L.; Miki, H.; Nicol, M. The dissolution of chalcopyrite in chloride solutions: Part 2: Effect of various parameters on the rate. *Hydrometallurgy* **2010**, *103*, 80–85.
10. Veloso, T.C.; Peixoto, J.J.; Pereira, M.S.; Leao, V.A. Kinetics of chalcopyrite leaching in either ferric sulphate or cupric sulphate media in the presence of NaCl. *Int. J. Miner. Process.* **2016**, *148*, 147–154. [CrossRef]
11. Dutrizac, J. The leaching of sulphide minerals in chloride media. *Hydrometallurgy* **1992**, *29*, 1–45. [CrossRef]
12. Nicol, M.; Miki, H.; Velásquez-Yévenes, L. The dissolution of chalcopyrite in chloride solutions: Part 3. Mechanisms. *Hydrometallurgy* **2010**, *103*, 86–95. [CrossRef]
13. Herreros, O.; Quiroz, R.; Restovic, A.; Viñals, J. Dissolution kinetics of metallic copper with $CuSO_4$–NaCl–HCl. *Hydrometallurgy* **2005**, *77*, 183–190.
14. Jamett, N.E.; Hernández, P.C.; Casas, J.M.; Taboada, M.E. Speciation in the Fe (III)-Cl (I)-H_2O System at 298.15 K, 313.15 K, and 333.15 K (25 °C, 40 °C and 60 °C). *Metall. Mater. Trans. B.* **2017**, *49*, 1–9. [CrossRef]

15. Carneiro, M.F.C.; Leão, V.A. The role of sodium chloride on surface properties of chalcopyrite leached with ferric sulphate. *Hydrometallurgy* **2007**, *87*, 73–82. [CrossRef]

16. Lu, Z.Y.; Jeffrey, M.I.; Lawson, F. The effect of chloride ions on the dissolution of chalcopyrite in acidic solutions. *Hydrometallurgy* **2000**, *56*, 189–202. [CrossRef]

17. Hernández, P.; Taboada, M.; Herreros, O.; Torres, C.; Ghorbani, Y. Chalcopyrite dissolution using seawater-based acidic media in the presence of oxidants. *Hydrometallurgy* **2015**, *157*, 325–332. [CrossRef]

18. Watling, H.; Shiers, D.; Li, J.; Chapman, N.; Douglas, G. Effect of water quality on the leaching of a low-grade copper sulfide ore. *Miner. Eng.* **2014**, *58*, 39–51. [CrossRef]

19. Torres, C.; Taboada, M.; Graber, T.; Herreros, O.; Ghorbani, Y.; Watling, H. The effect of seawater based media on copper dissolution from low-grade copper ore. *Miner. Eng.* **2015**, *71*, 139–145. [CrossRef]

20. Cisternas, L.; Moreno, L. *El Agua De Mar En La Minería: Fundamentos Y Aplicaciones*; Ril editors: Santiago, Chile, 2014.

21. Velásquez-Yévenes, L.; Quezada-Reyes, V. Influence of seawater and discard brine on the dissolution of copper ore and copper concentrate. *Hydrometallurgy* **2018**, *180*, 88–95. [CrossRef]

22. Cisternas, L.A.; Gálvez, E.D. The use of seawater in mining. *Miner. Process. Extr. Metall. Rev.* **2018**, *39*, 18–33. [CrossRef]

23. Padilla, R.; Vega, D.; Ruiz, M.C. Pressure leaching of sulfidized chalcopyrite in sulfuric acid–oxygen media. *Hydrometallurgy* **2007**, *86*, 80–88. [CrossRef]

24. Herreros, O.; Quiroz, R.; Campos, I.; Rojas, J.; Viñals, J. Lixiviación de concentrados calcopiriticos con cloro. *Innovación* **2001**, *13*, 7–15.

25. Córdoba, E.M.; Muñoz, J.A.; Blázquez, M.L.; González, F.; Ballester, A. Leaching of chalcopyrite with ferric ion. Part I: General aspects. *Hydrometallurgy* **2008**, *93*, 81–87. [CrossRef]

26. Lu, J.; Dreisinger, D. Copper leaching from chalcopyrite concentrate in Cu (II)/Fe (III) chloride system. *Miner. Eng.* **2013**, *45*, 185–190. [CrossRef]

27. Sokić, M.D.; Marković, B.; Živković, D. Kinetics of chalcopyrite leaching by sodium nitrate in sulphuric acid. *Hydrometallurgy* **2009**, *95*, 273–279. [CrossRef]

28. Habashi, F. Nitric acid in the hydrometallurgy of sulfides. In *EPD Congress*; The Minerals, Metals & Materials Society: McCandless, PA, USA, 1999.

29. Habashi, F. Action of nitric acid on chalcopyrite. *Trans. Soc. Min. Eng Aime. Aime.* **1973**, *254*, 224–228.

30. Gok, O.; Anderson, C.G. Dissolution of low-grade chalcopyrite concentrate in acidified nitrite electrolyte. *Hydrometallurgy* **2013**, *134–135*, 40–46. [CrossRef]

31. Anderson, C.G. Treatment of copper ores and concentrates with industrial nitrogen species catalyzed pressure leaching and non-cyanide precious metals recovery. *JOM.* **2003**, *55*, 32–36. [CrossRef]

32. *Anuario de la Minería de Chile 2016*; G.d.C. Sernageomin Ministerio de Minería: Santiago, Chile, 2016.

33. Shiers, D.; Collinson, D.; Kelly, N.; Watling, H. Copper extraction from chalcopyrite: Comparison of three non-sulfate oxidants, hypochlorous acid, sodium chlorate and potassium nitrate, with ferric sulfate. *Miner. Eng.* **2016**, *85*, 55–65. [CrossRef]

34. Tsogtkhangai, D.; Mamyachenkov, S.V.; Anisimova, O.S.; Naboichenko, S.S. Kinetics of leaching of copper concentrates by nitric acid. *Russ. J. Non-Ferrous Met.* **2011**, *52*, 469–472. [CrossRef]

35. Castellón, C.; Taboada, M.E.; Hernández, P.C. *Lixiviación de Concentrado de Calcopirita por Nitrato de Sodio Con Agua de Mar en un Medio Ácido*; XXXI Convención Internacional de Minería Acapulco: México, 2015; pp. 227–234.

36. Queneau, P.; Prater, J. Nitric acid process for recovering metal values from sulfide ore materials containing iron sulfides. U.S. Patent 3,793,429, 1974.

37. Arias, J.A. Heap leaching copper ore using sodium nitrate. U.S. Patent US 6.569.391 B1, 2003.

38. Hard, R.A. Process for in-situ mining. US Patent 3.910.636, 1975.

39. Carnahan, T.G.; Heinen, H.J. Chemical mining of copper porphyry ores. US 3.912.330, 1975.

40. Lueders, R.E.; Frankiewicz, T.C. Metal leaching from concentrates using nitrogen dioxide in acids. US 4.189.461, 1980.

41. Anderson, C.G.; Krys, L.E.; Harrison, K.D. Treatment of metal bearing mineral material. US 5.096.486, 1992.

42. Hernández, P.C.; Taboada, M.E.; Herreros, O.O.; Graber, T.A.; Ghorbani, Y. Leaching of chalcopyrite in acidified nitrate using seawater-based media. *Minerals* **2018**, *8*, 238. [CrossRef]

43. Bouffard, S.C. Review of agglomeration practice and fundamentals in heap leaching. *Miner. Process. Extr. Metall. Rev.* **2005**, *26*, 233–294. [CrossRef]

44. Schlesinger, M.; King, M.; Sole, K.; Davenport, W. *Hydrometallurgical copper extraction: introduction and leaching, In Extractive Metallurgy of Copper*; Elsevier: Amsterdam, The Netherlands, 2011; pp. 281–322.

45. Lu, J.; Dreisinger, D.; West-Sells, P. Acid curing and agglomeration for heap leaching. *Hydrometallurgy* **2017**, *167*, 30–35. [CrossRef]

46. Herreros, O.; Quiroz, R.; Gutierrez, M. Estimación de equipos de aglomeración en lixiviación en pilas. *Innovación* **1991**, *4*, 17–21.

47. Kodali, P.; Depci, T.; Dhawan, N.; Wang, X.; Lin, C.; Miller, J.D. Evaluation of stucco binder for agglomeration in the heap leaching of copper ore. *Miner. Eng.* **2011**, *24*, 886–893. [CrossRef]

48. Cerda, C.P.; Taboada, M.E.; Jamett, N.E.; Ghorbani, Y.; Hernández, P.C. Effect of pretreatment on leaching primary copper sulfide in acid-chloride media. *Minerals* **2017**, *8*, 1. [CrossRef]

49. Bahamonde, F.; Gómez, M.; Navarro, P. Pre-treatment with sodium chloride and sulfuric acid of a bornitic concentrate and later leaching in chloride solution. In *Leaching and Bioleaching of Sulfide Concentrates and Minerals*; Hydroprocess-ICMSE 2017: Santiago, Chile, 2017.

50. Taboada, M.E.; Quiroz, R.; Hernández, P.C.; Padilla, A.; Herreros, O.O.; Graber, T.A. Copper ore leaching with pre-treatment. In *Innovation, Development and Process Improvement*; Hydroprocess ICMSE 2017: Santiago, Chile, 2017; pp. 54–60.

51. Velásquez-Yévenes, L.; Torres, D.; Toro, N. Leaching of chalcopyrite ore agglomerated with high chloride concentration and high curing periods. *Hydrometallurgy* **2018**, *181*, 215–220. [CrossRef]

52. Senanayake, G. A review of chloride assisted copper sulfide leaching by oxygenated sulfuric acid and mechanistic considerations. *Hydrometallurgy* **2009**, *98*, 21–32. [CrossRef]

53. *HSC-Chemistry*, version 6; Outokumpu Researcher Oy Antti Roine: Piori, Finland, 2006.

Article

Leaching of White Metal in a NaCl-H$_2$SO$_4$ System under Environmental Conditions

Jonathan Castillo [1,*], Rossana Sepúlveda [1], Giselle Araya [1], Danny Guzmán [1], Norman Toro [2], Kevin Pérez [2], Marcelo Rodríguez [2] and Alessandro Navarra [3]

[1] Department of Metallurgy Engineering, University of Atacama, Av. Copayapu 485, Copiapó 1531772, Chile; rossana.sepulveda@uda.cl (R.S.); giselle.araya@alumnos.uda.cl (G.A.); danny.guzman@uda.cl (D.G.)

[2] Department of Metallurgical and Mining Engineering, Catholic University of the North, Av. Angamos 610, Antofagasta 1270709, Chile; ntoro@ucn.cl (N.T.); kps003@alumnos.ucn.cl (K.P.); mra039@alumnos.ucn.cl (M.R.)

[3] Department of Mining and Materials Engineering, McGill University, University Street 3610, Montreal, QC H3A 0C5, Canada; alessandro.navarra@mcgill.ca

* Correspondence: jonathan.castillo@uda.cl; Tel.: +56-52-225-5622 or +56-52-225-5614

Received: 24 April 2019; Accepted: 22 May 2019; Published: 24 May 2019

Abstract: The effect of NaCl on the leaching of white metal from a Teniente Converter was investigated in NaCl-H$_2$SO$_4$ media under environmental conditions. The copper dissolution from white metal was studied using ferric ions in the range of 1–10 g/L, NaCl in the range of 30–210 g/L, and sulfuric acid in the range of 10–50 g/L. The test without NaCl produced a dissolution of 55%; through the addition of NaCl, the dissolution increased to nearly 90%. The effect of sulfuric acid on the copper dissolution was not significant in the studied range, as the excess sulfuric acid simply increased the iron precipitation. The positive effect of NaCl seems to be related to the action of chloro-complex oxidizing agents in relation to the Cu^{+2}/Cu$^+$ couple. A simplified two-stage mechanism is proposed for the leaching of white metal. In the first stage, the white metal produces covellite and Cu^{2+}, and in the second stage it produces elemental sulfur and Cu^{2+}. The first stage is very rapidly compared to the second stage.

Keywords: metal extraction; acid leaching; white metal; ferric ion; chloride ion

---

1. Introduction

In the pyrometallurgical processing of copper concentrates, iron is eliminated through the formation of two immiscible phases called slag and matte; the iron reports to the slag as oxide, and the copper remains in the matte as sulfide. In particular, a Teniente Converter (TC) produces high-grade matte called "white metal" (74–76% Cu) that is further processed in a Peirce-Smith converter or similar furnace, to produce so-called blister copper (99% Cu) [1–3]. There are alternatives to Peirce-Smith converting that can greatly reduce the apollution in copper smelters [2,3]. Indeed, the hydrometallurgical treatment of copper matte is supported by a series of experimental tests carried out by different authors, who have achieved the total dissolution of copper under different operating conditions. In most cases, these conditions are chemically aggressive or very energetic, achieving the total dissolution in a few hours by agitation leaching, under a high pressure and high temperature [4–6].

Studies in copper matte leaching (also applicable to studies with chalcocite) are divided into two main areas: high temperature and pressure with an oxidizing agent, and the use of molecular chlorine [6–8].

In the literature, there are several cases of white metal leaching that apply a high temperature, high pressure, and oxidizing agents, to obtain a high recovery within a few hours; these systems are very effective even for other sulfides like chalcopyrite. Some variables considered in the studies are the temperature, oxygen pressure, concentration of sulfuric acid and iron, and particle size [4,6,9,10]. A high temperature white metal leaching has a faster rate of oxidation of Fe^{2+} to Fe^{3+} than a low temperature oxidation (Equations (1) and (2)) and tends to promote sulfate formation instead of elemental sulfur (Equation (3)) [4,11].

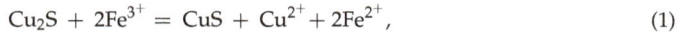

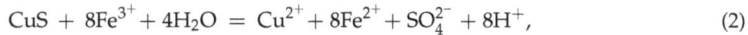

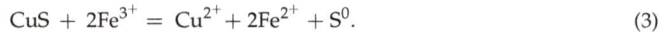

$$Cu_2S + 2Fe^{3+} = CuS + Cu^{2+} + 2Fe^{2+}, \tag{1}$$

$$CuS + 8Fe^{3+} + 4H_2O = Cu^{2+} + 8Fe^{2+} + SO_4^{2-} + 8H^+, \tag{2}$$

$$CuS + 2Fe^{3+} = Cu^{2+} + 2Fe^{2+} + S^0. \tag{3}$$

In contrast, the oxidative leaching of white metal occurs in two stages, with CuS as an intermediate product (Equations (4) and (5)), with the possible presence of nonstoichiometric sulfides, such as djurleite or digenite [4,9,12]. Although this mechanism changes in atmospheric conditions, to produce elemental sulfur instead of sulfate (Equation (6)) the intermediate product CuS remains the same [4,13].

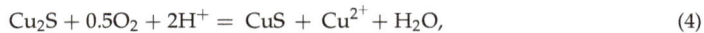

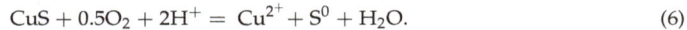

$$Cu_2S + 0.5O_2 + 2H^+ = CuS + Cu^{2+} + H_2O, \tag{4}$$

$$CuS + 2O_2 = Cu^{2+} + SO_4^{2-}, \tag{5}$$

$$CuS + 0.5O_2 + 2H^+ = Cu^{2+} + S^0 + H_2O. \tag{6}$$

The literature suggests that molecular chlorine may be an appropriately strong leaching agent for several complex ores, including refractory gold and platinum ores, as well as copper sulfides. However, chlorine leaching requires extreme care, in order to avoid the health risks of chlorine gas. In some tests, the molecular chlorine was provided by chlorine gas; other tests applied an in-situ generation. The in-situ generation consists in producing molecular chlorine through a reaction between sodium hypochlorite and hydrochloric acid (Equation (7)), the addition of MnO_2 in the presence of hydrochloric acid (Equation (8)), or the electro-generation of Cl_2 (Equation (9)) [7,14–22].

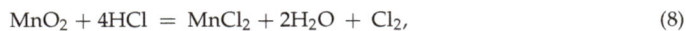

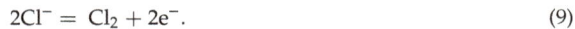

$$NaClO + 2HCl = NaCl + Cl_2 + H_2O, \tag{7}$$

$$MnO_2 + 4HCl = MnCl_2 + 2H_2O + Cl_2, \tag{8}$$

$$2Cl^- = Cl_2 + 2e^-. \tag{9}$$

Several authors propose the following mechanism for Cu_2S leaching with chlorine [7,15,23]:

$$Cu_2S + 5Cl_{2(g)} + 4H_2O = 2Cu_{(aq)}^{2+} + SO_4^{2-} + 10Cl_{(aq)}^- + 8H^+. \tag{10}$$

This method for copper recovery is effective but is very sensitive to the particle size and initial chlorine concentration.

The mechanism for the leaching of copper matte in strong chlorine media can be approximated to have two stages: the first is the transformation of chalcocite into covellite and Cu^{2+}, and the second is the formation of Cu^{2+} in the presence of elemental sulfur or sulfate. However, the real mechanism may be more complex, involving several transformations of nonstoichiometric compounds [24,25] through the following scheme:

$$Cu_2S \rightarrow Cu_{1.93}S \rightarrow Cu_{1.8}S \rightarrow Cu_{1.6}S \rightarrow Cu_{1-1.2}S. \tag{11}$$

This paper presents the experimental results of white metal leaching in $NaCl$-H_2SO_4 media, because the hydrometallurgical processing of white metal may be an attractive alternative to pyrometallurgical processing, but most research efforts have used aggressive methods such as a high pressure and temperature with an oxidizing agent, as well as the use of chlorine gas. These aggressive methods are centered on chemical considerations for copper recovery, but not on the other problems faced by the mining industry, such as sustainability, strong environmental regulations, the scarcity of fresh water, high-energy costs, and low ore grades [26].

2. Materials and Methods

2.1. White Metal Characterization

Representative white metal from a Teniente Converter reactor was supplied by the Hernán Videla Lira smelter in Copiapó, Chile. Pieces weighing 250 g were collected, crushed, milled and sieved to produce a fine powder with an average size of 64 μm.

The chemical characterization by Atomic absorption spectroscopy (AAS) (PerkinElmer PinAAcle™ 900F, Waltham, MA, USA) and sulfur analyzer (LECO) of white metal shows 74.03% Cu, 2.01% Fe and 20.5% S.

In order to identify the mineralogical species, the sample was analyzed by X-ray diffraction (XRD), and the result is shown in Figure 1. The species in the white metal are chalcocite (Cu_2S with a tetragonal structure), Spinel ($Fe_{2.57}Si_{0.43}O_4$ with an orthorhombic structure) and Cristobalite (SiO_2 with a tetragonal structure).

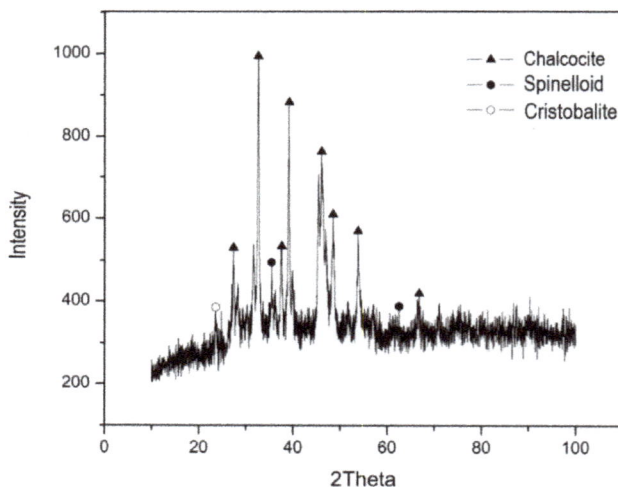

Figure 1. White metal diffractogram.

Scanning electron microscopy with energy dispersive spectroscopy (SEM-EDS) was performed in order to study the morphology of the particles and chemical composition. Mapping SEM-EDS suggests a majority presence of copper and sulfur, and small amounts of iron and oxygen (Figure 2), which reaffirms the XRD analysis. The morphology is similar to that of ceramic materials, with small irregular particles; crystallized structures or regular forms are not observed.

Finally, an optical microscopic observation of the white metal was carried out in order to corroborate the aforementioned results. In Figure 3, traces of metallic copper are observed by optical inspection in the white metal matrix. Metallic copper was not observed by SEM-EDS and XRF analysis, because these techniques are limited to trace levels of species.

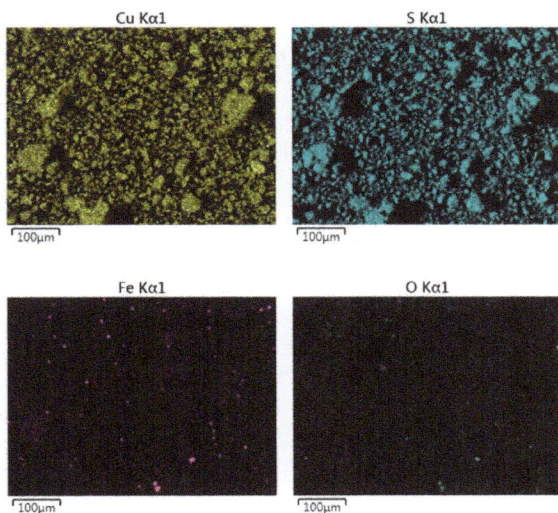

Figure 2. Analysis by scanning electron microscopy with energy dispersive spectroscopy SEM-EDS to the white metal with mapping to 200×.

Figure 3. Optical microscopy of white metal with magnification (**a**) 20× and (**b**) 100×.

2.2. Leaching Tests

The tests were conducted in shake flasks, at an ambient temperature, with magnetic stirring set to 200 rev/min. 500 mL of leach solution was put in contact with 2.5 g of white metal for 4 days. The pH, electrical potential and temperature were measured regularly. Aliquots of 10 mL of leached solution were collected and filtered to perform the copper and iron analyses. The concentration of copper and iron in the aqueous solution was quantified by Atomic absorption spectroscopy. After the leaching test, the residues were filtered, dried, and analyzed by SEM-EDS, XRD, and AAS.

The aqueous solutions were prepared using High performance liquid chromatography (HPLC) water from Merck. All chemical reagents that were used were of analytical grade, supplied by Merck.

3. Results and Discussion

The leaching tests were performed to evaluate the influence of the ferric ion, sodium chloride and sulfuric acid concentration on the copper dissolution. The tests considered the concentration ranges of 0–10 g/L, 0–210 g/L and 10–50 g/L, for Fe^{3+}, NaCl and H_2SO_4, respectively.

3.1. Baseline

The first test was performed to establish a baseline, evaluating the dissolution of white metal using a solution containing only water and sulfuric acid. The experiments were carried out with a stirring speed of 400 rpm, and with 20 g/L H_2SO_4 at room temperature for 4 days. As shown in Figure 4, copper leaching occurs very fast on the first day, in comparison to the other three days. Copper dissolution in the first day is 37.8%, ending with a 54.9% extraction on the fourth day. This asymptotic trend is attributed to a limited quantity of oxygen dissolved in the leaching solution. Consider that white metal is artificial chalcocite; thus, if the dissolved oxygen is the only oxidizing agent that is present, the following mechanism is proposed [4,9,24].

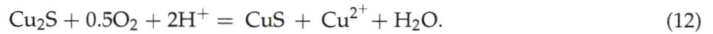

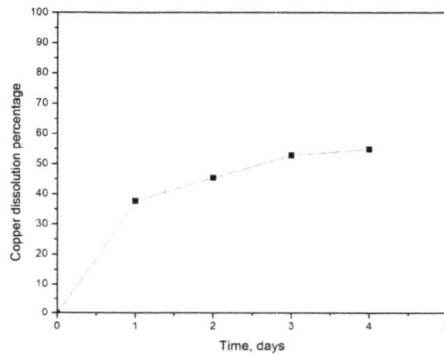

$$Cu_2S + 0.5O_2 + 2H^+ = CuS + Cu^{2+} + H_2O. \tag{12}$$

Figure 4. Copper leaching of white metal with a leaching solution containing 20 g/L H_2SO_4 at 1 atm, 22 °C and 400 rpm.

According to the mechanism, cupric ions are dissolved, and covellite is generated. Covellite is difficult to dissolve without a strong oxidizing agent; however, in this case, the copper dissolution is more than 50%, which may be due to the dissolved iron. The white metal contains 2% iron, which contributes to an iron concentration of 0.06 g/L in the final solution.

The residue characterization is shown in Figure 5, showing only covellite (hexagonal crystalline system). This is evidence of the mechanism of white metal leaching of Equation (12), since elemental sulfur is not detected.

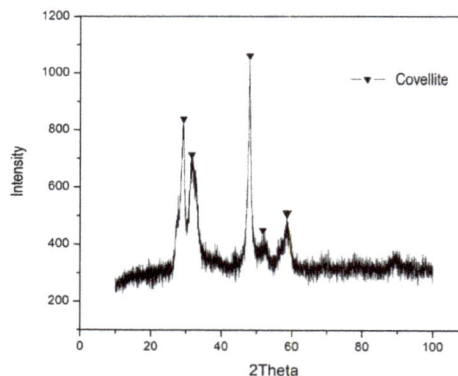

Figure 5. XRD analysis of residue for the baseline test.

3.2. Effect of Ferric Ion

The addition of ferric ions has a positive effect on the leaching in comparison to the baseline. The copper dissolution increased to a maximum of 75.3%, from the baseline value of 54.9%. As shown in Figure 6, the copper dissolution has a moderate increase with 1 g/L of Fe^{3+} (60.8%). With 4 and 10 g/L of Fe^{3+}, the effect is similar, at 73.5% and 75.3% respectively.

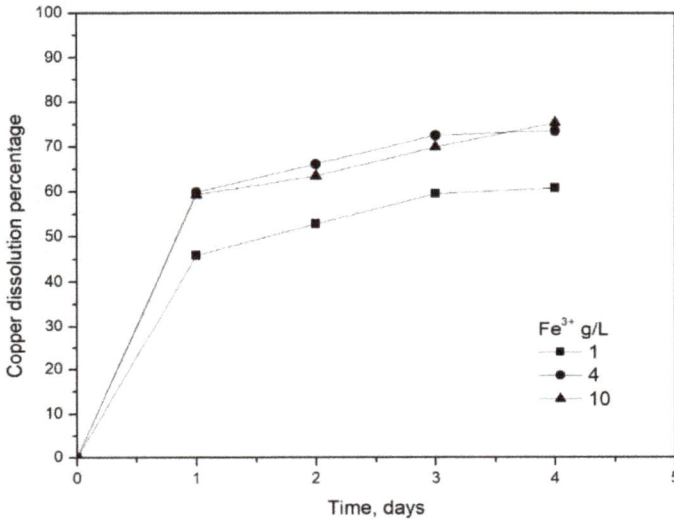

Figure 6. Copper recovery curves in leaching with a variation of Fe^{3+}, 20 g/L H_2SO_4 under conditions of 1 atm, 22 °C, and 400 rpm.

Ferric ions increase the kinetics and dissolution of copper in comparison to the baseline, but the curves show two distinct steps in the reaction. The copper dissolution is very fast within the first day; but from the second day onward, the copper dissolution is slower, tending toward an asymptote. The two-step behavior of the white metal dissolution is comparable to the leaching of chalcocite ores. Indeed, chalcocite is dissolved in the presence of Fe^{3+} in two steps; the first step is fast, but the second step is slow at room temperature. The mechanism is shown in the following equations:

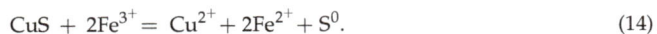

$$Cu_2S + 2Fe^{3+} = Cu^{2+} + 2Fe^{2+} + CuS, \tag{13}$$

$$CuS + 2Fe^{3+} = Cu^{2+} + 2Fe^{2+} + S^0. \tag{14}$$

Equation (13) shows the chemical dissolution of white metal (synthetic chalcocite) in the presence of Fe^{3+}, with a molar ratio of 1:2 between Cu_2S and Fe^{3+} to carry out a transformation from synthetic chalcocite to covellite and Cu^{2+}. The stoichiometry of Equation (13) explains the result for the slower kinetics and recovery at low Fe^{3+}, resulting from an insufficient amount of oxidizing agent. Indeed, a low concentration of ferric ion, added to the formation of elemental sulfur, is a barrier to copper leaching. The analysis of the residues by XRD indicates the presence of covellite and elemental sulfur, as shown in Figure 7, which supports the proposed mechanisms of Equations (13) and (14), which are similar to the leaching of natural chalcocite in the presence of ferric ions.

The additional ferric ions successfully increase the copper dissolution, but also lead to the contamination of the leaching solution with ferrous ions or precipitates; in an industrial setting, this would require auxiliary operations to purify the solution.

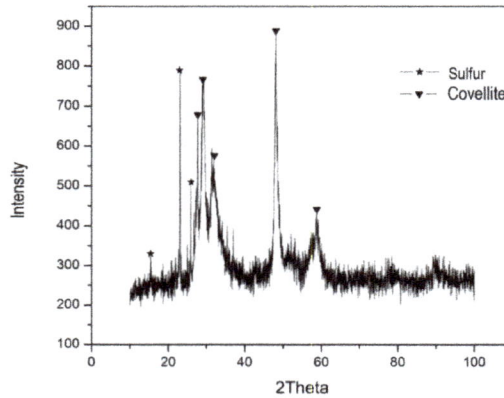

Figure 7. Residue diffractogram 4 days of leaching with 4 g/L of Fe^{3+}, 20 g/L of H_2SO_4 in conditions of 1 atm, 22 °C and 400 rpm.

3.3. Effect of Strong Chloride Media

The testing of a high chloride media similar to seawater is motivated by the scarcity of fresh water and the potential use of this resource in copper leaching [27–29]. There are no other studies in the literature that use a media similar to seawater for copper leaching, although there has been extensive research reported for copper ores with chloride, even for the leaching of concentrates and primary sulfides copper ores [26,30–35].

Within the current study, the effect of the NaCl concentration was tested from 30 to 210 g/L; these values consider seawater feed and high recirculating solutions that would be typical of copper leaching plants. As shown in Figure 8, the addition of chloride produces an important increase in copper dissolution compared to the test without NaCl. At concentrations over 30 g/L NaCl, the effect on the final recovery was not significant (although the rate of reaction seems to be affected). At 30 g/L of NaCl, the copper dissolution was 84.23%. Increasing this concentration to 210 g/L, the copper dissolution was 84.49%. This agrees with the observations of Miki et al. [13] who reported that the copper recovery from synthetic covellite increased marginally with an increasing chloride concentration in the range of 7–90 g/L.

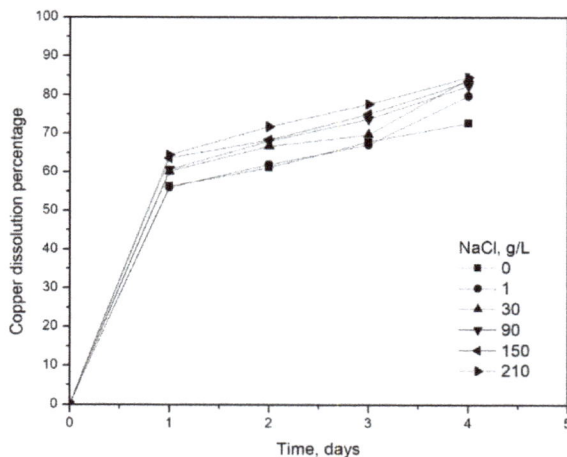

Figure 8. Copper recovery curves in leaching with a variation of NaCl, 20 g/L H_2SO_4, 4 g/L Fe^{3+} under conditions of 1 atm, 22 °C and 400 rpm.

The positive effects of strong chloride media in white metal leaching are shown in Figure 8. Interestingly, white metal is a high purity synthetic chalcocite, that has a similar leaching behaviour to chalcocite ores. The differences between both products (natural and synthetic) are the structure and morphology, which show a rough and slightly crystalline character (Figures 1–3), which could be significant for the copper recovery.

The efficiency of the chloride system for copper leaching from white metal is possibly attributed to the action of several chloro-complexes. As listed in Table 1, the leaching of copper sulfides in chloride media involves several stables species. Nonetheless, the speciation of the system is strongly conditioned by the pH, temperature and chloride concentration.

Table 1. Ion distribution as a function of the chloride concentration [36–38].

	Low Cl⁻					High Cl⁻
Cu(II)	Cu^{2+}	$CuCl^+$		$CuCl_2$	$CuCl_3^-$	$CuCl_4^{2-}$
Cu(I)	$CuCl_2^-$		$CuCl_3^{2-}$		$CuCl_4^{3-}$	
Fe(III)	Fe^{3+}			$FeCl^{2+}$		$FeCl_2^+$
Fe(II)	Fe^{2+}					$FeCl^+$

Although there are numerous stable species within the chloride medium, as shown in Table 1, the main ions that are present in strong chloride solutions are Cu^{2+} and $CuCl^+$. In the literature, it is proposed that $CuCl^+$ is generated constantly from reoxidized CuCl (formed by Cu_2S leach) to $CuCl^+$, thus continuously regenerating the oxidizing agent [27,39–41].

On the other hand, another possible beneficial effect of chloride is located at the residue surface. Reports in the literature suggest that chloride increases the surface area and porosity of residues, which tends to favor a higher copper dissolution. A high porosity sulfur layer facilitates the diffusion of leachants and products to and from the reaction surface [42,43].

The identification of species that are present in the residue is shown in Figure 9. According to the XRD analysis, the residue is composed of covellite and elemental sulfur. The diffraction pattern is similar to the leaching system in the absence of chloride ions; however, in this case, the intensities of the species are higher, especially the elemental sulfur. This result agrees with the higher recoveries obtained in the strong chloride system, since more elemental sulfur is produced as more copper is leached.

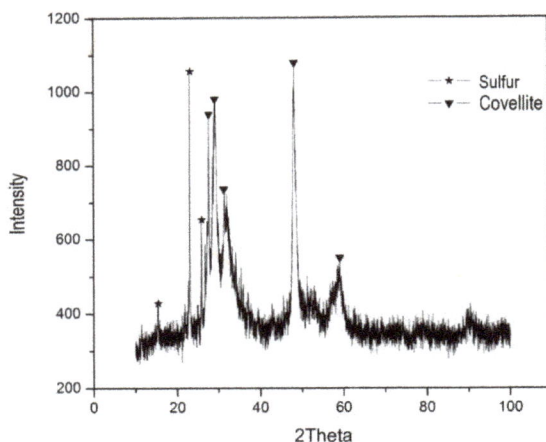

Figure 9. Residue diffractogram after 4 days of leaching with 4 g/L of Fe^{3+}, 20 g/L H_2SO_4, and 30 g/L NaCl in conditions of 1 atm, 22 °C and 400 rpm.

Based on the literature reports and the results obtained in this study, the leaching of white metal in strong chloride media at ambient conditions involves several chloride species, but the mechanism can be adequately described by:

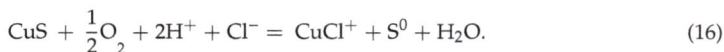

$$Cu_2S + \frac{1}{2}O_2 + 2H^+ + Cl^- = CuS + CuCl^+ + H_2O, \tag{15}$$

$$CuS + \frac{1}{2}O_2 + 2H^+ + Cl^- = CuCl^+ + S^0 + H_2O. \tag{16}$$

For the system studied in this research, the mechanism for the first stage (Equation (15)) is valid only at the initial stage (in the absence of Cu^{2+} ions); later, when the copper is dissolved, the cupric and chloride ions are oxidizing agents of white metal, as shown by the following equations:

$$Cu_2S + Cu^{2+} + 2Cl^- = CuS + 2CuCl, \tag{17}$$

$$Cu_2S + CuCl^+ + Cl^- = CuS + 2CuCl. \tag{18}$$

3.4. Effect of Sulfuric Acid

The effect of the concentration of sulfuric acid from 10 to 50 g/L is shown in Figure 10. It is observed that a substantial increase in the concentration of H_2SO_4 does not imply a significant increase in the dissolution of the copper; rather, there is a slight decrease at the end of the leaching test. Indeed, the increase in sulfuric acid causes a decrease in the oxygen solubility. This agrees with the observations of Ruiz et al. [9] who reported that the copper recovery from white metal increased marginally with an increasing sulfuric acid concentration in the range of 0.05–0.5 M.

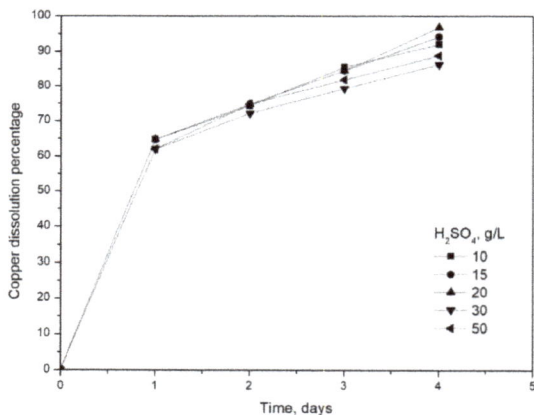

Figure 10. Copper recovery curves in leaching with a variation of H_2SO_4, 30 g/L NaCl and 4 g/L Fe^{3+} under conditions of 1 atm, 22 °C and 400 rpm.

Another important observation is that, for low levels of sulfuric acid, the dissolution of copper decreases slightly, while a significant iron hydroxide precipitation occurs (Table 2). This is a promising result from an industrial perspective, since a lower iron content in the leaching solution is favorable for solvent extraction [44].

Table 2. Total iron concentration in leach residues at different concentrations of H_2SO_4.

H_2SO_4, g/L	10	15	20	30	50
%Fe	3.14	2.99	2.67	2.35	2.11

A similar result was reported by Lu et al. [44], who in a study of chalcopyrite leaching in an acid medium with chloride ions, obtained more iron precipitation at low concentrations of acid. The iron precipitation is shown by the following equation:

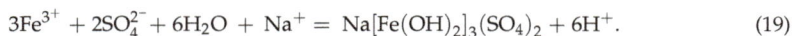

$$3Fe^{3+} + 2SO_4^{2-} + 6H_2O + Na^+ = Na[Fe(OH)_2]_3(SO_4)_2 + 6H^+. \tag{19}$$

The pregnant solution with cupric ions can be purified by solvent-extraction, to then obtain metallic copper by electrowinning [45,46]. The raffinate can be recycled to the leaching process.

4. Conclusions

The results of white metal leaching in the NaCl-H$_2$SO$_4$ system under environmental conditions indicate that white metal leaches in two stages. The first stage consists of the transformation of chalcocite into covellite and Cu^{2+}, and the second stage consists in the transformation of covellite into Cu^{2+} and elemental sulfur; on average, the first stage is about 5 times faster than the second.

In the baseline test, only a partial dissolution of white metal was observed, with a relatively fast first stage and a much slower second stage, with an asymptotic tendency close to a 55% dissolution. The residue showed no evidence of an elemental sulfur formation; this can be attributed to the presence of naturally dissolved oxygen in the solution, which acted as an oxidizing agent. The low concentration of oxygen in the solution and the room temperature limited the efficiency of the second dissolution stage.

The inclusion of chloride ions strongly increases the copper dissolution, approaching a dissolution of approximately 90%. The positive effect of the chloride is attributed to the Cu^{2+}/Cu$^+$ redox pair and the action of the oxidizing agents Cu^{2+}, CuCl$^+$, CuCl$_2$, and CuCl$_3$.

The effect of the sulfuric acid addition is not significant; in fact, the high sulfuric acid concentration causes an iron precipitation rather than increasing the copper dissolution.

Author Contributions: Conceptualization, J.C. and R.S.; formal analysis, D.G.; investigation, G.A.; methodology, K.P. and M.R.; validation, J.C. and N.T.; supervision, R.S.; writing—original draft, J.C. and R.S; writing—review and editing, A.N.

Funding: This research received no external funding.

Acknowledgments: The authors would like to thank CONICYT and GORE Atacama for funding the equipment used in this study (FONDEQUIP EQUR-160001) and the contribution of the Unit of Scientific Equipment - MAINI, from Catholic University of the North for facilitating the chemical tests of the solutions.

Conflicts of Interest: The authors declare no conflict of interest.

References

1. Wang, Q.-M.; Guo, X.-Y.; Wang, S.-S.; Liao, L.-L.; Tian, Q.-H. Multiphase equilibrium modeling of oxygen bottom-blown copper smelting process. *Trans. Nonferrous Met. Soc. China* **2017**, *27*, 2503–2511. [CrossRef]
2. Wang, S.; Guo, X. Thermodynamic Modeling of Oxygen Bottom-Blowing Continuous Converting Process. In *Extraction 2018*; Davis, B.R., Moats, M.S., Wang, S., Gregurek, D., Kapusta, J., Battle, T.P., Schlesinger, M.E., Alvear Flores, G.R., Jak, E., Goodall, G., et al., Eds.; Springer International Publishing: Cham, Switzerland, 2018; pp. 573–583.
3. Hogg, B.; Nikolic, S.; Voigt, P.; Telford, P. ISASMELT Technology for Sulfide Smelting. In *Extraction 2018*; Springer International Publishing: Cham, Switzerland, 2018; pp. 149–158.
4. Ruiz, M.C.; Gallardo, E.; Padilla, R. Copper extraction from white metal by pressure leaching in H$_2$SO$_4$-FeSO$_4$-O$_2$. *Hydrometallurgy* **2009**, *100*, 50–55. [CrossRef]
5. Park, K.H.; Mohapatra, D.; Hong-In, K.; Xueyi, G. Dissolution behavior of a complex Cu-Ni-Co-Fe matte in CuCl2-NaCl-HCl leaching medium. *Sep. Purif. Technol.* **2007**, *56*, 303–310. [CrossRef]
6. Anand, S.; Das, R.P.; Jena, P.K. Sulphuric acid pressure leaching of CuNiCo matte obtained from copper converter slag—Optimisation through factorial design. *Hydrometallurgy* **1991**, *26*, 379–388. [CrossRef]

7. Herreros, O.; Quiroz, R.; Viñals, J. Dissolution kinetics of copper, white metal and natural chalcocite in Cl2/Cl-media. *Hydrometallurgy* **1999**, *51*, 345–357. [CrossRef]
8. Neustroev, V.I.; Karimov, K.A.; Naboichenko, S.S.; Kovyazin, A.A. Autoclave leaching of arsenic from copper concentrate and matte. *Metallurgist* **2015**, *59*, 177–179. [CrossRef]
9. Ruiz, M.C.; Abarzúa, E.; Padilla, R. Oxygen pressure leaching of white metal. *Hydrometallurgy* **2007**, *86*, 131–139. [CrossRef]
10. Padilla, R.; Pavez, P.; Ruiz, M.C. Kinetics of copper dissolution from sulfidized chalcopyrite at high pressures in H_2SO_4-O_2. *Hydrometallurgy* **2008**, *91*, 113–120. [CrossRef]
11. Watling, H.R.; Shiers, D.W.; Li, J.; Chapman, N.M.; Douglas, G.B. Effect of water quality on the leaching of a low-grade copper sulfide ore. *Miner. Eng.* **2014**, *58*, 39–51. [CrossRef]
12. Arce, E.M.; González, I. A comparative study of electrochemical behavior of chalcopyrite, chalcocite and bornite in sulfuric acid solution. *Int. J. Miner. Process.* **2002**, *67*, 17–28. [CrossRef]
13. Miki, H.; Nicol, M.; Velásquez-Yévenes, L. The kinetics of dissolution of synthetic covellite, chalcocite and digenite in dilute chloride solutions at ambient temperatures. *Hydrometallurgy* **2011**, *105*, 321–327. [CrossRef]
14. Kim, M.S.; Lee, J.C.; Park, S.W.; Jeong, J.; Kumar, V. Dissolution behaviour of platinum by electro-generated chlorine in hydrochloric acid solution. *J. Chem. Technol. Biotechnol.* **2013**, *88*, 1212–1219. [CrossRef]
15. Herreros, O.; Quiroz, R.; Manzano, E.; Bou, C.; Viñals, J. Copper extraction from reverberatory and flash furnace slags by chlorine leaching. *Hydrometallurgy* **1998**, *49*, 87–101. [CrossRef]
16. Hilson, G.; Monhemius, A.J. Alternatives to cyanide in the gold mining industry: What prospects for the future? *J. Clean. Prod.* **2006**, *14*, 1158–1167. [CrossRef]
17. Pilone, D.; Kelsall, G.H. Prediction and measurement of multi-metal electrodeposition rates and efficiencies in aqueous acidic chloride media. *Electrochim. Acta* **2006**, *51*, 3802–3808. [CrossRef]
18. Kim, E.; Kim, M.-S.; Lee, J.-C.; Yoo, K.; Jeong, J. Leaching behavior of copper using electro-generated chlorine in hydrochloric acid solution. *Hydrometallurgy* **2010**, *100*, 95–102. [CrossRef]
19. Padilla, R.; Girón, D.; Ruiz, M.C. Leaching of enargite in H_2SO_4-NaCl-O_2 media. *Hydrometallurgy* **2005**, *80*, 272–279. [CrossRef]
20. Kim, E.Y.; Kim, M.S.; Lee, J.C.; Jeong, J.; Pandey, B.D. Leaching kinetics of copper from waste printed circuit boards by electro-generated chlorine in HCl solution. *Hydrometallurgy* **2011**, *107*, 124–132. [CrossRef]
21. Kim, M.S.; Park, S.W.; Lee, J.C.; Choubey, P.K. A novel zero emission concept for electrogenerated chlorine leaching and its application to extraction of platinum group metals from spent automotive catalyst. *Hydrometallurgy* **2016**, *159*, 19–27. [CrossRef]
22. Kleiv, R.; Aasly, K.; Kowalczuk, P.; Snook, B.; Drivenes, K.; Manaig, D. Galvanic leaching of seafloor massive sulphides using MnO_2 in H_2SO_4-NaCl media. *Minerals* **2018**, *8*, 235.
23. Beşe, A.V.; Ata, O.N.; Çelik, C.; Çolak, S. Determination of the optimum conditions of dissolution of copper in converter slag with chlorine gas in aqueous media. *Chem. Eng. Process.* **2003**, *42*, 291–298. [CrossRef]
24. Fisher, W.W.; Flores, F.A.; Henderson, J.A. Comparison of chalcocite dissolution in the oxygenated, aqueous sulfate and chloride systems. *Miner. Eng.* **1992**, *5*, 817–834. [CrossRef]
25. Zeng, W.; Qiu, G.; Chen, M. Investigation of Cu-S intermediate species during electrochemical dissolution and bioleaching of chalcopyrite concentrate. *Hydrometallurgy* **2013**, *134–135*, 158–165. [CrossRef]
26. Hernández, P.C.; Taboada, M.E.; Herreros, O.O.; Torres, C.M.; Ghorbani, Y. Chalcopyrite dissolution using seawater-based acidic media in the presence of oxidants. *Hydrometallurgy* **2015**, *157*, 325–332. [CrossRef]
27. Torres, C.M.; Taboada, M.E.; Graber, T.A.; Herreros, O.O.; Ghorbani, Y.; Watling, H.R. The effect of seawater based media on copper dissolution from low-grade copper ore. *Miner. Eng.* **2015**, *71*, 139–145. [CrossRef]
28. Hernández, P.; Taboada, M.; Herreros, O.; Graber, T.; Ghorbani, Y. Leaching of chalcopyrite in acidified nitrate using seawater-based media. *Minerals* **2018**, *8*, 238. [CrossRef]
29. Cisternas, L.A.; Gálvez, E.D. The use of seawater in mining. *Miner. Process. Extr. Metall. Rev.* **2018**, *39*, 18–33. [CrossRef]
30. Salinas, K.; Herreros, O.; Torres, C. Leaching of primary copper sulfide ore in chloride-ferrous media. *Minerals* **2018**, *8*, 312. [CrossRef]
31. Ruiz, M.C.; Montes, K.S.; Padilla, R. Chalcopyrite leaching in sulfate-chloride media at ambient pressure. *Hydrometallurgy* **2011**, *109*, 37–42. [CrossRef]
32. Veloso, T.C.; Peixoto, J.J.M.; Pereira, M.S.; Leao, V.A. Kinetics of chalcopyrite leaching in either ferric sulphate or cupric sulphate media in the presence of NaCl. *Int. J. Miner. Process.* **2016**, *148*, 147–154. [CrossRef]

33. Velásquez-Yévenes, L.; Torres, D.; Toro, N. Leaching of chalcopyrite ore agglomerated with high chloride concentration and high curing periods. *Hydrometallurgy* **2018**, *181*, 215–220. [CrossRef]

34. Velásquez-Yévenes, L.; Quezada-Reyes, V. Influence of seawater and discard brine on the dissolution of copper ore and copper concentrate. *Hydrometallurgy* **2018**, *180*, 88–95. [CrossRef]

35. Deniz Turan, M.; Boyrazlı, M.; Soner Altundoğan, H. Improving of copper extraction from chalcopyrite by using NaCl. *J. Cent. South Univ.* **2018**, *25*, 21–28.

36. Senanayake, G. Chloride assisted leaching of chalcocite by oxygenated sulphuric acid via Cu(II)-OH-Cl. *Miner. Eng.* **2007**, *20*, 1075–1088. [CrossRef]

37. Winand, R. Chloride hydrometallurgy. *Hydrometallurgy* **1991**, *27*, 285–316. [CrossRef]

38. Zhou, K.; Pan, L.; Peng, C.; He, D.; Chen, W. Selective precipitation of Cu in manganese-copper chloride leaching liquor. *Hydrometallurgy* **2018**, *175*, 319–325. [CrossRef]

39. Herreros, O.; Viñals, J. Leaching of sulfide copper ore in a NaCl-H_2SO_4-O_2 media with acid pre-treatment. *Hydrometallurgy* **2007**, *89*, 260–268. [CrossRef]

40. Senanayake, G. A review of chloride assisted copper sulfide leaching by oxygenated sulfuric acid and mechanistic considerations. *Hydrometallurgy* **2009**, *98*, 21–32. [CrossRef]

41. Xing, W.D.; Lee, M.S.; Senanayake, G. Recovery of metals from chloride leach solutions of anode slimes by solvent extraction. Part II: Recovery of silver and copper with LIX 63 and Alamine 336. *Hydrometallurgy* **2018**, *180*, 49–57. [CrossRef]

42. Lawson, F.; Chu-Yong, C.; Ying Lee, S. Leaching of copper sulphides and copper mattes in oxygenated chloride/sulphate leachants. *Miner. Process. Extr. Metall. Rev.* **1992**, *8*, 183–203. [CrossRef]

43. Carneiro, M.F.C.; Leão, V.A. The role of sodium chloride on surface properties of chalcopyrite leached with ferric sulphate. *Hydrometallurgy* **2007**, *87*, 73–82. [CrossRef]

44. Lu, Z.Y.; Jeffrey, M.I.; Lawson, F. Effect of chloride ions on the dissolution of chalcopyrite in acidic solutions. *Hydrometallurgy* **2000**, *56*, 189–202. [CrossRef]

45. Zhu, Z.; Zhang, W.; Cheng, C.Y. A synergistic solvent extraction system for separating copper from iron in high chloride concentration solutions. *Hydrometallurgy* **2012**, *113–114*, 155–159. [CrossRef]

46. Lu, J.; Dreisinger, D. Two-stage countercurrent solvent extraction of copper from cuprous chloride solution: Cu(II) loading coupled with Cu(I) oxidation by oxygen and iron scrubbing. *Hydrometallurgy* **2014**, *150*, 41–46. [CrossRef]

minerals

Article

Corrosion Behavior of a Pyrite and Arsenopyrite Galvanic Pair in the Presence of Sulfuric Acid, Ferric Ions and HQ0211 Bacterial Strain

Jia-Ning Xu [1],*, Wen-Ge Shi [2], Peng-Cheng Ma [2], Liang-Shan Lu [2], Gui-Min Chen [3] and Hong-Ying Yang [1],*

[1] School of Metallurgy, Northeastern University, Wenhua Road NO. 3-11, Heping District, Shenyang 100819, China
[2] Shandong Zhaojin Group Co., Ltd., Shengtai Road NO. 108, Zhaoyuan 265400, China; 13853566318@163.com (W.-G.S.); aust.pengcheng@163.com (P.-C.M.); jcllls@126.com (L.-S.L.)
[3] Hainan Shanjin Mining Co., Ltd., Ledong 572500, China; tongtong4858@163.com
* Correspondence: xuyaoiii@163.com (J.-N.X.); yanghy@smm.neu.edu.cn (H.-Y.Y.); Tel.: +86-024-8367-3932 (H.-Y.Y.)

Received: 4 January 2019; Accepted: 1 March 2019; Published: 9 March 2019

Abstract: In this paper, the galvanic effect of pyrite and arsenopyrite during the leaching pretreatment of gold ores was determined with the use of electrochemical testing (open circuit potential, linear sweep voltammetry, Tafel, and electrochemical impedance spectroscopy (EIS)) and frontier orbit calculations. The results show that (i) the linear sweep voltammetry curve and Tafel curve of the galvanic pair are similar to those of arsenopyrite, (ii) the corrosion behavior of the galvanic pair is consistent with that of arsenopyrite, and (iii) the galvanic effect promotes the corrosion of arsenopyrite by simultaneously increasing the cathode and anode currents and reducing oxidation resistance. The frontier orbit calculation explains the principle of the galvanic effect of pyrite and arsenopyrite from the view of quantum mechanics.

Keywords: galvanic effect; pyrite–arsenopyrite galvanic pair; electrochemical; frontier orbital methods

1. Introduction

The chemical formulas of pyrite and arsenopyrite are FeS_2 and FeAsS, respectively. Pyrite and arsenopyrite are the two most important gold-containing minerals in gold ores [1–4]. Fine-grained or submicroscopic gold is disseminated within pyrite and arsenopyrite, thereby rendering gold inaccessible to cyanide solutions [5]. Dissolving pyrite and arsenopyrite is the main objective of the biological pretreatment of gold ores [6–8]. Arsenic present in arsenopyrite is harmful to the bacteria used in bioleaching. A decision whether to leach run-of-mine ore or concentrate can only be made after determining the percentage of gold which is trapped in both sulfides and the free gold as well, to minimize the operating cost of the process [5]. In the case where gold is preferentially associated with arsenopyrite, only partial oxidation of this arsenopyrite is required for an increased gold liberation [9]. The oxidation reactions of pyrite and arsenopyrite during bioleaching have been widely discussed. Komnitsas et al. [5,9] and Smith et al. [10] described the aqueous oxidation of pyrite using stoichiometric chemical reactions:

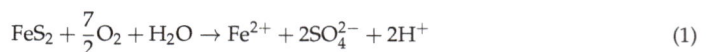

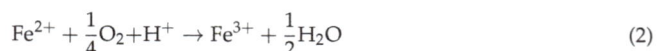

$$FeS_2 + \frac{7}{2}O_2 + H_2O \rightarrow Fe^{2+} + 2SO_4^{2-} + 2H^+ \tag{1}$$

$$Fe^{2+} + \frac{1}{4}O_2 + H^+ \rightarrow Fe^{3+} + \frac{1}{2}H_2O \tag{2}$$

$$FeS_2 + 14Fe^{3+} + 8H_2O \rightarrow 15Fe^{2+} + 2SO_4^{2-} + 16H^+ \tag{3}$$

$$FeS_2 + 6Fe^{3+} + 3H_2O \rightarrow 7Fe^{2+} + S_2O_3^{2-} + 6H^+ \tag{4}$$

$$S_2O_3^{2-} + 8Fe^{3+} + 5H_2O \rightarrow 8Fe^{2+} + 2SO_4^{2-} + 10H^+ \tag{5}$$

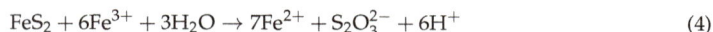

The reactions involved in the aqueous oxidation of arsenopyrite are shown in Equations (6) and (7) [11,12]:

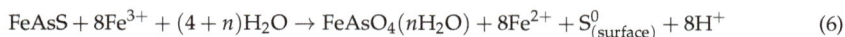

$$FeAsS + 8Fe^{3+} + (4+n)H_2O \rightarrow FeAsO_4(nH_2O) + 8Fe^{2+} + S^0_{(surface)} + 8H^+ \tag{6}$$

$$4FeAsS + 11O_2 + 6H_2O \rightarrow 4Fe^{2+} + 4H_3AsO_3 + 4SO_4^{2-} \tag{7}$$

The role of ferric ions, which are produced by the bacterial oxidation of ferrous ions, accelerates oxidation of sulfide phases and raises solution Eh [9]. In fact, ferric ions can be regarded as "catalysts" for the reactions involved in the oxidation of minerals. Ferrous ions are oxidized to ferric ions by oxygen, which then oxidizes minerals and produces more ferrous ions in a perpetual process which can be significantly accelerated by bacteria. Bioleaching takes place in an acidic solution, and the hydrogen ions are either the reactants or products of the bioleaching reaction. In a ferric ion solution (pH < 4), the oxidants of pyrite and arsenopyrite are ferric ions, rather than dissolved oxygen [12,13]. Thus, the galvanic effect of pyrite/arsenopyrite in the presence of sulfuric acid, ferric ion, and HQ0211 bacteria strain solutions was assessed in this paper.

Biological pretreatment of gold ores is associated with galvanic effects [14–16]. The oxidation potential of pyrite is around 630 mV, while arsenopyrite oxidation is initiated at much lower Eh values, ranging between 390 and 430 mV [5,9]. The rest potential of arsenopyrite is lower than that of pyrite. Under acidic conditions, a galvanic effect occurs when they contact each other, after which arsenopyrite dissolves as an anode, and pyrite acts as a cathode [17,18]. In addition, it is well known that the galvanic effect promotes the corrosion of arsenopyrite in the process of arsenical gold sulfide concentrate bio-pretreatment. In other words, the presence of pyrite accelerates oxidation of arsenopyrite [5–9]. In recent years, the galvanic mechanism of pyrite and arsenopyrite in different media has been discussed [14–18]. Urbano et al. [15] stated that the electrochemical reactivity of pyrite in contact with arsenopyrite mineral was delayed and shifted to more positive potentials with respect to the high-purity pyrite mineral electrochemical response due to the galvanic effect. This reaction process was described according to the results of scanning electron microscopy (in ferric-free solution with a pH of 6.5). Santos et al. [17] calculated the stability, structure, and electronic properties of the pyrite/arsenopyrite solid–solid interface by DFT (density functional theory) and stated that the valence band of the pyrite/arsenopyrite interface has large contributions from the pyrite phase, while the conduction band has large contributions from the arsenopyrite. This is consistent with the role of pyrite as the cathode and arsenopyrite as the anode in galvanic contact with unfavorable miscibility. Deng et al. [18] studied the galvanic pair in a 9 K culture medium and sulfuric acid (pH = 1.6), which revealed that the presence of pyrite increased the conductivity of the electrodes and electrolytes. Thus, this verified the catalytic effect of the galvanic interactions on the process of arsenopyrite leaching. In addition, the galvanic effect of other minerals has also been studied, such as pyrite/chalcopyrite [14], chalcopyrite/magnetite [16], pyrite/gold [19], and pyrite/galena [20]. Pyrite and arsenopyrite are contained in almost all types of gold ores, as they make contact naturally and form a galvanic pair during leaching. In the bioleaching process, the effective components in the solution are sulfuric acid, ferric ions, and the bio-leaching bacteria. However, the promoting mechanism of the galvanic effect of pyrite and arsenopyrite in the leaching process has been seldom discussed. The novelty of this paper is the study of the corrosion behavior of pyrite, arsenopyrite, and the galvanic pair in the presence of sulfuric acid, ferric ions, and the HQ0211 strain (a mixture of strains protected by a patent in our lab) by electrochemical methods.

The study is an in-depth attempt to interprete by means of electrochemical measurements and DFT calculations a well-known phenomenon of practical importance for gold ore pretreatment: the galvanic corrosion of arsenopyrite in contact with pyrite. The electrochemical corrosion behavior of pyrite, arsenopyrite, and galvanic pairs is compared in detail. The electrochemical impedance spectroscopy (EIS) models of pyrite/arsenopyrite galvanic pair are established and explained. Furthermore, an EIS model of arsenopyrite oxidation with less fitting error (absence of ferric ions, pH = 1.5) is proposed. The mechanism of galvanic effect accelerating arsenopyrite corrosion is discussed. Then the electrokinetic properties of pyrite and arsenopyrite are calculated according to the frontier orbital theory based on the density functional theory (DFT) which provided a quantum mechanics perspective of pyrite/arsenopyrite galvanic effect.

2. Experimental Methods

2.1. Electrochemical Methods

The pyrite and arsenopyrite specimens were obtained from Shandong Zhaojin Group Co., Ltd., Zhaoyuan, China. The results of X-ray diffraction (XRD) analysis (model of the instrument: D8 ADVANCE; wavelength radiation: 1.5405 Angstrom; kV and mA values during operation: 30 KV and 40 mA; type of detector: LYNXEYE, Bruker Corporation, Billerica, MA, USA) of the pyrite and arsenopyrite are shown in Figure 1. To avoid the influence of mineral oxidation on electrochemical behavior during grinding, block electrodes were used in all electrochemical experiments instead of carbon paste electrodes. The specimens were cut into rectangular slices with dimensions of approximately $10 \times 10 \times 5$ mm and mounted using epoxy resin. The slides were polished on both sides. An insulated copper wire was attached to the underside of the specimen using silver conductive epoxy, which was subsequently coated with epoxy resin. A fresh pyrite surface was created before each experiment by abrading with successively finer grades (600, 800, 1200, 1600, and 2000 grit) of silicon carbide paper, which was followed by successive polishing with 3, 1, and 0.5 μm diamond suspensions and rinsing with Milli-Q water in a nitrogen-filled vessel for 20 s. The exposed surfaces of pyrite, arsenopyrite, and the pyrite–arsenopyrite galvanic pair electrode are shown in Figure 2a–c, respectively. Milli-Q water with a specific resistance of 18.2 MΩ·cm^{-1} was used in all experiments, and all reagents used in this study were of analytical purity. The sulfuric acid solution (pH = 1.5) and ferric sulfate solution (pH = 1.5, 9 g·L^{-1} Fe^{3+}) were prepared from sulfuric acid and ferric sulfate. The HQ0211 bacteria strain was obtained (solution potential >600 mV, number of viable bacteria >10^7/mL, pH = 1.5) and 0.01 mol/L Na$_2$SO$_4$ was used as the supporting electrolyte in all electrochemical solutions. All electrochemical experiments were conducted on a CHI660e electrochemical workstation, (CH Instruments, Inc., Austin, TX, USA). Electrochemical measurements were performed using a conventional three-electrode system. The experimental conditions of the open circuit potential test (Table 1), linear sweep voltammetry test (Figure 4), polarization curve test (Figure 4), and AC impedance test (Figure 5) are shown in the annotations of the figures of each test. To simulate a real leaching environment, the dissolved oxygen in the solution was not removed during the experiment. All potential values in this paper are relative to a standard hydrogen electrode (SHE).

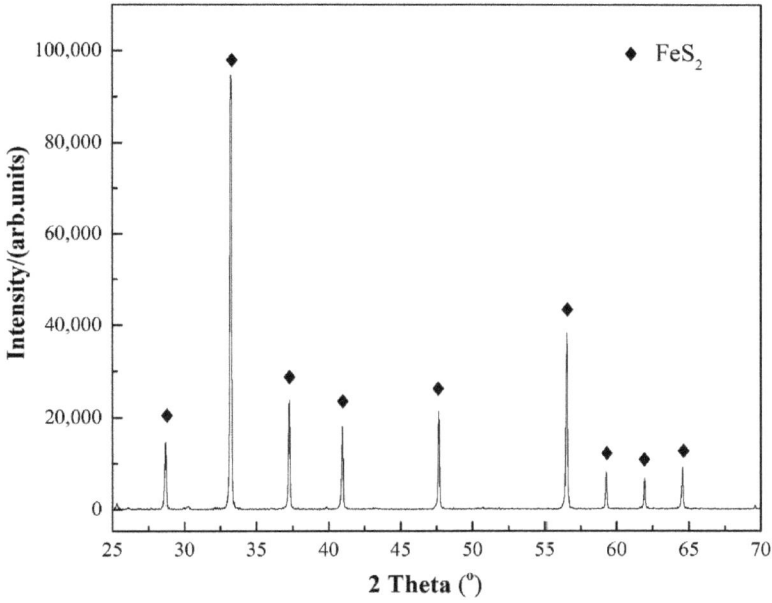

(a)

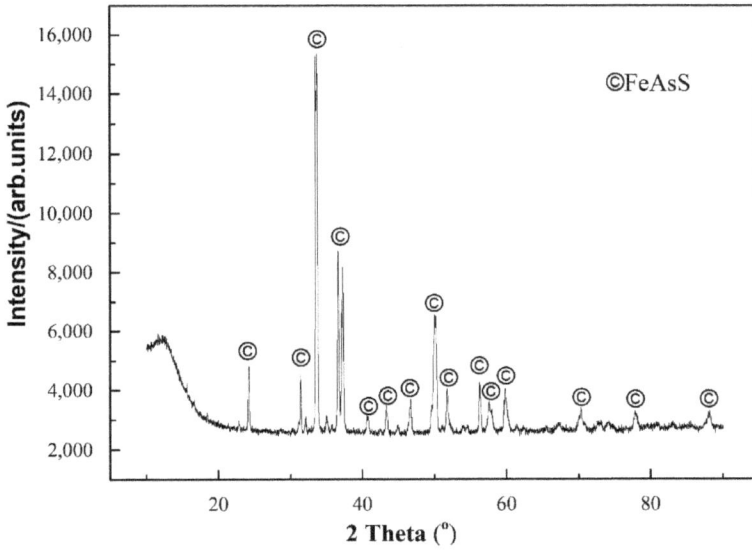

(b)

Figure 1. X-ray diffraction (XRD) patterns of pyrite (**a**) and arsenopyrite (**b**).

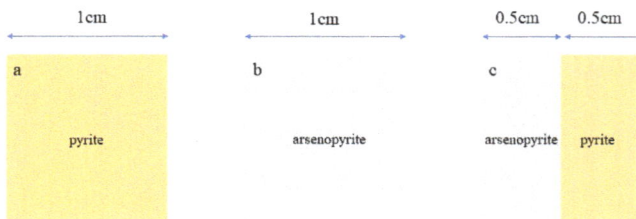

Figure 2. The exposed surfaces of pyrite (**a**), arsenopyrite (**b**), and the pyrite–arsenopyrite galvanic pair electrode (**c**).

Table 1. The mean open circuit potentials (OCPs) of pyrite, arsenopyrite, and the galvanic pair in the investigated solutions.

Solution Type	Pyrite (mV)	Arsenopyrite (mV)	Galvanic Pair (mV)
Sulfuric acid	704	316	354
Ferric ion	729	435	479
HQ0211 strain	768	446	614

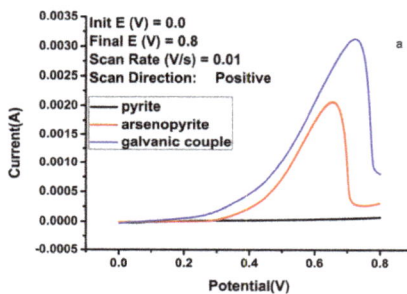

(**a**)

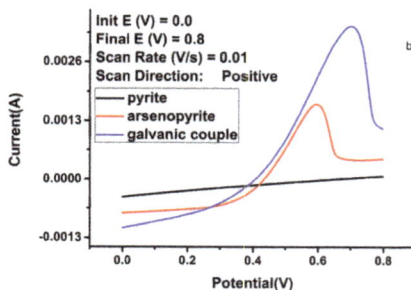

(**b**)

Figure 3. *Cont.*

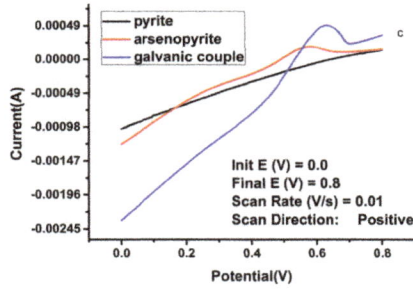

(**c**)

Figure 3. Linear sweep voltammetry curves of the three investigated electrodes in sulfuric acid (**a**), ferric ion (**b**), and HQ0211 strain solutions (**c**) (Init E: Initial potential of linear sweep voltammetry test, Final E: Final potential of linear sweep voltammetry test).

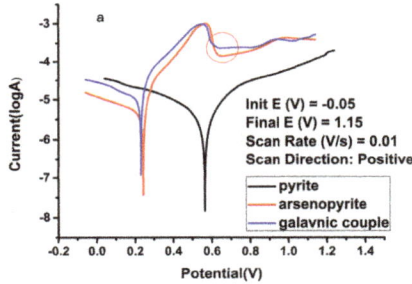

(**a**)

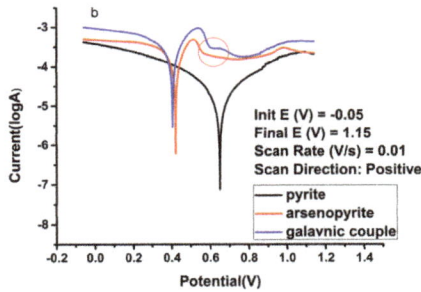

(**b**)

Figure 4. *Cont.*

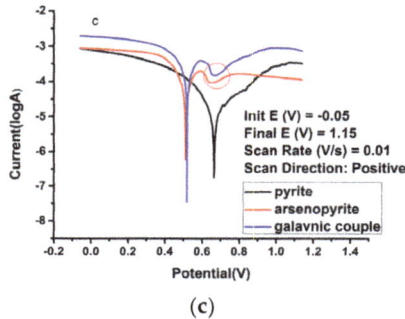

(c)

Figure 4. Tafel curves of the three investigated electrodes in sulfuric acid (**a**), ferric ion (**b**), and HQ0211 strain solutions (**c**).

2.2. DFT Calculational Details

All calculations were performed using DMOL3 (Version: 5.0, Accelrys, San Diego, CA, USA) [21] and the generalized gradient approximation (GGA) PW91 density functional [22]. Only the valence electrons (Fe $3d^64s^2$ and S $3s^23p^4$ As $4s^24p^3$) were considered by using ultra soft pseudopotentials [23] and a plane wave cut-off energy of 300 eV after testing. The Monkhorst–Pack k-point sampling density for pyrite was $4 \times 4 \times 4$ for a mesh and $2 \times 3 \times 2$ for arsenopyrite. The self-consistent field (SCF) convergence tolerance was set to 1.0×10^{-6} eV·atom^{-1}. Because of the existence of iron atoms in pyrite and arsenopyrite, the electron spin polarization was considered in calculations.

Pyrite (cubic symmetry) with a space group Pa-3 and cell parameter of 5.417 Å [24], and arsenopyrite with a space group B-1 (a = 9.51 Å; b = 5.65 Å; c = 6.42 Å; $\alpha = \beta = \gamma = 90°$) were used in our calculations [25].

3. Results and Discussion

3.1. Surface Activity

Galvanic interaction is closely related to the surface activity of two electrically connected materials [18]. The surface activity of pyrite and arsenopyrite in a solution was tested via open circuit potential (OCP) testing. The mean OCPs of pyrite, arsenopyrite, and the galvanic pair in a solution are listed in Table 1.

Due to the adsorption of ferric ions and bacteria onto the mineral surface and the previously mentioned oxidation reactions, the OCPs of pyrite and galvanic pair in the sulfuric acid, ferric ion, and HQ0211 strain solutions increased successively. The pyrite OCP value in HQ0211 strain is higher than that in the ferric ion solution. Ferric iron is produced by the oxidation of ferrous iron (reaction (3)) while the action of bacteria produces more ferric iron [9]. The mechanism of the enrichment of mineral surfaces with ferric ions produced by bacterial action has been mainly discussed elsewhere [26,27]. Due to the toxicity of arsenopyrite to bacteria, the OCP of arsenopyrite in the bacterial solution is only slightly higher (435 mV to 446 mV) than that in the ferric ion solution [28–30].

Pyrite shows higher OCP values than arsenopyrite in the three tested solutions (Table 1), which indicates that it is more stable than arsenopyrite. Since the electrode potential determines the intrinsic polarity of a galvanic pair, pyrite with the higher OCP would serve as the cathode while arsenopyrite with the lower OCP would act as the anode. The galvanic potentials of the galvanic pair in a two-electrode system in the three tested solutions were 388 mV, 294 mV, and 322 mV, respectively, meaning that the galvanic interaction between them truly exists.

Table 1 also shows that the OCP values of the galvanic pair were higher than those of arsenopyrite in all tested solutions. As the OCP of the galvanic pair reflects the new equilibrium point of pyrite and arsenopyrite, and the OCP of the galvanic pair is higher than the OCP of arsenopyrite, it can, therefore, polarize the arsenopyrite electrode positively and provide a positive over-potential to shift the overall reaction to the anodic region. This leads to an increase in the oxidation rate of arsenopyrite in all three solutions investigated in this paper.

3.2. Dissolution Behavior of the Galvanic Pair

Linear sweep voltammetry was performed to explain the dissolution process of the three electrodes. The linear sweep voltammetry curves of the three investigated electrodes in the three investigated solutions are shown in Figure 4. Compared with those of arsenopyrite and the galvanic pair, the oxidation peak of pyrite is insignificant in the three tested solutions, and the oxidation curve of arsenopyrite is similar to that of the galvanic pair. The results show that the oxidation behavior of the galvanic pair is similar to that of arsenopyrite within the sweep range, in agreement with the role of Fe^{3+} in acceleration oxidation (indirect mechanism). At the same time, in the three types of solution, the oxidation peak area of the galvanic pair is larger than that of arsenopyrite, which indicates that the galvanic effect accelerates the oxidation of arsenopyrite.

The corrosion behavior of the three electrodes in all investigated solutions was further studied using the Tafel curve (Figure 4). The calculated Tafel curve data are listed in Table 2.

Table 2. The calculated Tafel curve data of pyrite, arsenopyrite, and the galvanic pair in the investigated solutions (C.C.: Cathode Current; A.C.: Anode Current; i_{corr}: Corrosion current; E_{corr}: Corrosion potential).

Solution	Pyrite				Arsenopyrite				Galvanic Pair			
	C.C (lg i)	A.C (lg i)	i_{corr} $(10^{-6}A)$	E_{corr} (V)	C.C (lg i)	A.C (lg i)	i_{corr} $(10^{-6}A)$	E_{corr} (V)	C.C (lg i)	A.C (lg i)	i_{corr} $(10^{-6}A)$	E_{corr} (V)
Sulfuric acid	−5.61	−5.59	3.83	0.56	−5.37	−4.81	14.1	0.22	−5.00	−4.54	33.8	0.20
Ferric ion	−4.93	−4.96	1.46	0.63	−3.85	−3.71	260	0.40	−3.67	−3.57	390	0.39
HQ0211 strain	−4.83	−4.91	18.8	0.64	−3.92	−3.40	160	0.46	−3.40	−3.23	467	0.45

As shown in Figure 4, the corrosion potential of the galvanic pair is similar to that of arsenopyrite (0.22, 0.40, 0.46) but very different from that of pyrite (0.56, 0.63, 0.64 in three investigated solutions respectively), which indicates that the corrosion of the galvanic pair occurs at a potential similar to that of arsenopyrite. The slight decrease in the current marked in Figure 4 by a red circle is due to the oxidation of iron in the arsenopyrite, which is characteristic of arsenopyrite oxidation (pyrite does not have this characteristic) [26,28]. This is the same as the characteristic of the galvanic effect (an anodic reaction occurs first without changing the reaction process). It is also consistent with the results of the linear voltammetry analysis.

The acceleration of the anodic reaction due to the galvanic effect is reflected in the corrosion current. Table 2 shows that in the sulfuric acid, ferric ion, and HQ0211 strain solutions, the corrosion current of the galvanic pair was 2.40 times, 1.50 times, and 2.92 times that of arsenopyrite, respectively. The galvanic effect observably accelerates the oxidation of arsenopyrite. The limiting steps of the oxidation of arsenopyrite by the cathodic process can also be seen in Table 2. The increase in the cathodic current of arsenopyrite is larger than that in the anodic current when coupled with pyrite, which is due to the enhancement of the electron donation capability of the slightly negative pyrite.

The rate determining step and the reaction mechanism of the electrode system can be illustrated by electrochemical impedance spectroscopy (EIS). Nyquist plots of the galvanic pair and arsenopyrite in the three investigated solutions are shown in Figure 5. The shape of the Nyquist plot of the galvanic pair is very similar to that of arsenopyrite in the three investigated solutions with the same initial E. This shows that the reaction mechanisms of the galvanic pair and arsenopyrite alone are similar under the same potential.

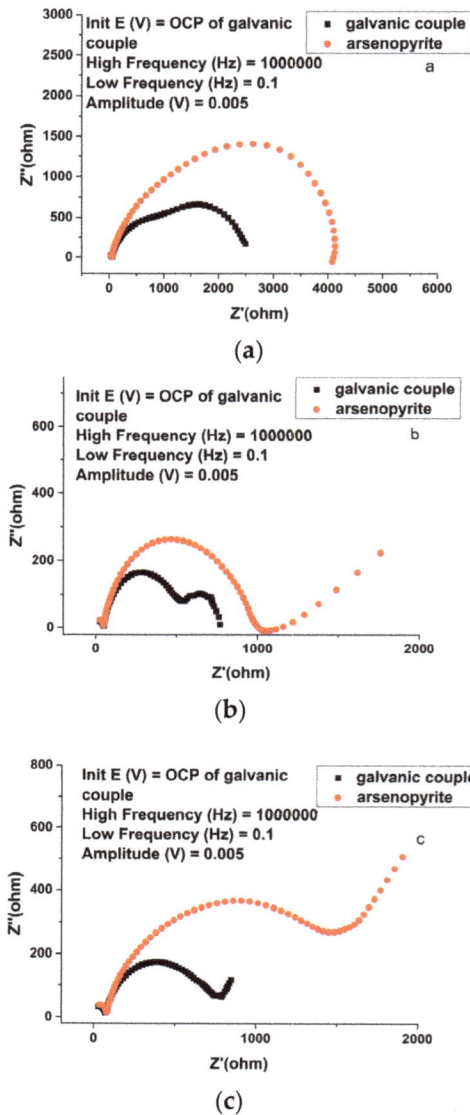

Figure 5. Nyquist plots of the galvanic pair and arsenopyrite in sulfuric acid (**a**), ferric ion (**b**), and HQ0211 strain solutions (**c**).

To further fit the data, this paper improved upon the analog circuit created by Deng et al. [31] in the sulfuric acid solution. The analog circuit and theory model are shown in Figure 6. A new oxidation resistance R_3 and interface Q_3 were added to the circuit. The resistance of the electrochemical reaction is parallel to the capacitance of the double layer. The solution resistance is connected in series with the parallel circuit of the electrode–electrolyte double-layer capacitance and the electrochemical reaction resistance. The initial oxidation resistance is connected in series with the parallel circuit of the surface film–electrolyte double-layer capacitance and the further oxidation resistance of the surface film. Furthermore, the second oxidation resistance is connected in series with the parallel circuit of the surface film–electrode capacitance and the further oxidation resistance of the surface film. Since the first layer of oxidation resistance disappears, we use the analog circuits from Deng et al. [31] in ferric ion and HQ0211 strain solutions.

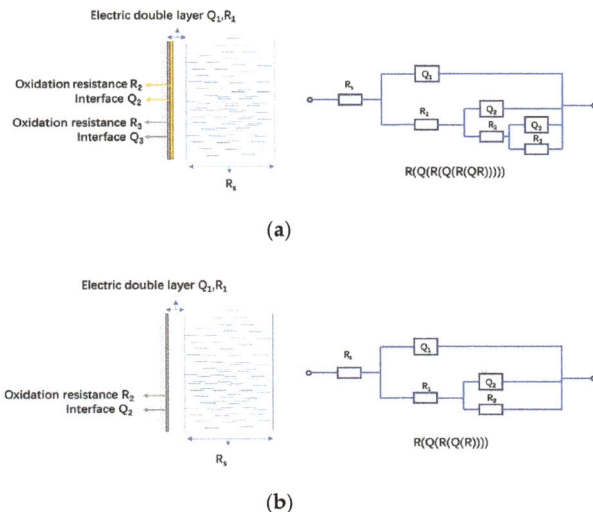

(a)

(b)

Figure 6. The theoretical model and analog circuit from electrochemical impedance spectroscopy (EIS) in the three investigated solutions: (**a**) analog circuit from EIS of the galvanic pair and arsenopyrite in sulfuric acid solution; and (**b**) the galvanic pair and arsenopyrite in ferric ion and HQ0211 strain solutions.

In the ferric ion solution and HQ0211 strain, Rs and R1 represent the solution resistance and the charge transfer resistance during arsenopyrite oxidation, respectively. Q1 and Q2 correspond to the double-layer capacitances between electrode/electrolyte interfaces and surface layer (S^0 layer)/electrolyte interfaces, respectively. R2 represents the charge transfer resistance during the oxidation of the surface layer. In the sulfuric acid solution, an additional passivation layer (As_2S_2) is formed after the reaction [27]. Rs and R1 represent the solution resistance and the charge transfer resistance during arsenopyrite oxidation. Q1, Q2, and Q3 correspond to the double-layer capacitances between electrode/electrolyte interfaces, surface layer (As_2S_2 layer) (Reaction (8))/surface layer (S^0 layer) interface, and surface layer (S^0 layer)/electrolyte interfaces, respectively. R2 and R3 represent the charge transfer resistance during the oxidation of the S layer and surface layer, respectively.

$$FeAsS \rightarrow Fe^{2+} + 0.5As_2S_2 + 2e^- \tag{8}$$

The fitting data and fitting errors are listed in Table 3. As shown in Table 3, the limiting step of the arsenopyrite oxidation reaction in a sulfuric acid solution is R_3 (charge transfer resistance during the oxidation of surface layer) and depends on whether arsenopyrite alone or the galvanic pair was used. The R_3 values of arsenopyrite and the galvanic pair are 2398 and 1475 $\Omega \cdot cm^2$, respectively. The galvanic effect can reduce the R_3 value to accelerate the oxidation of arsenopyrite. In ferric ion and HQ0211 strain solutions, the limiting step of arsenopyrite oxidation is R_2 (charge transfer resistance during the oxidation of the surface layer) for both arsenopyrite and the galvanic pair. The R_2 values of arsenopyrite and the galvanic pair are 898 and 498 $\Omega \cdot cm^2$, respectively, in the ferric ion solution. The R_2 values of arsenopyrite and the galvanic pair are 1153 and 809 $\Omega \cdot cm^2$, respectively, in the bacterial solution. The higher R_2 value in the bacterial solution may be due to the toxic effect of arsenopyrite on bacteria. This results in bacteria carrying ferric ions, which tend not to adsorb on the surface of arsenopyrite or the galvanic pair. Apart from this, the galvanic effect can accelerate arsenopyrite oxidation by reducing the value of R_2. Therefore, the galvanic effect can accelerate the oxidation of arsenopyrite and determine the rate-limiting step of the reaction.

Table 3. Fitting data and fitting errors of the investigated electrochemical impedance spectroscopy (EIS) data.

Solution	Electrode	$R_s/\Omega \cdot cm^2$	$Y_{o,1}/10^{-9}$ $S \cdot s^n \cdot cm^{-2}$ (n)	$R_1/\Omega \cdot cm^2$	$Y_{o,2}/10^{-5}$ $S \cdot s^n \cdot cm^{-2}$ (n)	$R_2/\Omega \cdot cm^2$	$Y_{o,3}/10^5$ $S \cdot s^n \cdot cm^{-2}$	$R_3/\Omega \cdot cm^2$	Fitting Error $\chi^2/10^{-3}$
Sulfuric	Arsenopyrite	7.958	5.33	42.44	1.059	1769	2.674	2398	0.23
acid	Galvanic pair	6.408	8.49	48.21	1.635	1002	12.97	1475	0.69
Ferric	Arsenopyrite	1.209	3.65	45.93	1.183	898	-	-	0.98
ion	Galvanic pair	9.284	6.5	39.41	1.369	483	-	-	0.74
HQ0211	Arsenopyrite	2.337	2.69	75.23	1.25	1153	-	-	0.81
strain	Galvanic pair	2.93	3.06	64.66	3.31	809	-	-	2.09

3.3. DFT Calculations

The calculation errors are listed in Table 4. Among all lattice parameters, the maximum calculation error of pyrite was 0.01%, and that of arsenopyrite was 0.05%, which meets the error requirement of Generalized Gradient Approximation (GGA)-Perdew-Wang (PW91). The crystal structure and highest occupied molecular orbital (HOMO) (green and yellow iso surface) of pyrite, arsenopyrite, and Fe^{2+} and the lowest unoccupied molecular orbital (LUMO) of Fe^{3+} are shown in Figure 7a–d, respectively. The HOMO orbitals of the arsenopyrite and pyrite crystals are distributed around the iron atoms, and the Fe^{2+} and Fe^{3+} orbital hybridization modes are different. The process of chemical reaction is actually the same as the process of electron migration from the HOMO of the reductant to the LUMO of the oxidant. The transfer of electrons during the oxidation of pyrite by ferric ions is shown in Reaction (3). The HOMO value of arsenopyrite (-0.180) is higher than that of pyrite (-0.220). When arsenopyrite and pyrite co-exist, the final electron donor is arsenopyrite. This is why the oxidation behavior of the galvanic pair is characterized by arsenopyrite oxidation behavior. However, because the HOMO of pyrite is closer to the LUMO of Fe^{3+} (-0.241) than to the HOMO of arsenopyrite, it is easier for electrons to migrate from arsenopyrite to the pyrite surface than for electrons to migrate from the pyrite surface to Fe^{3+}. Therefore, the presence of pyrite accelerates the oxidation of arsenopyrite in the galvanic pair.

Table 4. Calculation errors for pyrite and arsenopyrite.

Lattice Parameter	A (Angstrom)	B (Angstrom)	C (Angstrom)	α (°)	β (°)	γ (°)
Calculated pyrite lattice parameter	5.411	5.411	5.411	90	90	90
Calculation errors for pyrite	0.01%	0.01%	0.01%	0%	0%	0%
Calculated arsenopyrite lattice parameter	9.46	5.63	6.39	89.9	89.7	89.9
Calculation errors for arsenopyrite	0.5%	0.5%	0.4%	0.1%	0.3%	0.1%

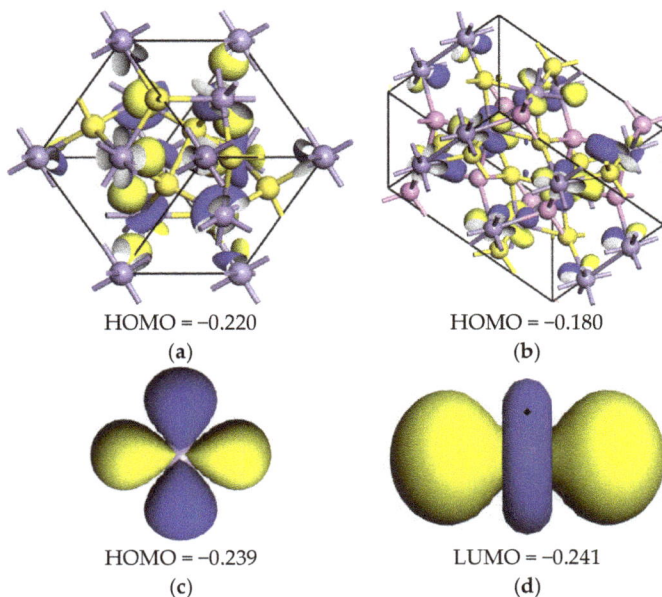

HOMO = −0.220
(a)

HOMO = −0.180
(b)

HOMO = −0.239
(c)

LUMO = −0.241
(d)

Figure 7. Crystal structure and highest occupied molecular orbital (HOMO) (green and yellow iso surface) of pyrite (**a**), arsenopyrite (**b**), and Fe^{2+} (**c**) and lowest unoccupied molecular orbital (LUMO) of Fe^{3+} (**d**) (blue ball for Fe atom, yellow ball for S atom, and purple ball for As atom).

4. Conclusions

By conducting electrochemical and DFT studies of three types of electrodes (pyrite, arsenopyrite, and a galvanic pair), the galvanic effect of pyrite and arsenopyrite in sulfuric acid, ferric ion, and HQ0211 strain solutions was determined to improve our understanding of the pyrite/arsenopyrite galvanic effect. The following conclusions were drawn:

1. After a detailed comparison, the corrosion process of the pyrite/arsenopyrite galvanic pair is similar to that of arsenopyrite, which is clearly supported by electrochemical data.
2. EIS models of pyrite/arsenopyrite galvanic pair were established as R(Q(R(Q(R(QR))))) and R(Q(R(Q(R)))) and the physical meaning was clearly indicated.
3. The mechanism of the galvanic effect that accelerates arsenopyrite corrosion by increasing the cathode and anode currents as well as the charge transfer resistance during the oxidation of the surface layer of the reaction was proposed.
4. The quantum mechanics perspective of pyrite/arsenopyrite galvanic effect was provided. The calculated results show that the oxidation behavior of the galvanic pair is characterized by arsenopyrite oxidation behavior, and the presence of pyrite accelerates the oxidation of arsenopyrite in the galvanic pair.

Author Contributions: J.-N.X. and H.-Y.Y. were responsible for experimental research and writing. J.-N.X., W.-G.S., P.-C.M., L.-S.L., G.-M.C. and H.-Y.Y. discussed the results and evaluated the data.

Funding: The authors acknowledge financial support from the National Key R&D Program of China (No. 2018YFC1902002) and the Special Funds for the National Natural Science Foundation of China (No. U1608254).

Acknowledgments: This research is supported by Northeastern University Testing Center and the Shandong Zhaojin Group Co., Ltd. The authors wish to acknowledge Xue-Min Qiu, Guo-Bao Chen, Linlin Tong, and Zhenan Jin for their technical and laboratory assistance. The assistance on Material Studio was provided by Jianhua Chen of Guangxi University.

Conflicts of Interest: The authors declare no conflict of interest.

References

1. Li, S.N.; Ni, P.; Bao, T.; Li, C.Z.; Xiang, H.L.; Wang, G.G.; Huang, B.; Chi, Z.; Dai, B.Z.; Ding, J.Y. Geology, fluid inclusion, and stable isotope systematics of the Dongyang epithermal gold deposit, Fujian Province, southeast China: Implications for ore genesis and mineral exploration. *J. Geochem. Explor.* **2018**, *195*, 16–30. [CrossRef]
2. Mansurov, Y.N.; Miklin, Y.A.; Miklin, N.A.; Nikol'Skii, A.V. Methods and Equipment for Breaking Down Gold-Containing Concentrates from Lean Ores and Mining Industry Waste. *Metallurgist* **2018**, *62*, 169–175. [CrossRef]
3. Kravtsova, R.G.; Tauson, V.L.; Nikitenko, E.M. Modes of Au, Pt, and Pd occurrence in arsenopyrite from the Natalkinskoe deposit, NE Russia. *Geochem. Int.* **2015**, *53*, 964–972. [CrossRef]
4. Khishgee, C.; Akasaka, M.; Ohira, H.; Sereenen, J. Gold Mineralization of the Gatsuurt Deposit in the North Khentei Gold Belt, Central Northern Mongolia. *Resour. Geol.* **2014**, *64*, 1–16. [CrossRef]
5. Komnitsas, C.; Pooley, F. Mineralogical characteristics and treatment of refractory gold ores. *Miner. Eng.* **1989**, *2*, 449–457. [CrossRef]
6. Rajasekar, A. Bio-oxidation and bio-cyanidation of refractory mineral ores for gold extraction: A review. *Crit. Rev. Environ. Sci. Technol.* **2015**, *45*, 1611–1643. [CrossRef]
7. Kaksonen, A.H.; Perrot, F.; Morris, C.; Rea, S.; Benvie, B.; Austin, P.; Hackl, R. Evaluation of submerged bio-oxidation concept for refractory gold ores. *Hydrometallurgy* **2014**, *141*, 117–125. [CrossRef]
8. Anjum, F.; Shahid, M.; Akcil, A. Biohydrometallurgy techniques of low grade ores: A review on black shale. *Hydrometallurgy* **2012**, *117–118*, 1–12. [CrossRef]
9. Komnitsas, C.; Pooley, F. Bacterial oxidation of an arsenical gold sulphide concentrate from Olympias, Greece. *Miner. Eng.* **1990**, *3*, 295–306. [CrossRef]
10. Smith, E.E.; Shumate, K.S.; Singer, P.C.; Stumm, W. Direct Oxidation by Adsorbed Oxygen during Acidic Mine Drainage. *Science* **1970**, *169*, 98. [CrossRef]
11. Breed, A.W.; Glatz, A.; Hansfor, G.S.; Harrison, S.T.L. The effect of As(III) and As(V) on the batch bioleaching of a pyrite-arsenopyrite concentrate. *Miner. Eng.* **1996**, *9*, 1235–1252. [CrossRef]
12. Yu, Y.; Zhu, Y.; Gao, Z.; Gammons, C.H.; Li, D. Rates of arsenopyrite oxidation by oxygen and Fe(III) at pH 1.8–12.6 and 15–45 °C. *Environ. Sci. Technol.* **2007**, *41*, 6460–6464. [CrossRef] [PubMed]
13. Moses, C.O.; Nordstrom, D.K.; Herman, J.S.; Aaron, L. Aqueous pyrite oxidation by dissolved oxygen and by ferric iron. *Geochim. Cosmochim. Acta* **1987**, *51*, 1561–1571. [CrossRef]
14. Mehta, A.P.; Murr, L.E. Fundamental studies of the contribution of galvanic interaction to acid-bacterial leaching of mixed metal sulfides. *Hydrometallurgy* **1983**, *9*, 235–256. [CrossRef]
15. Urbano, G.; Reyes, V.E.; Veloz, M.A.; GonzáLez, I. Pyrite–Arsenopyrite Galvanic Interaction and Electrochemical Reactivity. *J. Phys. Chem. C* **2008**, *112*, 10453–10461. [CrossRef]
16. Saavedra, A.; García-Meza, J.V.; Cortón, E.; González, I. Understanding galvanic interactions between chalcopyrite and magnetite in acid medium to improve copper (Bio)Leaching. *Electrochim. Acta* **2018**, *265*, 569–576. [CrossRef]
17. Santos, E.C.D.; Lourenço, M.P.; Pettersson, L.G.M.; Duarte, H.A. Stability, Structure, and Electronic Properties of the Pyrite/Arsenopyrite Solid-Solid Interface—A DFT Study. *J. Phys. Chem. C* **2017**, *121*, 8042–8051. [CrossRef]
18. Deng, S.; Gu, G.; He, G.; Li, L. Catalytic effect of pyrite on the leaching of arsenopyrite in sulfuric acid and acid culture medium. *Electrochim. Acta* **2018**, *263*, 8–16. [CrossRef]
19. Huai, Y.; Plackowski, C.; Peng, Y. The galvanic interaction between gold and pyrite in the presence of ferric ions. *Miner. Eng.* **2018**, *119*, 236–243. [CrossRef]
20. Qin, W.Q.; Wang, X.J.; Li-Yuan, M.A.; Jiao, F.; Liu, R.Z.; Gao, K. Effects of galvanic interaction between galena and pyrite on their flotation in the presence of butyl xanthate. *Trans. Nonferrous Met. Soc. China* **2015**, *25*, 3111–3118. [CrossRef]
21. Delley, B. From molecules to solids with the DMol^3 approach. *J. Chem. Phys.* **2000**, *113*, 7756–7764. [CrossRef]

22. Perdew, J.P.; Chevary, J.A.; Vosko, S.H.; Jackson, K.A.; Pederson, M.R.; Singh, D.J.; Fiolhais, C. Atoms, molecules, solids, and surfaces: Applications of the generalized gradient approximation for exchange and correlation. *Phys. Rev. B* **1993**, *46*, 6671–6687. [CrossRef]

23. Vanderbilt, D. Soft self-consistent pseudopotentials in a generalized eigenvalue formalism. *Phys. Rev. B* **1990**, *41*, 7892. [CrossRef]

24. Bayliss, P. Crystal chemistry and crystallography of some minerals within the pyrite group. *Am. Miner.* **1989**, *74*, 1168–1176.

25. Buerger, M. The symmetry and crystal structure of the minerals of the arsenopyrite group. *Z. Krist. Cryst. Mater.* **1936**, *95*, 83–113. [CrossRef]

26. Rohwerder, T.; Gehrke, T.; Kinzler, K.; Sand, W. Bioleaching review part A: Progress in bioleaching: Fundamentals and mechanisms of bacterial metal sulfide oxidation. *Appl. Microbiol. Biotechnol.* **2003**, *63*, 239. [CrossRef] [PubMed]

27. Sand, W.; Gehrke, T. Extracellular polymeric substances mediate bioleaching/biocorrosion via interfacial processes involving iron(III) ions and acidophilic bacteria. *Res. Microbiol.* **2006**, *157*, 49–56. [CrossRef]

28. Deng, S.; Gu, G.; Xu, B.; Li, L.; Wu, B. Surface characterization of arsenopyrite during chemical and biological oxidation. *Sci. Total Environ.* **2018**, *626*, 349. [CrossRef]

29. Corkhill, C.L.; Vaughan, D.J. Arsenopyrite oxidation—A review. *Appl. Geochem.* **2009**, *24*, 2342–2361. [CrossRef]

30. Yang, H.Y.; Liu, Q.; Chen, G.B.; Tong, L.L.; Ali, A. Bio-dissolution of pyrite by *Phanerochaete chrysosporium*. *Trans. Nonferrous Met. Soc. China* **2018**, *28*, 766–774. [CrossRef]

31. Deng, S.; Gu, G. An electrochemical impedance spectroscopy study of arsenopyrite oxidation in the presence of *Sulfobacillus thermosulfidooxidans*. *Electrochim. Acta* **2018**, *287*, 106–114. [CrossRef]

minerals

MDPI

Article

Initial Investigation into the Leaching of Manganese from Nodules at Room Temperature with the Use of Sulfuric Acid and the Addition of Foundry Slag—Part I

Norman Toro [1], Nelson Herrera [1], Jonathan Castillo [2], Cynthia M. Torres [1,* and Rossana Sepúlveda [2]

[1] Departamento de Ingeniería en Metalurgia y Minas, Universidad Católica del Norte, Antofagasta 1270709, Chile; ntoro@ucn.cl (N.T.); nelson.herrera@ucn.cl (N.H.)
[2] Departamento de Ingeniería en Metalurgia, Universidad de Atacama, Copiapó 1531772, Chile; jonathan.castillo@uda.cl (J.C.); rossana.sepulveda@uda.cl (R.S.)
* Correspondence: cynthia.torres@ucn.cl; Tel: +56-552651022

Received: 24 October 2018; Accepted: 28 November 2018; Published: 3 December 2018

Abstract: In this study, the surface optimization methodology was used to assess the effect of three independent variables—time, particle size and sulfuric acid concentration—on Mn extraction from marine nodules during leaching with H_2SO_4 in the presence of foundry slag. The effect of the MnO_2/Fe ratio and particle size (MnO_2) was also investigated. The maximum Mn extraction rate was obtained when a MnO_2 to Fe molar ratio of 0.5, 1 M of H_2SO_4, −320 + 400 Tyler mesh (−47 + 38 μm) nodule particle size and a leaching time of 30 min were used.

Keywords: manganese nodules; leaching; secondary mining; slag

1. Introduction

Oceans cover almost three-quarters of the Earth's surface and contain nine-tenths of its water, while being the habitat for 97% of the living things on the planet. Oceans are an essential part of the biosphere, influencing climate, health and wellbeing. Ocean sea beds comprise more than 60% of the earth's surface and contain great wealth, either in the form of Fe-Mn crust or Mn nodules [1].

Polymetallic nodules, also called manganese nodules, are rock concretions formed by concentric layers of Fe and Mn hydroxides. These polymetallic ores are a suitable alternative source of base metals for the growing manganese demand for steel production since high-grade ores are being depleted [2]. They were first discovered in the Siberian Arctic Ocean in 1968 [3]. Since then, new hydrometallurgical methods have been developed to extract valuable metals from nodules, including the use of sulfuric acid as an oxidation agent and other additives as reducing agents to extract manganese. Reductants such as iron from pyrite ore [4,5], ferrous ions [6] and wastewater from molasses-based alcohol production have been used [7]. Based on previous investigations, the advantages of iron as a reducing agent are its abundance, low cost and apparent efficiency [4–6,8].

Bafghi et al. [8] investigated the effect of elemental Fe (sponge iron at μm −600 + 250, −250 + 150) as a reducing agent at different iron to MnO_2 molar ratios (0.67, 0.80, 1.0, 1.2), and different acid to MnO_2 molar ratios (2.0, 2.4, 3.0) with a particle size of −600 + 250 μm and −250 + 150 μm of manganese ore. Considerable Mn extraction rates (98%) were obtained at room temperature and for short leaching periods (20 min). They concluded that the most important variables for extracting Mn from nodules are the Fe concentration and nodule particle size. They also compared their results with those of Zakeri et al. [6], noting that sponge iron performs better as a reducing agent than ferrous ions.

Kanungo and Das [9] conducted leaching tests of marine nodules in different acidic media. They obtained the maximum Mn extraction rates (100%) with concentrated HCl (11 mol·dm^{-3}) at 90–100 °C. Other studies report positive results for Mn extraction from marine nodules during leaching with HCl, with the co-dissolution of considerable amounts of Cu (II), Ni (II) and Co (II) [10,11].

Han et al. [12] conducted reactor-based leaching tests of marine nodules with sulfuric acid, and observed that low levels of manganese recovery (1%) were obtained with H_2SO_4 at room temperature (25 °C). They concluded that high temperature is required during the leaching of marine nodules in order to improve recovery, selectivity and kinetics. This indicates the need to use a reducing agent to obtain good results, since manganese oxides like pyrolusite are relatively insoluble in conventional leaching media [13].

Reducing agents such as SO_2 [14], pyrite iron [4,5], ferrous ions [6], molasses-based alcohol wastewater [7], H_2O_2 [15] and hydroxylammonium chloride [16] have been used to increase leaching kinetics.

Several authors have investigated Mn extraction from nodules during leaching with the use of sulfuric acid at different temperature in the presence of magnetite (1) and illite (2) marine nodules [17–21]. The most important reactions are shown below. These reactions indicate the important role of iron in extracting manganese in acidic environments.

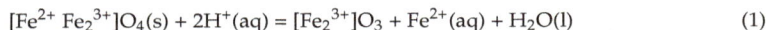

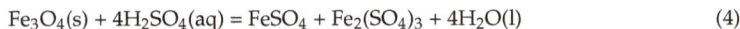

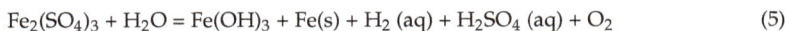

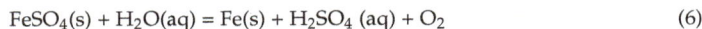

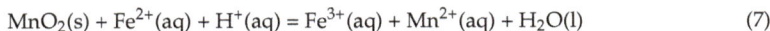

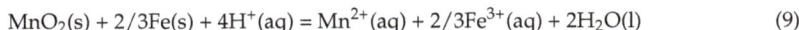

$$[Fe^{2+} Fe_2^{3+}]O_4(s) + 2H^+(aq) = [Fe_2^{3+}]O_3 + Fe^{2+}(aq) + H_2O(l) \tag{1}$$

$$3[Fe^{2+} Ti]O_3 (s) + 6H^+(aq) = 3[Ti]O_2 + 3Fe^{2+} (aq) + 3H_2O(l) \tag{2}$$

$$Fe_2O_3 + H_2SO_4 = Fe_2(SO_4)_3 + H_2O(l) \tag{3}$$

$$Fe_3O_4(s) + 4H_2SO_4(aq) = FeSO_4 + Fe_2(SO_4)_3 + 4H_2O(l) \tag{4}$$

$$Fe_2(SO_4)_3 + H_2O = Fe(OH)_3 + Fe(s) + H_2 (aq) + H_2SO_4 (aq) + O_2 \tag{5}$$

$$FeSO_4(s) + H_2O(aq) = Fe(s) + H_2SO_4 (aq) + O_2 \tag{6}$$

$$MnO_2(s) + Fe^{2+}(aq) + H^+(aq) = Fe^{3+}(aq) + Mn^{2+}(aq) + H_2O(l) \tag{7}$$

$$MnO_2(s) + Fe(s) + 8H^+(aq) = 2Fe^{3+}(aq) + Mn^{2+}(aq) + 2H_2O(l) + 2H_2(g) \tag{8}$$

$$MnO_2(s) + 2/3Fe(s) + 4H^+(aq) = Mn^{2+}(aq) + 2/3Fe^{3+}(aq) + 2H_2O(l) \tag{9}$$

The present study investigates the extraction of manganese from marine nodules in an acid medium (H_2SO_4) at room temperature using smelter slag as a source of iron. Currently in Chile, approximately 80 t of tailings and 1.8 t of smelter slag are generated during the production of one ton of copper. According to the Chilean National Service of Geology and Mining [22], there are 740 tailing dams in the country, of which 469 are inactive and 170 are abandoned. The volume of generated tailings increased by 213.8% from 2000 to 2016 [23].

2. Materials and Methods

2.1. Manganese Nodule Sample

The marine nodules used in this research were collected in the 1970s from the Blake Plateau in the Atlantic Ocean. The nodules were ground in a porcelain mortar to sizes ranging from −140 to +100 µm. The ground samples were analyzed by atomic emission spectrometry via induction-coupled plasma (ICP-AES), in the applied geochemistry laboratory of the Department of Geological Sciences of the Universidad Católica del Norte. Table 1 shows the chemical composition of the samples.

Table 1. Chemical analysis of the manganese ore.

Component	Mn	Fe	Cu	Co
Mass (%)	15.96	0.45	0.12	0.29

Table 2 shows the results of the elemental characterization of the manganese–iron nodules. The sample material was analyzed using Bruker® M4-Tornado μ-FRX table-top equipment (Fremont, CA, USA). μ-XRF data interpretation shows that the nodules were composed of pre-existing nodule fragments that formed their core, with concentric layers that precipitated around the core at later stages. The experiments showed that pyrolusite (MnO_2) was the predominant phase.

Table 2. Mineralogical analysis of the manganese ore.

Component	MgO	Al_2O_3	SiO_2	P_2O_5	SO_3	K_2O	CaO	TiO_2	MnO_2	Fe_2O_3
Mass (%)	3.54	3.69	2.97	7.20	1.17	0.33	22.48	1.07	29.85	26.02

2.2. Smelter Slag

The reducing agent (iron) was obtained in the form of slag from the Altonorte smelting plant. The same methods were used to determine the chemical and mineralogical composition of the slag as those used with the manganese nodules. Figure 1 shows the chemical species using QEMSCAN, several iron-containing phases are present while the content of Fe is estimated at 37.52%.

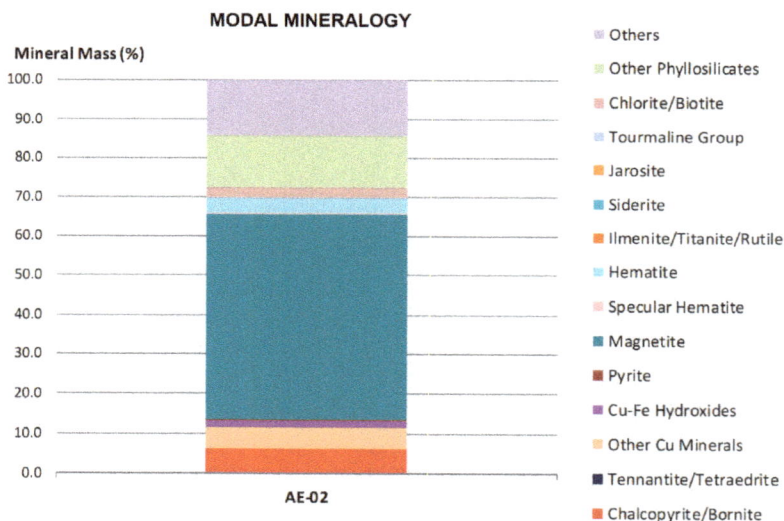

Figure 1. Detailed modal mineralogy.

Table 3 shows the mineralogical composition of the slag. The Fe in the slag was mainly in the form of magnetite.

Table 3. The mineralogical composition of the slag as determined by QEMSCAN.

Mineral	Amount % *w/w*
Chalcopyrite/bornite	6.05
Tennantite/tetraedrite	0.24
Other Cu minerals	5.22
Cu-Fe hydroxides	1.80
Pyrite	0.18
Magnetite	52.11
Specular hematite	0.47
Hematite	3.79
Ilminite/titanite/rutile	0.03
Siderite	0.07
Chlorite/biotite	2.55
Other phyllosilicates	13.14
Others	14.35
Total	100.00

2.3. Reagent and Leaching Test

The sulfuric acid used for the leaching tests was grade P.A., with a 95–97% purity, a density of 1.84 kg/L and a molecular weight of 98.8 g/mol.

Leaching tests were carried out in a 50 mL glass reactor with a 0.01 S/L ratio of leaching solution. A total of 200 mg of Mn nodules were maintained in agitation and suspension with the use of a 5-position magnetic stirrer (IKA ROS, CEP 13087-534, Campinas, Brasil) at a speed of 600 rpm. The tests were conducted at room temperature of 25 °C, with variations in additives, particle size and leaching time.

2.4. Experimental Design

The effects of independent variables on the Mn extraction rates from manganese nodules were studied using the response surface optimization method [24,25]. The central composite face (CCF) design and a quadratic model were applied to the experimental design.

Twenty-seven experimental tests were carried out to study the effects of H_2SO_4 concentration, particle size and time as dependent variables. Minitab 18 software was used for modeling and experimental design, which allowed for the study of the linear and quadratic effects of the independent variables. The experimental data were adjusted by a multiple regression analysis [26] to a quadratic model, considering only those factors that helped to explain the variability of the model.

Slag was used for all the tests and the experimental model, thus the MnO_2 and Fe ratio was 1 molar. Table 4 shows the experimental parameters for the central composite face design and data for Mn from H_2SO_4 extraction optimization.

The general form of the experimental model is represented by:

$$Y = (\text{overall constant}) + (\text{linear effects}) + (\text{interaction effects}) + (\text{curvature effects})$$
$$Y = b_0 + b_1x_1 + b_2x_2 + b_3x_3 + b_{12}x_1x_2 + b_{13}x_1x_3 + b_{23}x_2x_3 + b_{11}x_1^2 + b_{22}x_2^2 + b_{33}x_3^2 \tag{10}$$

where x_1 is time; x_2 is mesh size; x_3 is H_2SO_4 concentration, and b is the variable coefficients.

Table 4 presents the ranges of parameter values used in the experimental model. The variable values were codified in the model. Equation (11) transforms a real value (Z_i) into a coded value (X_i) according to the experimental design:

The Equation (11) coded value was found as follows:

$$X_i = \frac{Z_i - \frac{Z_{high} + Z_{low}}{2}}{\frac{Z_{high} - Z_{low}}{2}} \tag{11}$$

where Z_{high} and Z_{low} are respectively the highest and lowest level of a variable [27].

Table 4. Experimental configuration and Mn extraction data.

Exp. No.	Time (min)	Sieve Fraction (Tyler Mesh)	Particle Size (μm)	Sulphuric Acid (M)	Mn Extraction (%)
1	10	−200 + 270	−75 + 53	0.1	8.77
2	20	−100 + 140	−150 + 106	0.5	30.08
3	20	−200 + 270	−75 + 53	1.0	58.27
4	30	−200 + 270	−75 + 53	1.0	69.55
5	10	−320 + 400	−47 + 38	0.5	22.56
6	20	−100 + 140	−150 + 106	0.1	11.28
7	30	−100 + 140	−150 + 106	1.0	57.64
8	30	−320 + 400	−47 + 38	0.1	15.04
9	10	−100 + 140	−150 + 106	0.5	16.92
10	10	−100 + 140	−150 + 106	1.0	38.22
11	20	−200 + 270	−75 + 53	0.5	53.88
12	30	−100 + 140	−150 + 106	0.1	11.90
13	20	−200 + 270	−75 + 53	0.1	17.54
14	10	−100 + 140	−150 + 106	0.1	6.27
15	10	−200 + 270	−75 + 53	1.0	45.74
16	10	−320 + 400	−47 + 38	0.1	8.15
17	20	−320 + 400	−47 + 38	0.1	10.65
18	20	−320 + 400	−47 + 38	0.5	27.57
19	30	−200 + 270	−75 + 53	0.1	20.05
20	30	−200 + 270	−75 + 53	0.5	60.78
21	10	−320 + 400	−47 + 38	1.0	44.49
22	20	−320 + 400	−47 + 38	1.0	55.14
23	20	−100 + 140	−150 + 106	1.0	49.50
24	30	−320 + 400	−47 + 38	0.5	35.09
25	10	−200 + 270	−75 + 53	0.5	39.47
26	30	−320 + 400	−47 + 38	1.0	61.40
27	30	−100 + 140	−150 + 106	0.5	35.09

The statistical R^2, R^2 (pred), p values and Mallows's Cp indicate whether the model obtained is adequate to describe Mn extraction under a given domain. The R^2 coefficient is a measure of the goodness of fit, that is, it measures the proportion of total variability of the dependent variable with respect to its mean, which is explained by the regression model. The p values represent statistical significance, which indicates whether there is a statistically significant association between the response variable and the term. The predicted R^2 was used to determine how well the model predicts the response for new observations. Finally, Mallows's Cp is a precise measure in the model, estimating the true parameter regression [27].

2.5. MnO$_2$/Fe Ratio Effect

The experimental design was used to assess the interaction among the sulfuric acid concentration, manganese nodule particle size and leaching time, with foundry slag as an additive. Bafghi et al. [8] conducted experiments with sponge iron at different MnO_2/Fe ratios in acid media, and concluded that the amount of sponge iron is more crucial for manganese dioxide leaching than the sulfuric acid concentration. Zakeri et al. [6] concluded that the excess amounts of ferrous ions with reference to the Fe^{2+}/MnO_2 stoichiometric molar ratio of 3 was crucial for successful manganese dissolution.

In the present study, the effect of the MnO_2/Fe ratio was evaluated with the use of foundry slag over time. A particle size of −200 + 270 Tyler mesh (−75 + 53 μm), with a stirring speed of 600 rpm and 20 mL 1 M sulfuric acid concentration, and 200 mg of Mn nodules were used at a room temperature (25 °C).

2.6. The Effect of Particle Size

The effect of the manganese nodule particle size was evaluated by adding Fe slag at different sulfuric acid concentrations over time under the conditions shown in Table 5.

Table 5. Experimental conditions for the study of the effect of manganese nodule particle size.

Parameters	Values
Sieve fraction (Tyler mesh)	$-100 + 140$, $-200 + 270$, $-320 + 400$
Particle size (μm)	$-150 + 106$, $-75 + 53$, $-47 + 38$
Time (in min)	5, 10, 20, 30, 40
H_2SO_4 (M)	0.1, 0.5, 0.75, 1
MnO_2/Fe (slag)	1/1

3. Results and Discussion

3.1. Methodology

An ANOVA analysis (Table 6) showed no significant effect of the interactions (time, particle size) and {particle size, concentration} ($p > 0.05$) on the manganese extraction rate. However, the interaction {time, concentration} must also be considered ($p < 0.1$). The effects of the curvature of time and concentration are not significant in explaining the variability of the model.

Table 6. ANOVA Mn extraction.

Source	F-Value	p-Value
Regression	38.11	0.000
Time	36.29	0.000
Mesh size	26.95	0.000
H_2SO_4	269.22	0.000
Time × Mesh size	0.51	0.485
Time × H_2SO_4	3.89	0.065
Mesh size × H_2SO_4	0.37	0.549
Time × Time	0.62	0.443
Mesh size × Mesh size	3.28	0.088
H_2SO_4 × H_2SO_4	1.86	0.191

The linear effects of time and H_2SO_4 concentration contributed greatly to explaining the experimental model, as shown in the contour plot in Figure 2.

Figures 3–5 show that time, size range and H_2SO_4 concentration, as well as the interaction of time and H_2SO_4, and particle size curvature significantly affected the Mn extraction.

Equation (12) presents the Mn extraction model over the range of experimental conditions after eliminating the non-significant coefficients.

$$\%Extraction = 0.3112 + 0.0755X_1 + 0.0651X_2 + 0.2057X_3 + 0.0303X_1X_3 - 0.0393X_2^2 \quad (12)$$

where x_1, x_2 and x_3 are coded variables that respectively represent time, particle size and H_2SO_4 concentration.

Figure 6 graphically represents the order in which parameters were added to the model, with the contribution of each variable to explaining variability.

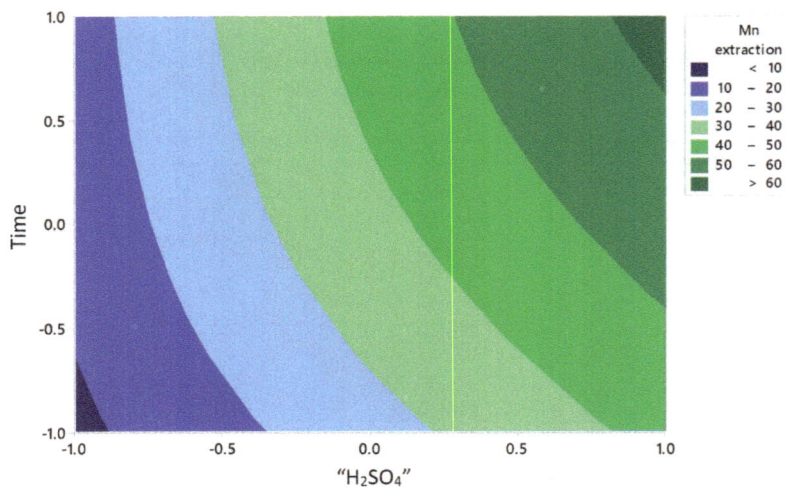

Figure 2. Experimental contour plot of Mn extraction (25 °C, −100 + 140, −200 + 270, −320 + 400 sieve fraction mesh, 5, 10, 20, 30, 40 min leaching time, H_2SO_4 0.1, 0.5, 0.75, 1 M and 1 MnO_2/Fe M ratio).

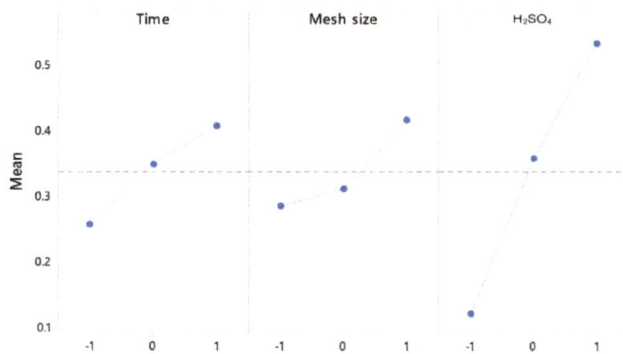

Figure 3. Linear effect plot for Mn extraction.

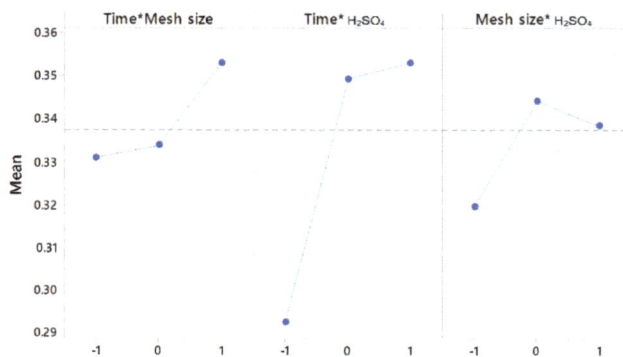

Figure 4. Interaction effect plot for Mn extraction.

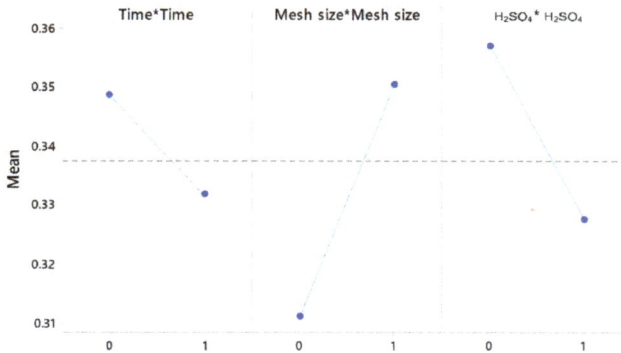

Figure 5. Curvature effect plot for Mn extraction.

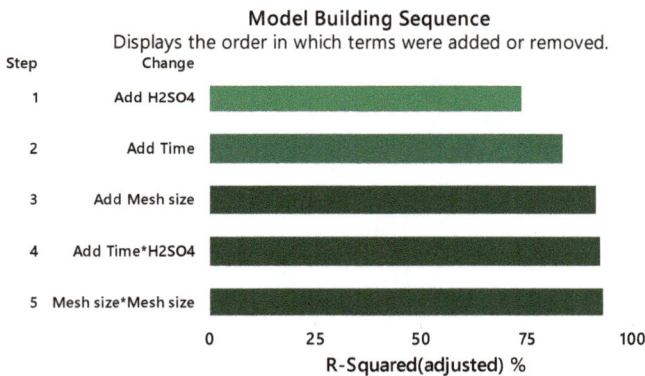

Figure 6. Construction sequence of the model.

An ANOVA test indicated that the quadratic model adequately represented Mn extraction under the established parameter ranges. The model did not require adjustment and it was validated by the R^2 value (94.34%) (Figure 7). The ANOVA analysis showed that the indicated factors influenced the manganese extraction from regression (70.07), 5% confidence level F4.22 (2.8167).

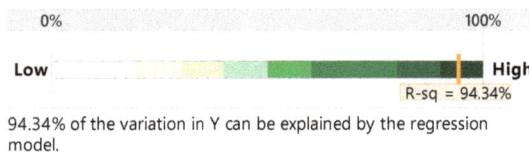

Figure 7. R^2 statistic with % of variation explained by the model.

The *p* value (Figure 8) of the model, as represented by Equation (12), indicated that the model was statistically significant.

The Mallows's Cp indicated that the model was relatively accurate and did not present bias in estimating the true regression coefficients. It also allows for prediction with an acceptable future forecast margin of error of R_{pred} = 91.13%. The response surface graphs in Figure 9A show that Mn extraction increased with a larger particle size and higher H_2SO_4 concentration. Figure 9B shows the effect of increased time and H_2SO_4 concentration, which significantly increased extraction. Finally,

Figure 9C shows that Mn extraction increased with increased mesh size and time, in the context of the size parameters used in the experiment.

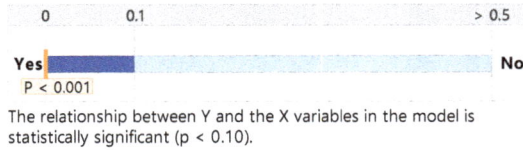

The relationship between Y and the X variables in the model is statistically significant (p < 0.10).

Figure 8. p statistic of the relationship between the Y and X variables in the model.

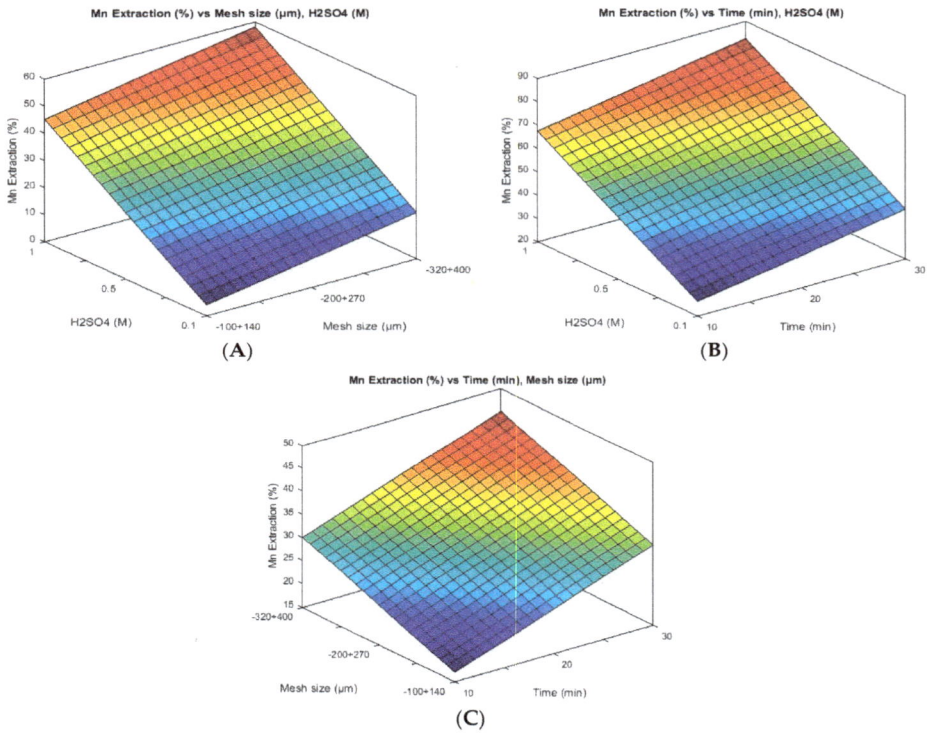

Figure 9. Response surface of the independent variables H_2SO_4 concentration and particle size (**A**), H_2SO_4 concentration and time (**B**), particle size and time, in the dependent variable Mn extraction (**C**).

Finally, from the adjustment of the ANOVA analysis, it was found that the factors considered, after analysis of the main components, explained the variation in the response. The difference between the predictive R^2 and R^2 of the model was minimal, thus reducing the risk that the model was over adjusted, that is, the probability that the model fits only the sample data is lower. The ANOVA analysis indicated that H_2SO_4, time, size and curvature of the mesh are the factors that explain to a greater extent the behavior of the system for the sampled data set.

3.2. Effect of the MnO_2/Fe Ratio

The results shown in Figure 10 indicate that the Mn extraction rates increased with higher Fe concentrations, which concurs with the conclusions of Zakeri et al. [6] and Bafghi et al. [8]. The highest Mn extraction rates were obtained with a MnO_2/Fe ratio of 1/2. However, the extraction did not tend to increase much with time. It was emphasized that this MnO_2/Fe ratio 1/2 resulted in high extraction rates in short periods of time, such as 68% Mn extraction in only 5 min; by decreasing the MnO_2/Fe ratio to 1/1, it was possible to obtain an Mn extraction rate of 70% in 30 min. The same tendency was noted with a 0.5 ratio, where there was a small extraction rate at 40 min, and the extraction rates were lower with shorter periods of time (5, 10 min). A low extraction rate of 47% in 40 min was obtained with an MnO_2/Fe ratio of 2/1. The leaching results support the principle of dissolution using two rate-balancing corrosion couples, MnO_2/Fe^{2+} and FeS_2/Fe^{3+}, and form the theoretical background [28]. Zakeri et al. [6] obtained better results during the leaching of Mn from marine nodules using iron instead of ferrous sulphate. This is because the iron in the system maintains ferrous ion regeneration, resulting in high levels of ferrous ion and ferric ion activity [8]. Under the ranges of pH (-2 to 0.1) and potential (-0.4 to 1.4), Mn ions remain in solution and do not precipitate through oxidation-reduction reactions, given the presence of ions Fe^{2+} and Fe^{3+} [29]. Based on the positive results shown in Figure 10, slag is a viable reducing agent for Mn dissolution from marine nodules.

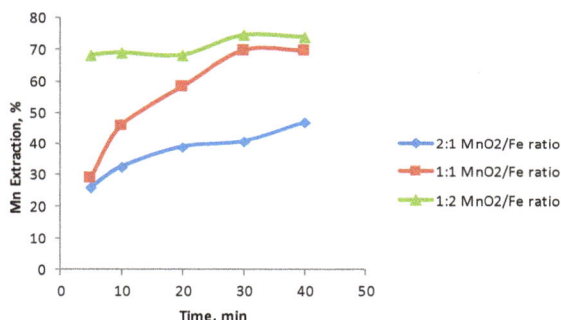

Figure 10. Effect of the MnO_2/Fe ratio on manganese extraction (25 °C, $-200 + 270$ Tyler mesh ($-75 + 53$ μm), H_2SO_4 1 M).

3.3. Effect of Particle Size

Figure 11 shows that the effect of particle size on Mn extraction was not as significant as the effect of sulfuric acid concentration. Particle size is important in the context of a sulfuric acid concentration of 1 M. The highest Mn extraction rate of 70% was obtained for particle sizes between $-320 + 400$ Tyler ($-47 + 38$ μm) mesh and an H_2SO_4 concentration of 1 M (Figure 11B). However, a similar result, with a 65% extraction rate, was obtained with the same parameters and sizes ranging between $-200 + 270$ Tyler mesh ($-75 + 53$ μm) (Figure 11C). The lowest extraction rate (60%) was obtained for particle sizes between $-150 + 106$ μm and 1M H_2SO_4 (Figure 11A).

Figure 11C shows that Mn extraction rates are lower with higher H_2SO_4 concentrations. Extraction did not exceed 3% with H_2SO_4 concentrations of 0.5, 0.75 and 1 M. The results obtained in this research (Figures 10 and 11) indicate the promising use of an industrial waste in dissolving Mn from marine nodules. There is a need for additional research to overcome production barriers and provide technological alternatives as a viable option for extracting metals from raw materials [30]. Reusing smelter slag in element extraction also produces considerable savings in disposal costs and reduces environmental impacts, which can result in greater social acceptance of the industry.

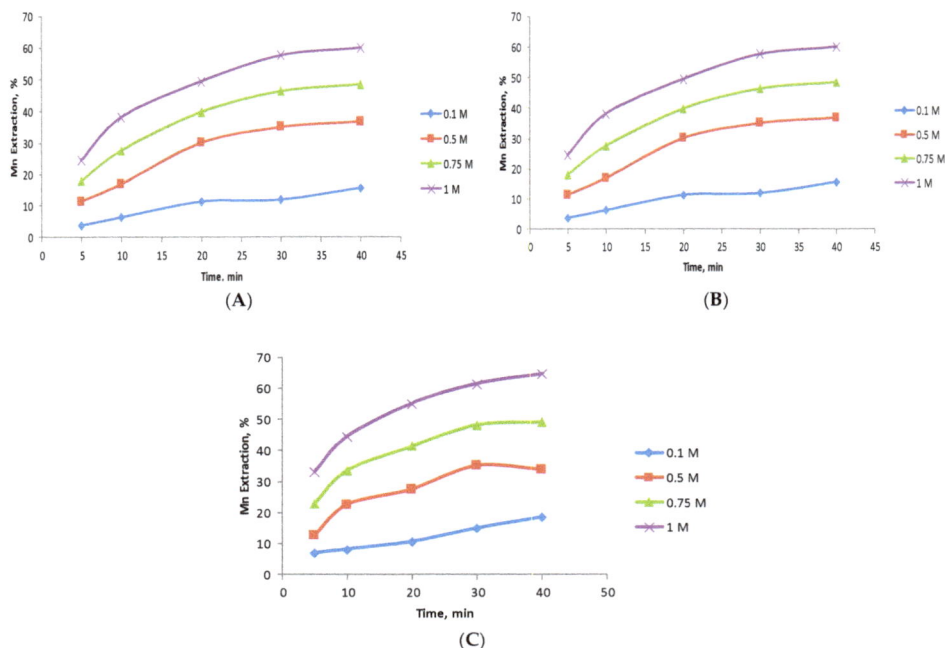

Figure 11. Effect of particle size (manganese nodule) on Mn extraction (25 °C, H_2SO_4: 0.1, 0.5, 0.75, 1 M). (**A**): $-100 + 140$ Tyler mesh ($-150 + 106$ μm), (**B**): $-200 + 270$ Tyler mesh ($-75 + 53$ μm), (**C**): $-320 + 400$ Tyler mesh ($-47 + 38$ μm).

4. Conclusions

The present investigation presents the laboratory results for dissolving Mn from marine nodules in an acid medium at room temperature (25 °C) with the use of foundry slag. The iron metal proved to be a good reducer when leaching MnO_2 in acid media. The findings of this study were:

(1) High ratios of MnO_2/Fe (0.5) and 1 M H_2SO_4 significantly shorten the manganese dissolution time (from 30 to 5 min).

(2) The MnO_2 particle size is not as significant for the extraction of Mn in solution as the concentration of H_2SO_4 in the presence of Fe from foundry slag.

(3) At low granulometries ($-47 + 38$ μm) the Mn extraction margins were narrower when higher H_2SO_4 concentrations were used.

(4) The highest Mn extraction obtained in this experimental study at a MnO_2 ratio of 0.5, 1 M H_2SO_4, $-320 + 400$ Tyler mesh ($-47 + 38$ μm), with 30 min of leaching was 61.4%.

Author Contributions: N.T., C.M.T., N.H. contributed in the methodology, conceived and designed the experiments; analyzed the data and wrote paper, J.C. and R.S. performed the experiments.

Funding: This research received no external funding.

Acknowledgments: The authors are grateful for the contribution of the Scientific Equipment Unit- MAINI of the Universidad Católica del Norte for aiding in generating data by automated electronic microscopy QEMSCAN® and for facilitating the chemical analysis of the solutions. We are also grateful to the Altonorte Mining Company for supporting this research and providing slag for this study, and we thank the students Reina Valdés Marangunic, Kevin Pérez Salinas and Manuel Saldaña Pino of the Universidad Católica del Norte for supporting the experimental tests. FCAC 2018-UCN contributed reagents/materials/analysis tools.

Conflicts of Interest: The authors declare no conflict of interest.

References

1. Somoza, L.; González, F.; León, R.; Medialdea, T.; De Torres, T.; Ortiz, J.; Lunar, R.; Martínez-Frías, J.; Merinero, R. Ferromanganese nodules and micro-hardgrounds associated with the Cadiz Contourite Channel (NE Atlantic): Palaeoenvironmental records of fluid venting and bottom currents. *Chem. Geol.* **2012**, *310–311*, 56–78.

2. Senanayake, G. Acid leaching of metals from deep-sea manganese nodules—A critical review of fundamentals and applications. *Miner. Eng.* **2011**, *24*, 1379–1396. [CrossRef]

3. Lenoble, J.P. *Polymetallic Nodules*; International Seabed Authority: Kingston, Jamaica, 2000; p. 8.

4. Kanungo, S.B. Rate process of the reduction leaching of manganese nodules in dilute HCl in presence of pyrite. Part I. Dissolution behavior of iron and sulphur species during leaching. *Hydrometallurgy* **1999**, *52*, 313–330. [CrossRef]

5. Kanungo, S.B. Rate process of the reduction leaching of manganese nodules in dilute HCl in presence of pyrite: Part II. Leaching behavior of manganese. *Hydrometallurgy* **1999**, *52*, 331–347. [CrossRef]

6. Zakeri, A.; Bafghi, M.S.; Shahriari, S. Dissolution of manganese dioxide ore in sulfuric acid in the presence of ferrous ion. *Iran. J. Mater. Sci. Engi.* **2007**, *4*, 22–27.

7. Su, H.; Liu, H.; Wang, F.; Lü, X.; Wen, Y. Kinetics of Reductive Leaching of Low-grade Pyrolusite with Molasses Alcohol Wastewater in H_2SO_4. *Chin. J. Chem. Eng.* **2010**, *18*, 730–735. [CrossRef]

8. Bafghi, M.; Zakeri, A.; Ghasemi, Z.; Adeli, M. Reductive dissolution of manganese ore sulfuric acid in the presence of iron metal. *Hydrometallurgy* **2008**, *90*, 207–212. [CrossRef]

9. Kanungo, S.B.; Das, R.P. Extraction of metals from manganese nodules of the Indian Ocean by leaching in aqueous solution of sulphur dioxide. *Hydrometallurgy* **1988**, *20*, 135–146. [CrossRef]

10. Charewicz, W.A.; Chaoyin, Z.; Chmielewski, T. The leaching behavior of ocean polymetallic nodules in chloride solutions. *Physicochem. Probl. Miner. Process.* **2001**, *35*, 55–56.

11. Kanungo, S.B.; Jena, P.K. Reduction leaching of manganese nodules of Indian Ocean origin in dilute hydrochloric acid. *Hydrometallurgy* **1998**, *2*, 41–58. [CrossRef]

12. Han, K.N.; Fuerstenau, D.W. Acid leaching of ocean floor manganese nodules at elevated temperature. *Int. J. Miner. Process.* **1975**, *2*, 163–171. [CrossRef]

13. Jiang, T.; Yang, Y.; Huang, Z.; Zhang, B.; Qiu, G. Leaching kinetics of pyrolusite from manganese–silver ores in the presence of hydrogen peroxide. *Hydrometallurgy* **2004**, *72*, 129–138. [CrossRef]

14. Petrie, L.M. Molecular interpretation for SO_2 dissolution kinetics of pyrolusite, manganite and hematite. *Appl. Geochem.* **1995**, *10*, 253–267. [CrossRef]

15. Nayl, A.A.; Ismail, I.M.; Aly, H.F. Recovery of pure $MnSO_4 \cdot H_2O$ by reductive leaching of manganese from pyrolusite ore by sulfuric acid and hydrogen peroxide. *Int. J. Miner. Process.* **2011**, *100*, 116–123. [CrossRef]

16. Hariprasad, D.; Mohapatra, M.; Anand, S. Non-isothermal self-sustained one pot dissolution of metal values from manganese nodules using NH_3OHCl as a novel reductant in sulfuric acid medium. *J. Chem. Technol. Biotechnol.* **2013**, *88*, 1114–1120. [CrossRef]

17. White, F.; Peterson, M.L.; Hochella, M.F. Electrochemistry and dissolution kinetics of magnetite and ilmenite. *Geochim. Cosmochim. Acta* **1994**, *58*, 1859–1875. [CrossRef]

18. Nijjer, S.; Thonstad, J.; Haarberg, G.M. Oxidation of manganese(II) and reduction of manganese dioxide in sulphuric acid. *Electrochim. Acta* **2000**, *46*, 395–399. [CrossRef]

19. Godunov, E.B.; Izotov, A.D.; Gorichev, I.G. Reactions of manganese oxides with sulfuric acid solutions studied by kinetic and electrochemical methods. *Inorg. Mater.* **2017**, *53*, 831–837. [CrossRef]

20. Anacleto, N.; Ostrovski, O.; Ganguly, S. Reduction of Manganese Oxides by Methane-containing Gas. *ISIJ Int.* **2004**, *44*, 1480–1487. [CrossRef]

21. Sesen, F.E. Practical reduction of manganese oxide. *J. Chem. Technol. Appl.* **2017**, *1*, 1–2.

22. COCHILCO. Análisis del Catastro de Depósitos de Relaves en Chile y guía de estructura de datos. Servicio Nacional de Geología y Minería. 2018. Available online: http://www.sernageomin.cl/wp-content/uploads/2018/05/An%C3%A1lisis-de-los-Dep%C3%B3sitos-de-Relaves-en-Chile_VF.pdf (accessed on 1 December 2018).

23. DGA 1998, Universidad de Chile. Informe País: Estado del Medio Ambiente en Chile 2012. 2013. Available online: http://www.repositorio.uchile.cl/handle/2250/123564 (accessed on 1 December 2018).

24. Dean, A.; Voss, D.; Draguljic, D. Response Surface Methodology. *Des. Anal. Exp.* **2017**, 565–614.

25. Bezerra, M.A.; Santelli, R.E.; Oliveira, E.P.; Villar, L.S.; Escaleira, L.A. Response surface methodology (RSM) as a tool for optimization in analytical chemistry. *Talanta* **2008**, *76*, 965–977. [CrossRef] [PubMed]
26. Berger, P.D.; Maurer, R.E.; Celli, G.B. Multiple Linear Regression. In *Experimental Design*; Springer International Publishing: Cham, Switzerland, 2018; pp. 505–532.
27. Montgomery, D.C. *Design and Analysis of Experiments*; Wiley: Hoboken, NJ, USA, 2012; Volume 8, pp. 3–10.
28. Paramguru, R.K.; Kanungu, S.B. Electrochemical phenomena in MnO_2–FeS_2 leaching in dilute HCl. Part 3. Manganese dissolution from indian ocean nodules. *Can. Metall. Q.* **1998**, *37*, 405–417. [CrossRef]
29. Komnitsas, K.; Bazdanis, G.; Bartzas, G.; Sahinkaya, E.; Zaharaki, D. Removal of heavy metals from leachates using organic/inorganic permeable reactive barriers. *Desalin. Water Treat.* **2013**, *51*, 3052–3059. [CrossRef]
30. Komnitsas, K.; Zaharaki, D.; Perdikatsis, V. Effect of synthesis parameters on the compressive strength of low-calcium ferronickel slag inorganic polymers. *J. Hazard. Mater.* **2009**, *161*, 760–768. [CrossRef] [PubMed]

minerals

MDPI

Article

Leaching of Manganese from Marine Nodules at Room Temperature with the Use of Sulfuric Acid and the Addition of Tailings

Norman Toro [1],*, Manuel Saldaña [2], Jonathan Castillo [3], Freddy Higuera [2] and Roxana Acosta [4]

[1]   Departamento de Ingeniería en Metalurgia y Minas, Universidad Católica del Norte, Antofagasta 1270709, Chile
[2]   Departamento de Ingeniería Industrial, Universidad Católica del Norte, Antofagasta 1270709, Chile; msp018@alumnos.ucn.cl (M.S.); fhiguera@ucn.cl (F.H.)
[3]   Departamento de Ingeniería en Metalurgia, Universidad de Atacama, Copiapó 1531772, Chile; jonathan.castillo@uda.cl
[4]   Departamento de Educación, Universidad de Antofagasta, Antofagasta 1270300, Chile; roxana.acosta@uantof.cl
*   Correspondence: ntoro@ucn.cl; Tel.: +56-552651021

Received: 21 March 2019; Accepted: 9 May 2019; Published: 11 May 2019

Abstract: Based on the results obtained from a previous study investigating the dissolution of Mn from marine nodules with the use of sulfuric acid and foundry slag, a second series of experiments was carried out using tailings produced from slag flotation. The proposed approach takes advantage of the Fe present in magnetite contained in these tailings and is believed to be cost-efficient. The surface optimization methodology was used to evaluate the independent variables of time, particle size, and sulfuric acid concentration in the Mn solution. Other tests evaluated the effect of agitation speed and the MnO_2/Fe_2O_3 ratio in an acid medium. The highest Mn extraction rate of 77% was obtained with an MnO_2/Fe_2O_3 ratio of 1/2 concentration of 1 mol/L of H_2SO_4, particle size of $-47 + 38$ μm, and 40 min of leaching. It is concluded that higher rates of Mn extraction were obtained when tailings instead of slag were used, while future research needs to focus on determination of the optimum Fe_2O_3/MnO_2 ratio to improve dissolution of Mn from marine nodules.

Keywords: secondary products; reducing agent; waste reuse; acid media

1. Introduction

Ferromanganese (Fe–Mn) deposits are present in the oceans across the world, marine ridges, and plateaus where the currents have delivered sediments for millions of years [1]. These deposits form through the accumulation of iron and manganese oxides in seawater, within either volcanic or sedimentary rocks that act as substrates, as observed in the central and northeastern ocean beds of the Pacific [2]. They may have economic potential [3], due to the high concentrations of Co, Ni, Te, Ti, Pt, and rare earth elements [4]. These Fe–Mn oceanic deposits include ferromanganese crusts, as well as cobalt-rich crusts, polymetallic nodules, and hydrothermal infusions [5]. Polymetallic nodules have a particular importance for the steel industry as an they may eventually become an alternate source of manganese [6].

In order to extract manganese and other metals from marine nodules, the use of a reducing agent is necessary [7]. Acid leaching of marine nodules, with the use of iron as a reducing agent, has shown good results [8–10]. In a previous study carried out by Toro et al. [11], several parameters were evaluated for dissolving Mn from marine nodules using slag at room temperature in an acid medium. This study established that high MnO_2/Fe_2O_3 ratios significantly shorten the manganese dissolution time from 30 to 5 min. They also conclude that MnO_2 particle size does not significantly affect the Mn extraction rate in an acid medium in the presence of Fe contained in ferrous slag.

The positive effect of Fe as a reducing agent for dissolving Mn from marine nodules was noted when lower Mn/Fe ratios were used [8–11]. Bafghi et al. [12] and Toro et al. [11] determined that sulfuric acid concentration is less important than Fe concentration in dissolving Mn.

The Mn extraction rate increases with a higher agitation speed [13–15]. Jiang et al. [13] evaluated the kinetic aspects of manganese and silver extraction during leaching of pyrolusite in sulfuric acid solutions in the presence of H_2O_2, and concluded that agitation speed was one of the most important variables affecting the Mn extraction rate. Su et al. [14] indicated that the Mn extraction rate increases significantly when the agitation speed increases from 100 to 700 rpm because high speed improves mixing and allows better contact between reagents and reactants. Jiang et al. [13] also reported that the extraction rate decreases slightly at 1000 rpm because excessive agitation can cause material to adhere to the walls of the reactor and prevent it from being leached. Velásquez et al. [16] indicated that it is only necessary to keep particles in suspension and prevent agglomeration.

The addition of Fe as a reducing agent in temperature-controlled acid media has already been studied [8,10,12]. In particular, Zakeri et al. [10] used ferrous ions with a Fe^{2+}/MnO_2 ratio of 2.4 and sulfuric acid as a leaching agent with a H_2SO_4/MnO_2 ratio of 2.0 over a temperature range of 20 to 60 °C, and found out that Mn extraction was notably higher at 60 °C and reached 96% after 60 min. Bafghi et al. [12] used Fe sponge with a molar ratio of 2, and H_2SO_4 with a molar ratio of 4 (both ratios with respect to MnO_2), under the same temperatures as Zakeri et al. [10]; at 60 °C, 100% of the Mn present within the nodules was dissolved in 3 min. Both cases demonstrate the positive impact of higher temperature on the extraction rate; however, the positive impact of the presence of iron indicated that effective processing may take place even at ambient temperatures. Furthermore, both studies demonstrate that the acid concentration is less significant than the Fe/MnO_2 ratio.

The present work investigates the effect of using of tailings, obtained after flotation of slag at the Altonorte Foundry Plant, on the dissolution of Mn from marine nodules. A report by SERNAGEOMIN [17] indicates that the production of copper concentrate in Chile has been increasing steadily, and is expected to almost double by 2026 from its 2014 level, from 3.9 to 5.4 million tons. For every ton of Cu concentrate obtained by flotation, 151 tons of tailings are generated [18], which are disposed of in tailing dams and have significant impacts on the environment [19]. Consequently, it is necessary to find new uses for tailings with the application of more environmentally friendly hydrometallurgical techniques [20]. This results in an attractive proposal given the quantities of waste generated in the country by flotation, providing an added value for this material while introducing a new initiative in the context of the need to overcome stagnation in the mining sector [21].

2. Materials and Methods

2.1. Manganese Nodule Sample

The marine nodules used in this work were the same as those used in Toro et al. [11]. They were composed of 15.96% Mn and 0.45% Fe. Table 1 shows the chemical composition. The sample material was analyzed with a Bruker®M4-Tornado μ-FRX tabletop device (Fremont, CA, USA). The μ-XRF data shows that the nodules were composed of fragments of preexisting nodules that formed their nuclei, with concentric layers that precipitated around the nuclei in later stages.

Table 1. Chemical analysis (in the form of oxides) of manganese nodules.

Component	MgO	Al_2O_3	SiO_2	P_2O_5	SO_3	K_2O	CaO	TiO_2	MnO_2	Fe_2O_3
Weight (%)	3.54	3.69	2.97	7.20	1.17	0.33	22.48	1.07	29.85	26.02

2.2. Tailings

The sample of tailings used in this study was obtained after flotation of slag during the production of copper concentrate at the Altonorte Smelting Plant. The methods used to determine the chemical and mineralogical composition of the tailings were the same as those used to determine marine nodule content. Chemical species were determined by QEMSCAN. Several iron-containing phases were present, while the Fe content was estimated at 41.9%. Table 2 shows the mineralogical composition of the tailings. As the Fe was mainly in the form of magnetite, the most appropriate method of extraction was the same as that used in Toro et al. [11].

Table 2. Mineralogical composition of tailings, as determined by QEMSCAN.

Mineral	Amount % (w/w)
Chalcopyrite/Bornite $(CuFeS_2/Cu_5FeS_4)$	0.47
Tennantite/Tetrahedrite $(Cu_{12}As_4S_{13}/Cu_{12}Sb_4S_{13})$	0.03
Other Cu Minerals	0.63
Cu–Fe Hydroxides	0.94
Pyrite (FeS_2)	0.12
Magnetite (Fe_3O_4)	58.52
Specular Hematite (Fe_2O_3)	0.89
Hematite (Fe_2O_3)	4.47
Ilmenite/Titanite/Rutile $(FeTiO_3/CaTiSiO_5/TiO_2)$	0.04
Siderite $(FeCO_3)$	0.22
Chlorite/Biotite $(Mg)_3(Si)_4O_{10}(OH)_2(Mg)_3(OH)_6/K(Mg)_3AlSi_3O_{10}(OH)_2$	3.13
Other Phyllosilicates	11.61
Fayalite (Fe_2SiO_4)	4.59
Dicalcium Silicate (Ca_2SiO_4)	8.30
Kirschsteinite $(CaFeSi\,O_4)$	3.40
Forsterita (Mg_2SiO_4)	2.30
Baritine $(BaSO_4)$	0.08
Zinc Oxide (ZnO)	0.02
Lead Oxide (PbO)	0.01
Sulfate (SO_4)	0.20
Others	0.03
Total	100.00

2.3. Reagents Used—Leaching Parameters

The sulfuric acid used for the leaching tests was grade P.A., with 95%–97% purity, a density of 1.84 kg/L, and a molecular weight of 98.8 g/mol. The leaching tests were carried out in a 50 mL glass reactor with a 0.01 solid/liquid ratio. A total of 200 mg of Mn nodules were maintained in suspension with the use of a 5-position magnetic stirrer (IKA ROS, CEP 13087-534, Campinas, Brazil) at a speed of 600 rpm. The tests were conducted at a room temperature of 25 °C, while the parameters studied were additives, particle size, and leaching time. Also, the tests were performed in duplicate, measurements (or analyses) were carried on 5 mL of undiluted samples using atomic absorption spectrometry with a coefficient of variation ≤5% and a relative error between 5% to 10%.

2.4. Experimental Design

The effect of the independent variables on the extraction rate of Mn from manganese nodules was studied using the response surface method [22,23], which helped in understanding and optimizing the response by refining the determinations of relevant factors using the model. An experiment was designed involving three factors that could influence the response variable, and with three levels for each factor for a total of 27 experimental tests (Table 3), the purpose of which was to study the effects of H_2SO_4 concentration, particle size, and time on the dependent variable. Minitab 18 software was used for modeling and experimental design, providing the same analytical approach as used in Toro et al. [11].

Table 3. Experimental configuration and Mn extraction data.

Exp. No.	Time (min)	Sieve Fraction (Tyler Mesh)	Particle Size (μm)	Sulfuric Acid Conc. (mol/L)	Mn Extraction (%)
1	10	−320 + 400	−47 + 38	0.1	8.12
2	20	−100 + 140	−150 + 106	0.5	29.10
3	20	−320 + 400	−47 + 38	1	55.51
4	30	−320 + 400	−47 + 38	1	71.00
5	10	−200 + 270	−75 + 53	0.5	19.12
6	20	−100 + 140	−150 + 106	0.1	7.63
7	30	−100 + 140	−150 + 106	1	49.8
8	30	−200 + 270	−75 + 53	0.1	17.79
9	10	−100 + 140	−150 + 106	0.5	13.98
10	10	−100 + 140	−150 + 106	1	41.22
11	20	−320 + 400	−47 + 38	0.5	52.51
12	30	−100 + 140	−150 + 106	0.1	10.89
13	20	−320 + 400	−47 + 38	0.1	19.12
14	10	−100 + 140	−150 + 106	0.1	5.24
15	10	−320 + 400	−47 + 38	1	46.23
16	10	−200 + 270	−75 + 53	0.1	9.54
17	20	−200 + 270	−75 + 53	0.1	11.11
18	20	−200 + 270	−75 + 53	0.5	29.41
19	30	−320 + 400	−47 + 38	0.1	19.43
20	30	−320 + 400	−47 + 38	0.5	59.16
21	10	−200 + 270	−75 + 53	1	46.77
22	20	−200 + 270	−75 + 53	1	54.00
23	20	−100 + 140	−150 + 106	1	47.24
24	30	−200 + 270	−75 + 53	0.5	33.67
25	10	−320 + 400	−47 + 38	0.5	38.23
26	30	−200 + 270	−75 + 53	1	63.50
27	30	−100 + 140	−150 + 106	0.5	30.00

The response variable can be expressed as showed in Equation (1):

$$Y = (\text{overall constant}) + (\text{linear effects}) + (\text{interaction effects}) + (\text{curvature effects}) \tag{1}$$

Table 4 shows the ranges for values of the parameters used for the experimental design.

Table 4. Experimental conditions.

Parameters/Values	Low	Medium	High
Sieve fraction (Tyler mesh)	−100 + 140	−200 + 270	−320 + 400
Particle size (μm)	−150 + 106	−75 + 53	−47 + 38
Time (in min)	10	20	30
H_2SO_4 (mol/L)	0.1	0.5	1

The levels of the factors are coded as (−1, 0, 1), where each number represents a particular value of the factor, with (−1) as the lowest value, (0) as the intermediate, and (1) as the highest. Equation (2) is used to transform a real value (Z_i) into a coded value (X_i) according to the experimental design:

$$X_i = \frac{Z_i - \frac{Z_{high} + Z_{low}}{2}}{\frac{Z_{high} - Z_{low}}{2}} \tag{2}$$

where Z_{high} and Z_{low} are, respectively, the highest and lowest values of a variable [22].

The statistics used to determine whether the model can adequately describe the extraction of Mn from marine nodules are similar with those used in the study of Toro et al. [11].

2.5. Effect of Stirring Speed

The effect of particle size was evaluated by Toro et al. [11]. It was concluded that this variable did not significantly influence the manganese solutions. Consequently, the present work assessed the effect of agitation speed on Mn dissolution kinetics.

This investigation determined the effect of increasing agitation speed (200, 400, 600, 800, and 1000 rpm) on leaching manganese nodules, using a particle size of $-75 + 53$ µm, MnO_2/Fe_2O_3 ratio of 1, leaching solution volume of 20 mL, 1 mol/L sulfuric acid, and room temperature (25 °C).

2.6. Effect of the MnO_2/Fe_2O_3 Ratio

The present study evaluated the effect of the MnO_2/Fe_2O_3 ratio on leaching time with the use of tailings, using a particle size of $-75 + 53$ µm, agitation speed of 600 rpm, leaching solution volume of 20 mL, 1 mol/L sulfuric acid, and room temperature (25 °C).

3. Results and Discussion

3.1. Effect of Variables

Based on the information obtained from the ANOVA analysis (Table 5), the linear effects of particle size, H_2SO_4, and time contribute greatly to explaining the experimental model, as shown in the contour plots (Figures 1 and 2), while there was no significant effect of any of the curvatures and interactions of the variables considered ($p >> 0.05$) on the manganese extraction rate.

Table 5. ANOVA of the Mn extraction rate.

Source	F-Value	p-Value
Regression	32.13	0.000
Time	27.12	0.000
Particle size	30.39	0.000
Sulfuric acid	226.50	0.000
Time × Time	0.43	0.522
Particle size × Particle size	0.67	0.423
Sulfuric acid × Sulfuric acid	0.39	0.542
Time × Particle size	1.81	0.196
Time × Sulfuric acid	1.57	0.228
Particle size × Sulfuric acid	0.34	0.568

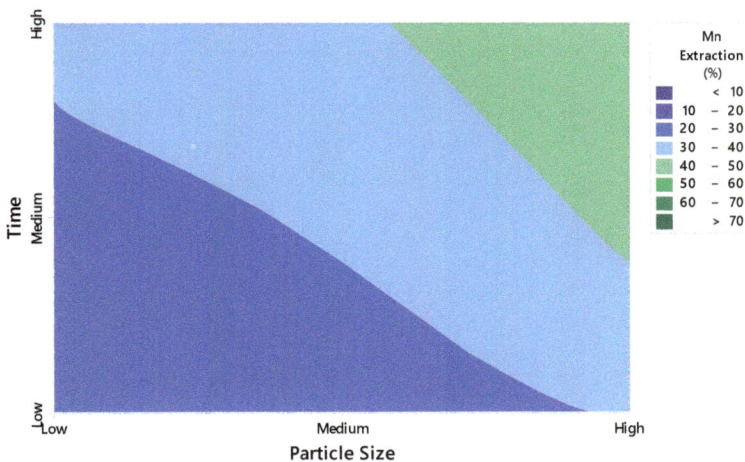

Figure 1. Experimental contour plot of Mn extraction (25 °C; $-150 + 106$, $-75 + 53$, $-47 + 38$ µm particle size; 10, 20, 30 min leaching time).

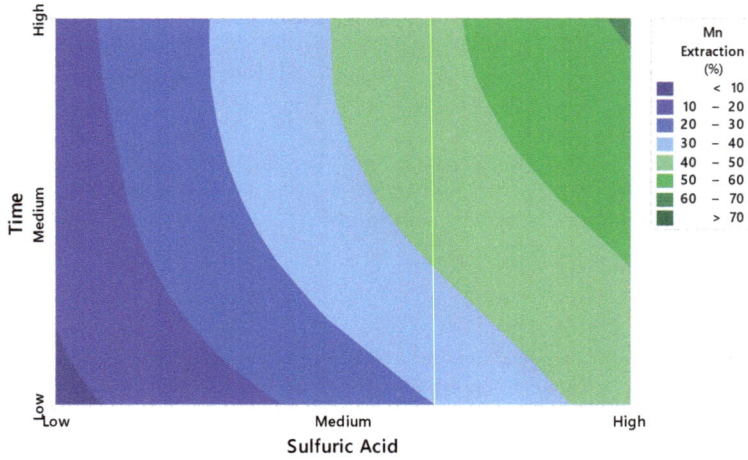

Figure 2. Experimental contour plot of Mn extraction (25 °C; 10, 20, 30 min leaching time; 0.1, 0.5, 1 mol/L H_2SO_4).

Figures 3–5 show that the linear effects of time, particle size, and H_2SO_4 concentration had the most significant impact on Mn extraction rates.

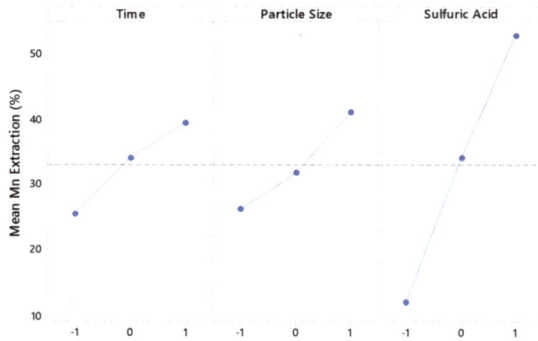

Figure 3. Linear effect plot for Mn extraction.

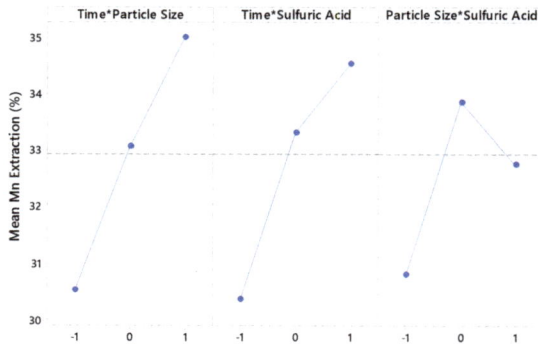

Figure 4. Interaction effect plot for Mn extraction.

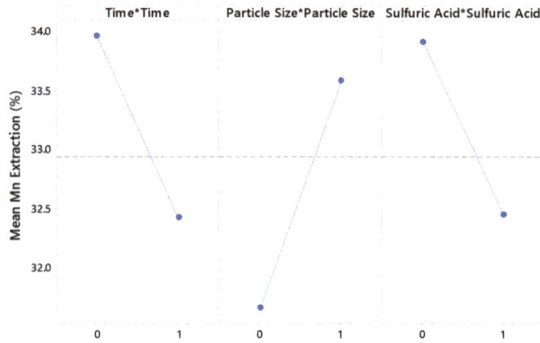

Figure 5. Curvature effect plot for Mn extraction.

After eliminating non-significant coefficients, the model developed to predict ore extraction over the range of experimental conditions is presented in Equation (3).

$$\text{Mn Extraction} = 0.3294 + 0.0704x_1 + 0.0746x_2 + 0.2036x_3 \tag{3}$$

where x_1, x_2, and x_3 are coded variables representing time, particle size, and H_2SO_4 concentration, respectively.

Figure 6 shows the order of adding parameters to the model, graphically showing the contribution to explaining the variability of each new parameter.

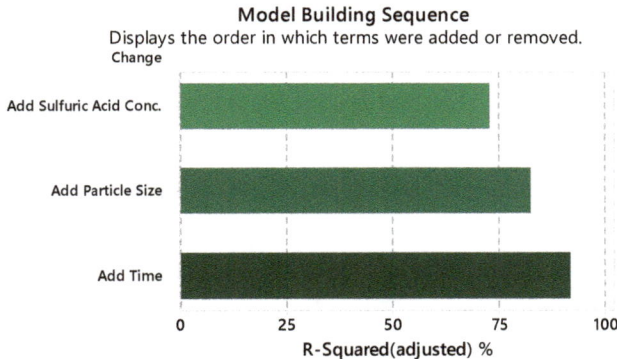

Figure 6. Construction sequence of the model.

The ANOVA indicates that the model adequately represents Mn extraction under the range of established parameters. The model does not require adjustment and is validated by the value of R^2 (0.9275) (Figure 7). The ANOVA shows that the effect of the indicated factors on manganese extraction is $F_{regression}$ (98.07) > F_{Table}, at the 95% confidence level $F_{4,22}$ (2.8167).

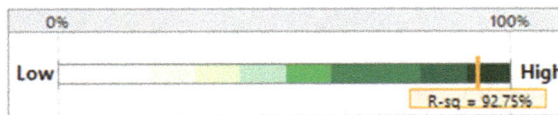

92.75% of the variation in Y can be explained by the regression model.

Figure 7. Statistic R^2 (% of variation explained by the model).

Additionally, the p-value (Figure 8) of the model represented by the Equation (3) indicates that the model is statistically significant.

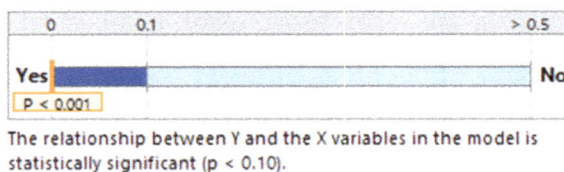

The relationship between Y and the X variables in the model is statistically significant (p < 0.10).

Figure 8. Statistic p.

In favor of the above analysis, the number of parameters plus the constant of the regression does not differ greatly from Mallows' Cp statistic, which indicates that the model is relatively accurate and does not present a bias in estimating the true coefficients of the regression, in addition to making predictions with an acceptable margin of error ($R_{pred} = 90.02\%$).

The data points from the normality test applied to the residuals resulting from the regression in Figure 9 are relatively close to the adjusted normal distribution line, and the p-value of the test is greater than the level of significance of 0.05, so it is not possible to reject the assumption of the regression model, that the residuals are distributed normally.

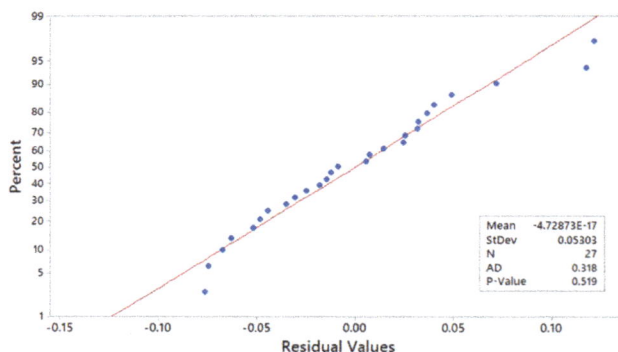

Figure 9. Probability plot of residual values.

Figure 10 shows that residuals do not correlate, indicating that they are independent of each other as there are no obvious trends or patterns.

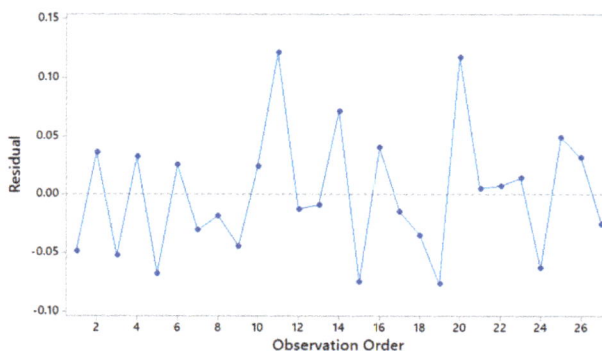

Figure 10. Residuals according to observations.

The response surface graphs in Figure 11A show that manganese extraction increases with time and particle size, while Figure 11B shows, graphically, that the effect of the variable H_2SO_4 concentration is greater than that of time, resulting in a more significant increase in extraction only when the acid concentration increases. The effect described above occurs analogously with variations in particle size and sulfuric acid concentration (Figure 11C).

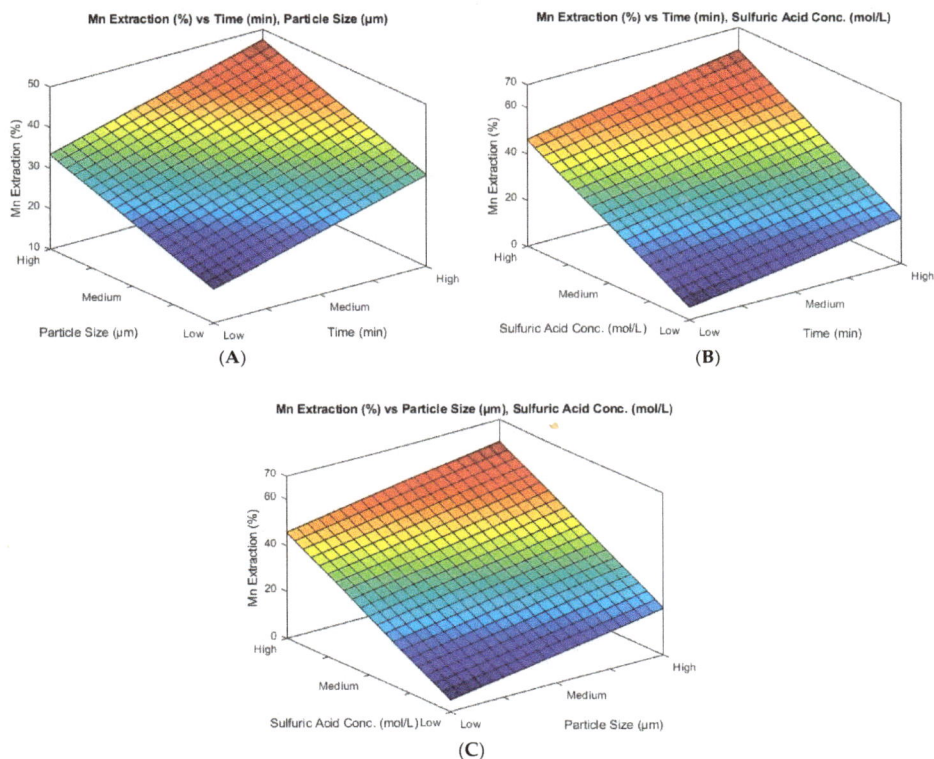

Figure 11. Response surface of the independent variables of time and particle size (**A**); time and H_2SO_4 concentration (**B**); and particle size and H_2SO_4 concentration (**C**) on the dependent variable of Mn extraction.

3.2. Effect of Agitation Speed

In Figure 12, it can be seen that higher Mn extraction rates are obtained at higher agitation speeds. In this study, the highest rate of 69% was obtained with a speed of 600 rpm and a time of 30 min. The extraction rate was lower at 800 and 1000 rpm because, at these speeds, some of the mineral breaks away and adheres to the reactor wall. Jiang et al. [13] had a similar observation at the speed of 1000 rpm. The extraction rate, at 400 rpm (58%), was not significantly different from what was obtained at 600 rpm, while at a low speed of 200 rpm, the Mn extraction rate was only 35% at 30 min. It was observed that not all the particles were in suspension at a stirring speed of 200 rpm, which explains why the extraction rate was so much lower. This is consistent with what Velásquez et al. [16] found in a study of leaching chalcopyrite mineral in chlorinated media. These authors concluded that agitation speed was not the most important factor in determining extraction rates as long as all the particles of the system are kept in suspension.

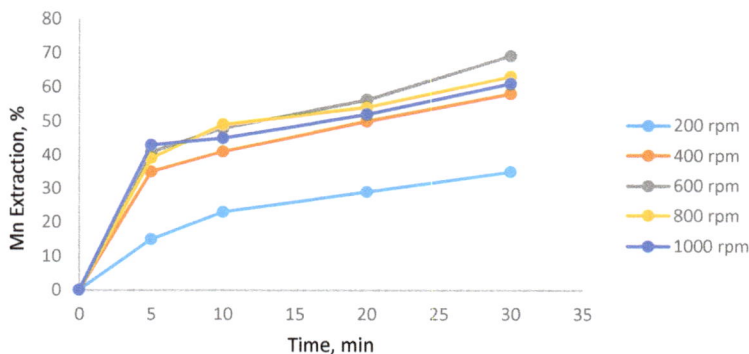

Figure 12. Effect of stirring speed on manganese extraction (25 °C, MnO_2/Fe_2O_3 ratio of 1, −75 + 53 μm, 1 mol/L H_2SO_4).

3.3. Effect of the MnO_2/Fe_2O_3 Ratio

The results presented in Figure 13 show the benefit of operating at high concentrations of reducing agent (Fe) in terms of shortening the dissolution time. The highest Mn extraction of 77% was obtained after 40 min with an MnO_2/Fe_2O_3 ratio of 1/2. Notably, at this MnO_2/Fe_2O_3 ratio, the leaching time required to reach a 70% extraction rate has been shortened significantly, while 67% extraction was reached in 5 min. However, the extraction graph shows asymptotic behavior, with no significant increase in the extraction rate vs. time. It can be observed that the extraction rate for 30 min with an MnO_2/Fe_2O_3 ratio of 1/1 is close to that obtained with a ratio of 1/2. However, the differences in dissolution rates are more significant for short periods of time (between 5 and 20 min). Finally, the Mn extraction rate was lower (maximum of 47% in 40 min) with an MnO_2/Fe_2O_3 ratio of 2/1 than with the ratios mentioned above. The tests conducted in this investigation were in pH ranges between −2 to 0.1, and potentials from −0.4 to 1.4 V, because the presence of Fe_2O_3 maintains the regeneration of ferrous ions, which results in high levels of ferrous ion concentration and activity, favoring the dissolution of Mn and avoiding the formation of precipitates through oxidation–reduction reactions.

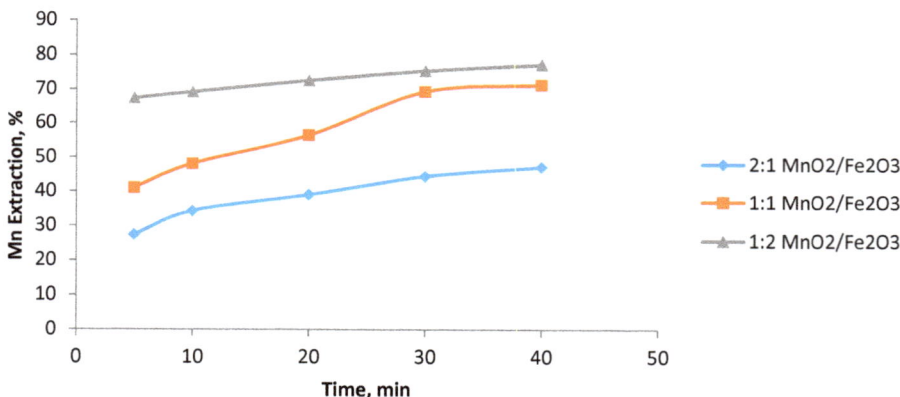

Figure 13. Effect of the MnO_2/Fe_2O_3 ratio on manganese extraction (25 °C, −75 + 53 μm, 1 mol/L H_2SO_4).

Table 6 compares the results using Fe present in slag and tailings as a reducing agent for Mn dissolution under the same operational conditions. In both cases, dissolution over a short period of time (5 min) immediately reached values close to 70%, with almost identical levels in the two investigations.

However, better results were obtained at 40 min using tailings instead of slag, although the difference is small (7%). This is possibly due to the presence of 13.07% Cu in the slag, which consumes protons. However, this issue requires additional study, since the reactivity of Cu in acid media is associated with slag mineralogy, and the presence of high silica often results in gel formation when leaching is carried out in low pH values [24]. In addition, the tailings are more reactive since they are derived from flotation, and have been attacked by chemicals, resulting in exposure of their surface [25]. These results are promising for future hydrometallurgical studies to investigate the use of slag and tailings as reducing agents for manganese ores. In future studies, it is proposed that this research be continued under the same operating parameters and by applying elemental iron (Fe^0) to determine if it is possible to achieve better results within short periods of time (5 min). In addition, when slag and tailings are used for Mn reduction, the effect of temperature should be evaluated to determine if it is possible to obtain 100% extraction in short periods of time. Finally, an optimal MnO_2/Fe_2O_3 ratio must be found.

Table 6. Comparison of the experimental results.

Experimental Conditions	Toro et al. [11] (2018)	Present Investigation
Temperature (°C)	25	25
Particle size of Mn nodules and slag/tailings (μm)	−75 + 53	−75 + 53
H_2SO_4 concentration (mol/L)	1	1
MnO_2/Fe_2O_3 ratio	1/2	1/2
Mn dissolution rate at 5 min (%)	68	67
Mn dissolution rate at 40 min (%)	70	77

For the recovery of Mn from the solution, the use of zerovalent iron (ZVI) is proposed. In a study by Bartzas et al. [26], the performance of a Fe^0 permeable reactive barrier (PRB) was evaluated for the treatment of acid leachates, where it was observed that metals, such as aluminum, manganese, nickel, cobalt, and zinc, were mainly removed from solution, as metal hydroxides, by precipitation. This can be an attractive proposal because zerovalent iron is a cheap byproduct obtained from the metal finishing industries.

Table 6 shows a comparison of the experimental results of Mn extraction from marine nodules with the use of slag and slag flotation tailings.

4. Conclusions

This investigation presents the results of dissolving Mn from marine nodules in an acid medium at room temperature (25 °C) with the use of tailings obtained from flotations of smelter slag. The Fe present in the tailings proved to be a good reducing agent, increasing MnO_2 dissolution kinetics. The findings of this study are as follows:

(1) The ANOVA test indicates that sulfuric acid is the factor that has the greatest impact on manganese extraction under the studied conditions.
(2) The manganese dissolution rate was generally higher when tailings were used instead of slag, possibly because tailings are more reactive to leaching.
(3) Increase of the agitation speed did not significantly increase Mn extraction.
(4) The highest Mn extraction rate of 77% was obtained at an MnO_2/Fe_2O_3 ratio 0.5, 1 mol/L H_2SO_4, particle size of −47 + 38 μm, and leaching time of 40 min.

In future work, the leaching of marine nodules should be studied using different Fe reducing agents but under the same operational conditions. It is also necessary to determine the optimal MnO_2/Fe_2O_3 ratio that improves dissolution of Mn. In addition, SEM studies need to be carried out on the tailings and manganese nodules after leaching, in order to observe their morphology and determine the possible formation of any iron precipitates.

Author Contributions: N.T. and M.S. contributed in the methodology, conceived and designed the experiments; analyzed the data and wrote paper, J.C. performed the experiments, F.H. contributed with resources and R.A. contributed with review and editing.

Funding: This research received no external funding.

Acknowledgments: The authors are grateful for the contribution of the Scientific Equipment Unit MAINI of the Universidad Católica del Norte for aiding in generating data by automated electronic microscopy QEMSCAN®and for facilitating the chemical analysis of the solutions. We are also grateful to the Altonorte Mining Company for supporting this research and providing slag for this study, and we thank to Marina Vargas Aleuy and María Barraza Bustos of the Universidad Católica del Norte for supporting the experimental tests. This research was supported by FCAC 2018-UCN.

Conflicts of Interest: The authors declare they have no conflict of interest.

References

1. Marino, E.; González, F.J.; Somoza, L.; Lunar, R.; Ortega, L.; Vázquez, J.T.; Reyes, J.; Bellido, E. Strategic and rare elements in Cretaceous-Cenozoic cobalt-rich ferromanganese crusts from seamounts in the Canary Island Seamount Province (northeastern tropical Atlantic). *Ore Geol. Rev.* **2017**, *87*, 41–61. [CrossRef]
2. Nishi, K.; Usui, A.; Nakasato, Y.; Yasuda, H. Formation age of the dual structure and environmental change recorded in hydrogenetic ferromanganese crusts from Northwest and Central Paci fi c seamounts. *Ore Geol. Rev.* **2017**, *87*, 62–70. [CrossRef]
3. Konstantinova, N.; Cherkashov, G.; Hein, J.R.; Mirão, J.; Dias, L.; Madureira, P.; Kuznetsov, V.; Maksimov, F. Composition and characteristics of the ferromanganese crusts from the western Arctic Ocean. *Ore Geol. Rev.* **2017**, *87*, 88–99. [CrossRef]
4. Usui, A.; Nishi, K.; Sato, H.; Nakasato, Y.; Thornton, B.; Kashiwabara, T. Continuous growth of hydrogenetic ferromanganese crusts since 17 Myr ago on Takuyo-Daigo Seamount, NW Pacific, at water depths of 800–5500 m. *Ore Geol. Rev.* **2017**, *87*, 71–87. [CrossRef]
5. Josso, P.; Pelleter, E.; Pourret, O.; Fouquet, Y.; Etoubleau, J.; Cheron, S.; Bollinger, C. A new discrimination scheme for oceanic ferromanganese deposits using high fi eld strength and rare earth elements. *Ore Geol. Rev.* **2017**, *87*, 3–15. [CrossRef]
6. Senanayake, G. Acid leaching of metals from deep-sea manganese nodules—A critical review of fundamentals and applications. *Miner. Eng.* **2011**, *24*, 1379–1396. [CrossRef]
7. Randhawa, N.S.; Hait, J.; Jana, R.K. A brief overview on manganese nodules processing signifying the detail in the Indian context highlighting the international scenario. *Hydrometallurgy* **2016**, *165*, 166–181. [CrossRef]
8. Kanungo, S.B. Rate process of the reduction leaching of manganese nodules in dilute HCl in presence of pyrite. Part I. Dissolution behaviour of iron and sulphur species during leaching. *Hydrometallurgy* **1999**, *52*, 313–330. [CrossRef]
9. Kanungo, S.B. Rate process of the reduction leaching of manganese nodules in dilute HCl in presence of pyrite. Part II: leaching behavior of manganese. *Hydrometallurgy* **1999**, *52*, 331–347. [CrossRef]
10. Zakeri, A.; Bafghi, M.S.; Shahriari, S.; Das, S.C.; Sahoo, P.K.; Rao, P.K. Dissolution kinetics of manganese dioxide ore in sulfuric acid in the presence of ferrous ion. *Hydrometallurgy* **2007**, *8*, 22–27.
11. Toro, N.; Herrera, N.; Castillo, J.; Torres, C.; Sepúlveda, R. Initial Investigation into the Leaching of Manganese from Nodules at Room Temperature with the Use of Sulfuric Acid and the Addition of Foundry Slag—Part I. *Minerals* **2018**, *8*, 565. [CrossRef]
12. Bafghi, M.S.; Zakeri, A.; Ghasemi, Z.; Adeli, M. Reductive dissolution of manganese ore in sulfuric acid in the presence of iron metal. *Hydrometallurgy* **2008**, *90*, 207–212. [CrossRef]
13. Jiang, T.; Yang, Y.; Huang, Z.; Zhang, B.; Qiu, G. Leaching kinetics of pyrolusite from manganese-silver ores in the presence of hydrogen peroxide. *Hydrometallurgy* **2004**, *72*, 129–138. [CrossRef]
14. Su, H.; Liu, H.; Wang, F.; Lü, X.; Wen, Y. Kinetics of reductive leaching of low-grade pyrolusite with molasses alcohol wastewater in H2SO4. *Chin. J. Chem. Eng.* **2010**, *18*, 730–735. [CrossRef]
15. Zhang, Y.; You, Z.; Li, G.; Jiang, T. Manganese extraction by sulfur-based reduction roasting-acid leaching from low-grade manganese oxide ores. *Hydrometallurgy* **2013**, *133*, 126–132. [CrossRef]
16. Velásquez Yévenes, L.; Miki, H.; Nicol, M. The dissolution of chalcopyrite in chloride solutions: Part 2: Effect of various parameters on the rate. *Hydrometallurgy* **2010**, *103*, 80–85. [CrossRef]
17. SERNAGEOMIN. *Anuario de la mineria de Chile 2017*; SERNAGEOMIN: Santiago, Chile, 2017.

18. COCHILCO. *Sulfuros primarios: desafíos y oportunidades I Comisión Chilena del Cobre*; COCHILCO: Santiago, Chile, 2017.

19. Oyarzun, R.; Oyarzún, J.; Lillo, J.; Maturana, H.; Higueras, P. Mineral deposits and Cu-Zn-As dispersion-contamination in stream sediments from the semiarid Coquimbo Region, Chile. *Environ. Geol.* **2007**, *53*, 283–294. [CrossRef]

20. Baba, A.A.; Ayinla, K.I.; Adekola, F.A.; Ghosh, M.K.; Ayanda, O.S.; Bale, R.B.; Sheik, A.R.; Pradhan, S.R. A Review on Novel Techniques for Chalcopyrite Ore Processing. *Int. J. Min. Eng. Miner. Process.* **2012**, *1*, 1–16. [CrossRef]

21. Centro de Estudios del Cobre y la Minería (CESCO). Available online: http://www.cesco.cl/en/home-en/ (accessed on 11 May 2019).

22. Montgomery, D.C. *Design and Analysis of Experiments*; Wiley: Hoboken, NJ, USA, 2012.

23. Bezerra, M.A.; Santelli, R.E.; Oliveira, E.P.; Villar, L.S.; Escaleira, L.A. Response surface methodology (RSM) as a tool for optimization in analytical chemistry. *Talanta* **2008**, *76*, 965–977. [CrossRef] [PubMed]

24. Komnitsas, K.; Zaharaki, D.; Perdikatsis, V. Effect of synthesis parameters on the compressive strength of low-calcium ferronickel slag inorganic polymers. *J. Hazard. Mater.* **2009**, *161*, 760–768. [CrossRef] [PubMed]

25. Komnitsas, K.; Manousaki, K.; Zaharaki, D. Assessment of reactivity of sulphidic tailings and river sludges. *Geochemistry Explor. Environ. Anal.* **2009**, *9*, 313–318. [CrossRef]

26. Bartzas, G.; Komnitsas, K.; Paspaliaris, I. Laboratory evaluation of Fe0 barriers to treat acidic leachates. *Miner. Eng.* **2006**, *19*, 505–514. [CrossRef]

![minerals logo] *minerals*

MDPI

Article

Optimization of Parameters for the Dissolution of Mn from Manganese Nodules with the Use of Tailings in An Acid Medium

Norman Toro [1,2,*], Manuel Saldaña [3], Edelmira Gálvez [1], Manuel Cánovas [1], Emilio Trigueros [2], Jonathan Castillo [4] and Pía C. Hernández [5]

1   Departamento de Ingeniería en Metalurgia y Minas, Universidad Católica del Norte, Av. Angamos 610, Antofagasta 1270709, Chile
2   Departamento de Ingeniería Minera y Civil. Universidad Politécnica de Cartagena, Paseo Alfonso Xlll N°52, 30203 Cartagena, Spain
3   Departamento de Ingeniería Industrial, Universidad Católica del Norte, Av. Angamos 610, Antofagasta 1270709, Chile
4   Departamento de Ingeniería en Metalurgia, Universidad de Atacama, Av. Copayapu 485, Copiapó 1531772, Chile
5   Departamento de Ingeniería Química y Procesos de Minerales, Universidad de Antofagasta, Avda. Angamos 601, Antofagasta 1240000, Chile
*   Correspondence: ntoro@ucn.cl; Tel.: +56-552651021

Received: 25 May 2019; Accepted: 24 June 2019; Published: 26 June 2019

Abstract: Manganese nodules are an attractive source of base metals and critical and rare elements and are required to meet a high demand of today's industry. In previous studies, it has been shown that high concentrations of reducing agent (Fe) in the system are beneficial for the rapid extraction of manganese. However, it is necessary to optimize the operational parameters in order to maximize Mn recovery. In this study, a statistical analysis was carried out using factorial experimental design for the main parameters, including time, MnO_2/Fe_2O_3 ratio, and H_2SO_4 concentration. After this, Mn recovery tests were carried out over time at different ratios of MnO_2/Fe_2O_3 and H_2SO_4 concentrations, where the potential and pH of the system were measured. Finally, it is concluded that high concentrations of $FeSO_4$ in the system allow operating in potential and pH ranges (-0.2 to 1.2 V and -1.8 to 0.1) that favor the formation of Fe^{2+} and Fe^{3+}, which enable high extractions of Mn (73%) in short periods of time (5 to 20 min) operating with an optimum MnO_2/Fe_2O_3 ratio of 1:3 and a concentration of 0.1 mol/L of H_2SO_4.

Keywords: leaching; manganese nodules; optimization of parameters; tailings

1. Introduction

The oxides of Fe and Mn are formed by direct precipitation in ambient seawater and are mainly deposited on the flat parts and the flanks of seamounts, where ocean currents prevent sedimentation [1,2]. These deposits are found in the oceans around the world [3] and among these are the manganese nodules [4].

The economic interest in ferromanganese (Fe-Mn) nodules is due to high grades of base, critical, and rare metals [5]. These metals that provide mineral deposits on the seabed are necessary for the rapid development of high technology application. They also support the growth and quality of life of the middle class in densely populated countries with expanding markets and economies [6]. Manganese is the most abundant marine nodule metal, with an average content of around 24% [7].

In order to dissolve Mn present in marine nodules in acidic media, it is necessary to use a reducing agent [8]. Studies have reported that the acid leaching of manganese nodules with the use of Fe as the reducing agent is efficient at room temperature [8–11]. In a study conducted by Zakeri et al. [8], ferrous ions were added for the reductive dissolution of manganese nodules. The authors indicated that in a molar H_2SO_4/MnO_2 ratio of 2:1 and a molar Fe^{2+}/MnO_2 ratio of 3:1, it was possible to dissolve 90% of Mn in 20 min at 20 °C. Bafgui et al. [9] performed acid leaching of manganese nodules by adding Fe, comparing their results with those previously obtained by Zakeri et al. [8]. The authors concluded that when operating with high Fe/MnO_2 ratios, Fe^0 is a more efficient reducing agent compared to Fe^{2+} because it maintains high activity in the system through the regeneration of ferrous ions.

For the acidic leaching of marine nodules with the use of residues (tailings and slags) containing Fe_2O_3, only two studies have been presented [10,11]. In the studies carried out by Toro et al. [10] and Toro et al. [11], it was shown that variables, such as particle size and agitation speed, do not majorly influence the dissolution of MnO_2 and that the most important variable is the Fe_2O_3 concentration in the system. In the study carried out by Toro et al. [10] with the use of smelter slag, extraction of 70% of Mn was achieved in 40 min when operating at a MnO_2/Fe_2O_3 ratio of 1:2, a particle size of $-47 + 38$ μm, and a H_2SO_4 concentration of 1 mol/L. In the later study carried out by Toro et al. [11], involving the use of tailings, it was demonstrated that under the same operating conditions as in Toro et al. [10], greater extraction of Mn (77%) was achieved because tailings are more amenable to leaching. In both studies, the following Reactions (R1)–(R9) involving the use of Fe_2O_3 were proposed.

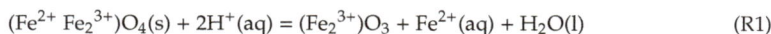

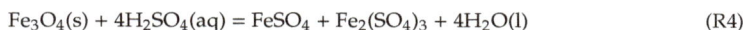

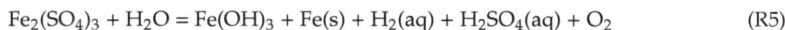

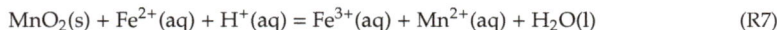

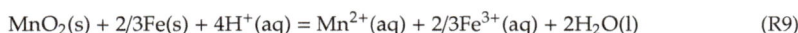

$$(Fe^{2+} Fe_2^{3+})O_4(s) + 2H^+(aq) = (Fe_2^{3+})O_3 + Fe^{2+}(aq) + H_2O(l) \tag{R1}$$

$$3(Fe^{2+} Ti)O_3(s) + 6H^+(aq) = 3TiO_2 + 3Fe^{2+}(aq) + 3H_2O(l) \tag{R2}$$

$$Fe_2O_3 + H_2SO_4 = Fe_2(SO_4)_3 + H_2O(l) \tag{R3}$$

$$Fe_3O_4(s) + 4H_2SO_4(aq) = FeSO_4 + Fe_2(SO_4)_3 + 4H_2O(l) \tag{R4}$$

$$Fe_2(SO_4)_3 + H_2O = Fe(OH)_3 + Fe(s) + H_2(aq) + H_2SO_4(aq) + O_2 \tag{R5}$$

$$FeSO_4(s) + H_2O(aq) = Fe(s) + H_2SO_4(aq) + O_2 \tag{R6}$$

$$MnO_2(s) + Fe^{2+}(aq) + H^+(aq) = Fe^{3+}(aq) + Mn^{2+}(aq) + H_2O(l) \tag{R7}$$

$$MnO_2(s) + Fe(s) + 8H^+(aq) = 2Fe^{3+}(aq) + Mn^{2+}(aq) + 2H_2O(l) + 2H_2(g) \tag{R8}$$

$$MnO_2(s) + 2/3Fe(s) + 4H^+(aq) = Mn^{2+}(aq) + 2/3Fe^{3+}(aq) + 2H_2O(l) \tag{R9}$$

However, in previous studies [8,9], thermodynamic aspects were not considered.

Table 1 reports the statistical information of the reactions of interest with iron as a reducing agent and its transformations during manganese leaching. It is emphasized that, unlike previous investigations [11], under these conditions, elemental iron (Fe^0) was not formed, since this reaction is not spontaneous (G = 744.22 kJ) and requires a lot of energy. On the other hand, the main reducing agent is ferrous sulfate ($FeSO_4$), which is produced from the reaction between magnetite (Fe_3O_4, mostly present in tailings) and sulfuric acid (Equation (2)). With this reducing agent, it is possible to reduce manganese present in pyrolusite (Mn^{4+}), obtaining a manganese sulphate (Mn^{2+}), as observed in Equation (5).

Table 1. Thermodynamic information of the reactions (based on HSC Chemistry 5.1).

Reaction	Equation	Reaction Coefficient (K) 25 °C	$\Delta G°$ (kJ) 25 °C
$Fe_2O_3(s) + 3 H_2SO_4(aq) = Fe_2(SO_4)_3(s) + 3 H_2O(l)$	(1)	4.21×10^{28}	−163,37
$Fe_3O_4(s) + 4H_2SO_4(l) = FeSO_4(aq) + Fe_2(SO_4)_3(s) + 4 H_2O(l)$	(2)	6.06×10^{45}	−261,30
$Fe_2(SO_4)_3(s) + 6 H_2O(l) = 2 Fe(OH)_3(s) + 3 H_2SO_4(l)$	(3)	2.14×10^{-35}	197,87
$2 FeSO_4(aq) + 2 H_2O(l) = 2 Fe(s) + 2 H_2SO_4(l) + O_2(g)$	(4)	4.02×10^{-131}	744,22
$2 FeSO_4(aq) + 2 H_2SO_4(aq) + MnO_2(s) = Fe_2(SO_4)_3(s) + 2 H_2O(l) + MnSO_4(aq)$	(5)	9.06×10^{34}	−199,52

The smelting slag is one of the main solid wastes of the copper industry and the produced volume increases day by day [12]. In Chile, the smelters produce 163 tons of slag per day [13] and companies such as Altonorte perform slag flotation for the recovery of Cu. During flotation for each ton of Cu obtained, 151 tons of tailings are generated [14], which are mainly disposed of in tailing dams and represent the most significant environmental liability according to their size and risk of a mining site [15]. Another example is what happened in Lavrio, Greece, due to the intensive mining and metallurgical activities in the last century. This generated huge amount of waste, including acid-generating sulfidic tailings, carbonaceous tailings, and slags. Quantification of the human health risks indicated that direct ingestion of contaminated particles is the most important exposure route for the intake of contaminants by humans [16]. Komnitsas et al. [17] conducted research on waste generated by intensive mining and mineral processing activities in Navodari and Baia, on the Romanian Black Sea coast. Analyzing the experimental results and the associated risks, the authors conclude that it is necessary to rehabilitate the affected areas through removal of toxic and heavy elements from sulphidic tailings and leachates with biosorption and biosolubilisation techniques and development of a vegetative cover on phosphogypsnm stacks and sulphidic tailings dumps. For this reason, it is important to highlight the importance of studying options to reuse waste generated from metallurgical processes.

In Chile, ocean mining is not regulated and is also under-exploited for security reasons [18]. Due to this, it is not possible to carry out a cost-effectiveness study on the extraction of nodules from sea depths. Mining technologies have been developed in the world for the extraction of polymetallic nodules [19]. However, there is no study indicating cost differences between the different methods available on the market. In spite of this, it is necessary to continue investigating processes for the extraction of elements from marine nodules, because technologies are being developed to collect minerals from sea beds and, in the near future, they could be considered as viable alternatives to meet the high demand for metals.

In this investigation, the use of Fe_2O_3, which is present in tailings, to facilitate reductive leaching of MnO_2 from marine nodules for the recovery of Mn is evaluated. The objective is to minimize these environmental liabilities and optimize the most important process variables (time, acid concentration, and MnO_2/Fe_2O_3 ratio).

2. Materials and Methods

2.1. Manganese Nodule Sample

The marine nodules used in this work were the same as those previously used in Toro et al. [11]. They were analyzed by means of atomic emission spectrometry by induction-coupled plasma (ICP-AES), developed in the applied geochemistry laboratory of the department of geological sciences of the Catholic University of the North. They contained 15.96% Mn and 0.45% Fe; Mn was present as MnO_2 (29.85%) and Fe as Fe_2O_3 (26.02%).

2.2. Tailings

The tailings used for the present investigation were the same as those used in Toro et al. [11]. The methods used to determine their chemical and mineralogical composition are the same as those used for the analysis of the manganese nodules. Figure 1 shows the chemical species determined by QEMSCAN. There were several phases that contained iron (mainly magnetite (58.52%) and hematite (4.47%), while the content of Fe was estimated at 41.90%.

MODAL MINERALOGY

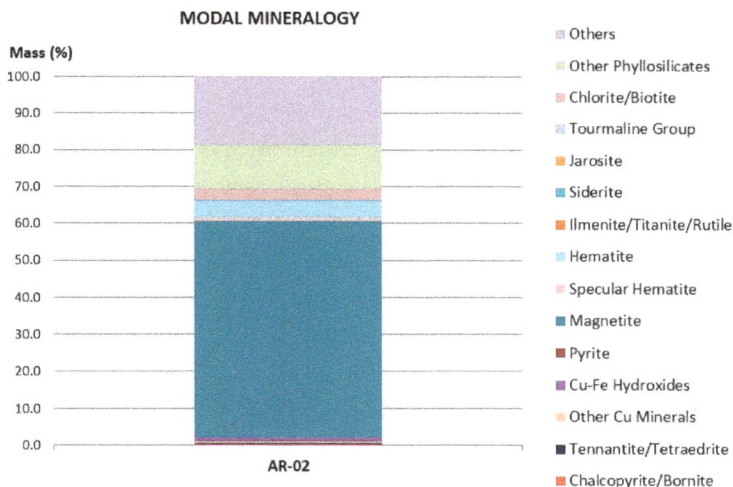

Figure 1. Detailed modal mineralogy.

2.3. Reagents Used—Leaching Parameters

The sulfuric acid used for the leaching tests was grade PA, with 95–97% purity, a density of 1.84 kg/L, and a molecular weight of 98.80 g/mol. The leaching tests were carried out in a 50 mL glass reactor with a 0.01 solid:liquid ratio. A total of 200 mg of Mn nodules were maintained in suspension with the use of a 5 position magnetic stirrer (IKA ROS, CEP 13087-534, Campinas, Brazil) at a speed of 600 rpm. The tests were conducted at a room temperature of 25 °C, while the studied variables were additives, particle size, and leaching time. Also, the tests were performed in duplicate and measurements (or analyses) were carried out on 5 mL undiluted samples using atomic absorption spectrometry with a coefficient of variation ≤5% and a relative error between 5% and 10%. Measurements of pH and oxidation-reduction potential (ORP) of leach solutions were made using a pH-ORP meter (HANNA HI-4222). The ORP solution was measured using a combination of an ORP electrode cell composed of a platinum operating electrode and a saturated Ag/AgCl reference electrode. The solid waste obtained was analyzed by XRD with the use of a Bruker brand diffractometer; the patterns of the main crystalline phases were obtained using Eva software.

2.4. Estimation of Linear and Interaction Coefficients for Complete Factorial Designs of Experiments of 3^3

In previous studies [8–11], in which the dissolution of Mn from marine nodules was investigated with the use of Fe as a reducing agent, it was demonstrated that for high concentrations of Fe in the system (ratios of Fe/MnO_2 greater than 1), quite high extractions were obtained (over 70%) in short periods of time (5 to 30 min). The studies conducted by Bafghi et al. [9] and Toro et al. [10] indicated that the concentration of Fe in the system is the most important variable in order to shorten MnO_2 dissolution times; it was also found that the concentration of H_2SO_4 is not an important parameter. However, in these studies, it was not possible to indicate an optimum MnO_2/Fe ratio and H_2SO_4 concentration in relation to time. In order to overcome this and elucidate Mn extraction from marine nodules, three independent variables were selected for the factorial design of 3^3 experiments, namely time, sulfuric acid concentration, and MnO_2/Fe_2O_3 ratio. This approach allows the determination of the effect of the most relevant factors, as well as their levels, and the development of an experimental model that allows through the determination of coefficients the optimization of the response variable [20–22]. A factorial design was applied involving three factors, each one having three levels; thus, 27 experimental tests were carried out. Minitab 18 software was used for the experimental design and development of a multiple regression equation [23].

Then, the response variable was expressed based on the linear effect of the variables of interest and considering the effects of interaction and curvature, as shown in Equation (6).

$$\text{Mn Extraction (\%)} = \alpha + \beta_1 \times x_1 + \beta_2 \times x_2 + \beta_3 \times x_3 + \beta_{12} \times x_1 \times x_2 + \beta_{13} \times x_1 \times x_3 +$$
$$\beta_{23} \times x_2 \times x_3 + \beta_{11} \times x_1{}^2 + \beta_{22} \times x_2{}^2 + \beta_{33} \times x_3{}^2, \tag{6}$$

where α is an overall constant, x_i is the value of the level "i" of the factor x, β_i is the coefficient of the linear factor x_i, β_{ij} is the coefficient of the interactions $x_i \times x_j$, n is the level of the factor, and Mn extraction is the dependent variable.

Table 2 shows the values of the levels for each factor, while Table 3 shows the recovery obtained for each configuration.

Table 2. Experimental conditions.

Parameters/Values	Low	Medium	High
Time (min)	10	20	30
MnO_2/Fe_2O_3	2/1	1/1	1/2
H_2SO_4 (mol/L)	0.1	0.5	1
Codifications	−1	0	1

Table 3. Experimental configuration and Mn extraction.

Exp. No.	Time (min)	MnO_2/Fe_2O_3 Ratio	Sulfuric Acid Conc. (mol/L)	Mn Extraction (%)
1	10	1/1	0.1	48.42
2	20	2/1	0.5	38.78
3	20	1/1	1	57.32
4	30	2/1	1	42.55
5	10	1/2	0.5	70.24
6	20	2/1	0.1	38.10
7	30	1/1	1	72.96
8	30	1/2	0.1	74.20
9	10	2/1	0.5	33.23
10	10	2/1	1	33.33
11	20	1/1	0.5	56.80
12	30	2/1	0.1	42.30
13	20	1/1	0.1	55.95
14	10	2/1	0.1	32.83
15	10	1/1	1	50.23
16	10	1/2	0.1	70.21
17	20	1/2	0.1	73.20
18	20	1/2	0.5	73.20
19	30	1/1	0.1	71.96
20	30	1/1	0.5	72.33
21	10	1/2	1	70.90
22	20	1/2	1	73.40
23	20	2/1	1	39.22
24	30	1/2	0.5	74.90
25	10	1/1	0.5	48.91
26	30	1/2	1	75.21
27	30	2/1	0.5	42.00

3. Results and Discussion

3.1. Effect of Variables

From the principal components analysis, it is seen that there is no main effect of the sulfuric acid concentration factor, which means that the average response is the same across all levels of the factor, while the time and MnO_2/Fe_2O_3 ratio factors have a main effect since the variation between the different levels affects the response differently, as shown in main effects plot for Mn extraction of Figure 2. Developing the ANOVA test and the multiple linear regression adjustment, the recovery according to the predictive variables of time and MnO_2/Fe_2O_3 ratio is given by Equation (7).

$$\text{Mn Extraction (\%)} = 53.90 + 6.12 \times x_1 - 17.40 \times x_2 - 4.00 \times x_2^2, \tag{7}$$

where x_1 represents the time factor and x_2 represents the MnO_2/Fe_2O_3 ratio (previous coding). Then, it is seen that the double and triple interaction factors, together with the curvature of time and H_2SO_4 concentration factors, do not contribute to the explanation of the variability of the model.

A gradient analysis of manganese extraction, ∇Mn Extraction $(x_1, x_2) = (6.12, -23.40)$, indicates an increase in the positive direction of the predictor variables. The response decreases faster with respect to the variable x_2 than with respect to the variable x_1, as shown in Figure 3.

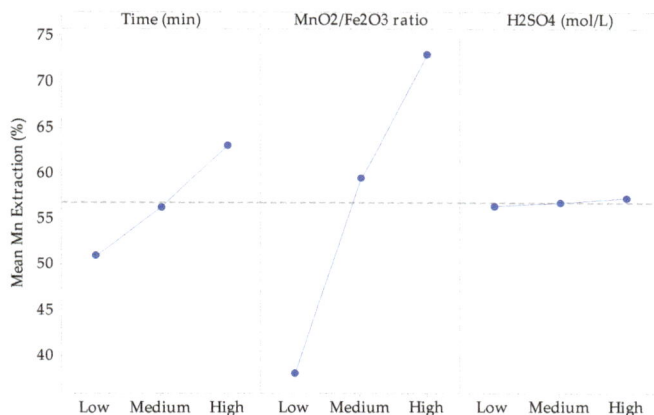

Figure 2. Linear effect plot for Mn extraction (%).

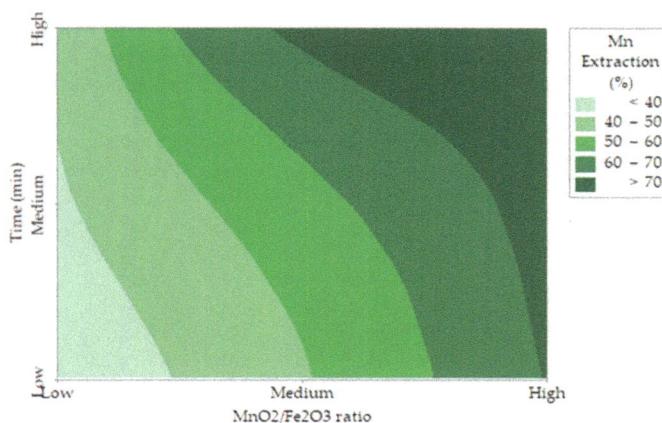

Figure 3. Contour plot of Mn extraction (%) versus MnO_2/Fe_2O_3, time (min).

The ANOVA test indicates that the model adequately represents manganese extraction for the set of sampled values. Also, the model does not require additional adjustments and is validated by the following goodness-of-fit statistics. The p-value of the model ($p < 0.05$) indicates that it is statistically significant. The value of the R^2 statistic is 94.94%, which indicates that approximately 95% of the total variability is explained by the model, while the predictive R^2 is 92.79%, indicating that the model can adequately predict the response to new observations. The F test indicates the significance of the model, given that $F_{Regression}(143.89) >> (F_{Table} = F_{3,23}(3.03))$, while the residual normality test indicates that these are distributed with media -7.41×10^{-7}, with a standard deviation of 3.5. The p-value of the Kolmogorov-Smirnov test is greater than the level of significance, so it is not possible to reject the assumption of the regression model, which is that the residuals are normally distributed.

3.2. Effect on Acid Concentration and MnO₂/Fe₂O₃ Ratio

In Figure 4, it can be seen that the largest extractions of Mn from marine nodules are obtained when operating at MnO_2/Fe_2O_3 ratios less than 1:1, which agrees with the theories proposed by Kanungo et al. [24], Zakeri et al. [8], Bafghi et al. [9], and Toro et al. [10], which mentioned that the presence of more Fe than MnO_2 in the system improved Mn dissolution in short periods of time. The highest Mn dissolution (76.10%) was obtained by operating at a MnO_2/Fe_2O_3 ratio of 1:3 with a H_2SO_4 concentration of 1 mol/L at 30 min. However, this extraction is not far from that obtained when operating at a MnO_2/Fe_2O_3 ratio of 1:2 (75.50%) at the same acid concentration. For the MnO_2/Fe_2O_3 ratios described in Figure 4b–d, it can be seen that, for leaching times of 30 min, very similar values are obtained, but much higher dissolution kinetics are seen in 1:2 and 1:3 ratios where higher than 65% recoveries of Mn are obtained after 5 min. For a MnO_2/Fe_2O_3 ratio of 2:1 (Figure 4a), much lower dissolution is obtained compared to the other cases under the same operating conditions; the maximum dissolution achieved is 48.30% after 30 min when the H_2SO_4 concentration is 1 mol/L.

The concentration of H_2SO_4 in the system is not significant when MnO_2/Fe_2O_3 ratios are 1:2 and 1:3; this finding agrees with those of Toro et al. [10], who mention that high concentrations of Fe_2O_3 in the system are independent of the acid concentration. However, it can be observed that this factor has a greater impact as long as there is a lower concentration of reducing agent ($FeSO_4$) in the system. For a MnO_2/Fe_2O_3 ratio of 2:1, there is a difference of 3.30% between 0.1 and 1 mol/L of H_2SO_4.

For these two variables analyzed under the exposed operational parameters, it can be observed that when operating at a MnO_2/Fe_2O_3 ratio of 1:2, at a concentration of H_2SO_4 of 0.5 mol/L at 20 min, similar results like those obtained when operating in a MnO_2/Fe_2O_3 ratio of 1:3 (73.50% approximately) are obtained. This is consistent with what was proposed by Toro et al. [11], who indicated that, when operating at high concentrations of Fe_2O_3 in the system, the dissolution kinetics of MnO_2 were drastically increased and significant differences were only observed in short periods of time (5 to 10 min). However, for the second case mentioned in Figure 4d, better results are obtained at low acid concentrations (0.1 mol/L). For this reason, it can be concluded that it is more convenient to operate at MnO_2/Fe_2O_3 ratios of 1:3 and low concentrations of acid (0.1 mol/L) at 20 min. This is because the tailings are wastes that do not have a commercial value and their reuse is beneficial, while the increase of the acid concentration in the system results in a direct increase in the cost of the process.

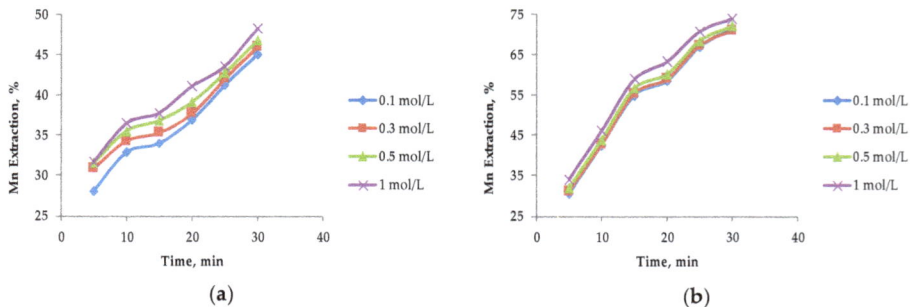

Figure 4. *Cont.*

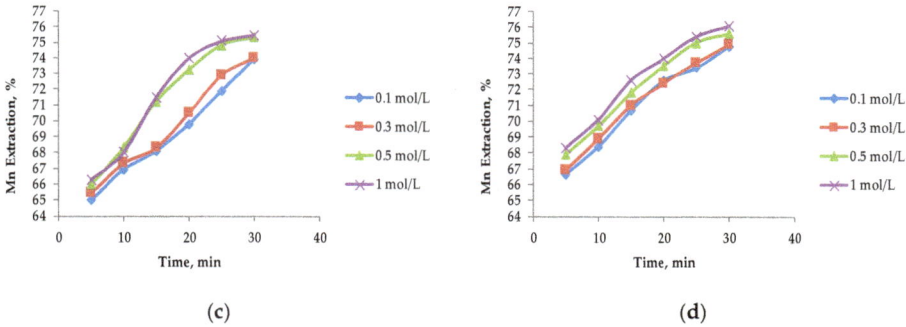

(c)   (d)

Figure 4. Effect on the acid concentration at different ratios of MnO$_2$/Fe$_2$O$_3$: (**a**) Ratio 2:1; (**b**) ratio 1:1; (**c**) ratio 1:2; and (**d**) ratio 1:3 (25 °C, 600 rpm, −75 + 53 µm).

Based on the results presented in Figure 4, Figure 5 shows that an optimum MnO$_2$/Fe$_2$O$_3$ ratio can be determined. It can be seen that there is no difference in manganese extractions when operating at 1:3 and 1:4 ratios. For short periods of time (5 to 20 min), it can be observed that there are greater extractions for ratios higher than 1:2, achieving dissolutions of Mn over 70% at 15 min. However, it can be seen that at 30 min the results converge in extractions of approximately 75%. Finally, it can be indicated that for times between 15 to 25 min, it is convenient to operate at a MnO$_2$/Fe$_2$O$_3$ ratio of 1:3, while at 30 min, the optimum ratio is 1:2. Figure 6 shows the potential and pH values obtained in the tests presented in Figure 5, which vary between (−0.2 V to 1.2 V) and (−1.8 to 0.1), respectively.

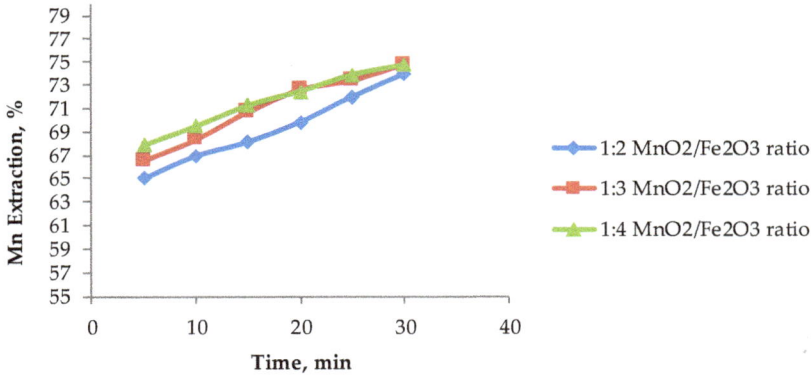

Figure 5. Effect of the MnO$_2$/Fe$_2$O$_3$ ratio on manganese extraction (25 °C, 600 rpm, −75 + 53 µm, acid concentration of 0.1 mol/L).

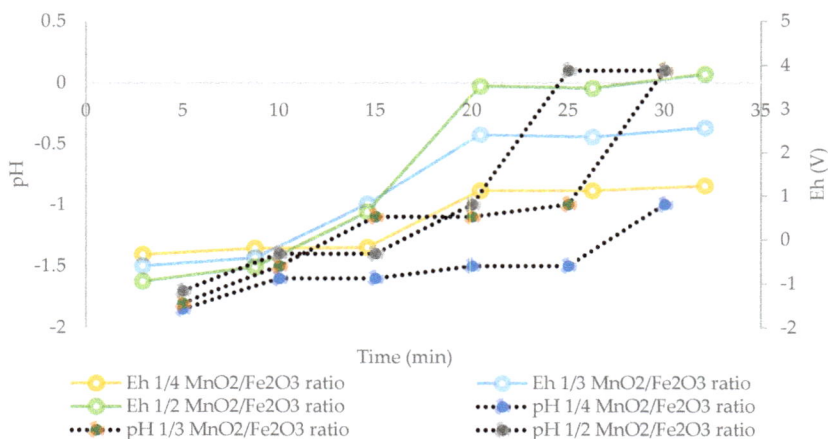

Figure 6. Effect of the potential and pH in solution at different MnO_2/Fe_2O_3 ratios (25 °C, 600 rpm, −75 + 53 μm, acid concentration of 0.1 mol/L).

Senanayake [25] stated that during reductive leaching of MnO_2 with the use of $FeSO_4$ as a reducing agent, the values of potential and pH must be in the range of −0.4 to 1.4 V and −2 to 0.1 in order to dissolve Mn. In addition, it is indicated that the divalent Fe (II), produced by the partial acid dissolution of Fe_3O_4, acts as a reducing agent for MnO_2. Under these conditions, Mn ions remain in solution and do not precipitate through oxidation-reduction reactions by the presence of Fe^{2+} and Fe^{3+} ions [26]. This can be seen in Figure 7, when analyzing the residues of the present study by XRD analysis and mainly the presence of fayalite ($Fe_2^{+2} SiO_4$), magnetite (Fe_3O_4), and gypsum ($CaSO_4 \cdot 2H_2O$) is observed. It is concluded that, in these residues, no Fe precipitates were generated from the solution when tailings were added. In future studies, it may be interesting to perform a kinetic study to elucidate the effect of temperature in order to determine the Mn dissolution mechanisms from marine nodules in very short periods of time (5 min).

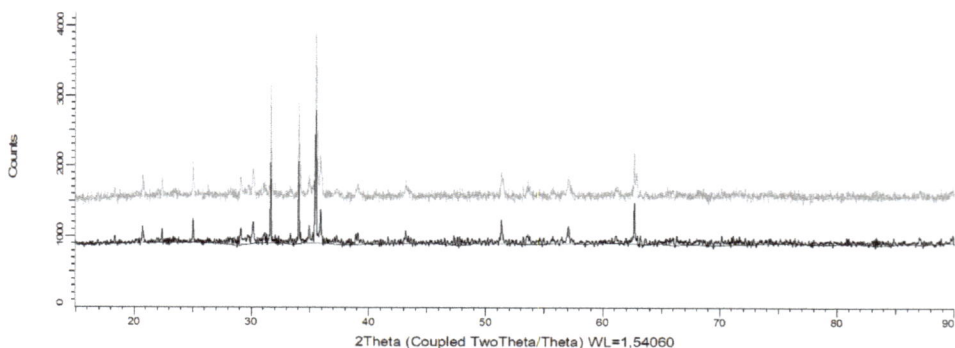

Figure 7. X-ray diffractogram of solid waste after leaching for 30 min at 25 °C using a MnO_2/Fe_2O_3 ratio of 1:2, 600 rpm, −75 + 53 μm, and an acid concentration of 0.1 mol/L.

4. Conclusions

The present study shows results by means of a statistical model as well as extraction curves versus time to investigate the extraction of Mn from MnO_2 present in manganese nodules using tailings obtained from slag flotation when operating in an acid medium and a room temperature of 25 °C.

Minerals **2019**, *9*, 387

$FeSO_4$ proves to be a good reducing agent, shortening the dissolution time of MnO_2. The main findings are the following:

1. At 30 min, the optimum MnO_2/Fe_2O_3 ratio is 1:2, with a H_2SO_4 concentration of 0.1 mol/L, achieving a Mn extraction of 74%.
2. For short periods of time (5 to 20 min), the optimum MnO_2/Fe_2O_3 ratio is 1:3, with a H_2SO_4 concentration of 0.1 mol/L, achieving a Mn extraction between 68% and 73%.
3. When operating at MnO_2/Fe_2O_3 ratios lower than 1:1, the concentration of acid in the system is not an important factor.
4. High concentrations of $FeSO_4$ in the system allow the operation in potential and pH ranges, which favor the generation of Fe^{2+} and Fe^{3+}; thus, the formation of Fe precipitates is avoided.

The reductive leaching of marine nodules in an acidic medium with the addition of tailings is an attractive and cost efficient alternative and results in high extraction of Mn in short periods of time with the use of low concentrations of acid. In the future, a study should be carried out to improve the economic viability of the process.

Author Contributions: N.T. contributed in project administration, investigation and wrote paper, M.S. and E.G. contributed in the data curation and software, M.C. and E.T. contributed in validation and supervision and J.C and P.C.H. performed the experiments, review and editing.

Funding: This research received no external funding.

Acknowledgments: The authors are grateful for the contribution of the Scientific Equipment Unit- MAINI of the Universidad Católica del Norte for aiding in generating data by automated electronic microscopy QEMSCAN®, and for facilitating the chemical analysis of the solutions. We are also grateful to the Altonorte Mining Company for supporting this research and providing slag for this study, and we thank Marina Vargas Aleuy, María Barraza Bustos and Carolina Ossandón Cortés of the Universidad Católica del Norte for supporting the experimental tests.

Conflicts of Interest: The authors declare they have no conflict of interest.

References

1. Konstantinova, N.; Cherkashov, G.; Hein, J.R.; Mirão, J.; Dias, L.; Madureira, P.; Kuznetsov, V.; Maksimov, F. Composition and characteristics of the ferromanganese crusts from the western Arctic Ocean. *Ore Geol. Rev.* **2017**, *87*, 88–99. [CrossRef]
2. Hein, J.R.; Koschinsky, A. Deep-Ocean Ferromanganese Crusts and Nodules. In *Treatise on Geochemistry*; Elsevier: Amsterdam, The Netherlands, 2014; Volume 13, pp. 273–291.
3. Marino, E.; González, F.J.; Somoza, L.; Lunar, R.; Ortega, L.; Vázquez, J.T.; Reyes, J.; Bellido, E. Strategic and rare elements in Cretaceous-Cenozoic cobalt-rich ferromanganese crusts from seamounts in the Canary Island Seamount Province (northeastern tropical Atlantic). *Ore Geol. Rev.* **2017**, *87*, 41–61. [CrossRef]
4. Josso, P.; Pelleter, E.; Pourret, O.; Fouquet, Y.; Etoubleau, J.; Cheron, S.; Bollinger, C. A new discrimination scheme for oceanic ferromanganese deposits using high fi eld strength and rare earth elements. *Ore Geol. Rev.* **2017**, *87*, 3–15. [CrossRef]
5. Cronan, D.S. Cobalt-rich ferromanganese crusts in the Pacific. In *Handbook of Marine Mineral Deposits*; CRC Press: Boca Raton, FL, USA, 2000; pp. 239–279.
6. Hein, J.R.; Cherkashov, G.A. Preface for Ore Geology Reviews Special Issue: Marine Mineral Deposits: New resources for base, precious, and critical metals. *Ore Geol. Rev.* **2017**. [CrossRef]
7. Sharma, R. Environmental Issues of Deep-Sea Mining. *Procedia Earth Planet. Sci.* **2015**, *11*, 204–211. [CrossRef]
8. Zakeri, A.; Bafghi, M.S.; Shahriari, S.; Das, S.C.; Sahoo, P.K.; Rao, P.K. Dissolution kinetics of manganese dioxide ore in sulfuric acid in the presence of ferrous ion. *Hydrometallurgy* **2007**, *8*, 22–27.
9. Bafghi, M.S.; Zakeri, A.; Ghasemi, Z.; Adeli, M. Reductive dissolution of manganese ore in sulfuric acid in the presence of iron metal. *Hydrometallurgy* **2008**, *90*, 207–212. [CrossRef]
10. Toro, N.; Herrera, N.; Castillo, J.; Torres, M.C.; Sepúlveda, R.S. Initial Investigation into the Leaching of Manganese from Nodules at Room Temperature with the Use of Sulfuric Acid and the Addition of Foundry Slag—Part I. *Minerals* **2018**, *8*, 565. [CrossRef]

11. Toro, N.; Saldaña, M.; Castillo, J.; Higuera, F.; Acosta, R. Leaching of Manganese from Marine Nodules at Room Temperature with the Use of Sulfuric Acid and the Addition of Tailings. *Minerals* **2019**, *9*, 289. [CrossRef]

12. Alejandra, C.; Estay, S. *Utilización de Escorias de Fundición para la Producción de Compuestos de Hierro*; Universidad de Chile: Santiago, Chile, 2006.

13. Vásquez, M. En Chile Diariamente se Desecha Cobre Avaluado en una Cifra Cercana a Los 450 Mil dólares. 2019. Available online: https://www.pucv.cl/uuaa/vriea/noticias/nuestros-investigadores/en-chile-diariamente-se-desecha-cobre-avaluado-en-una-cifra-cercana-a/2016-08-05/124009.html (accessed on 7 April 2019).

14. COCHILCO. *Sulfuros primarios: Desafíos y oportunidades I Comisión Chilena del Cobre*; COCHILCO: San Diego, Chile, 2017.

15. Medvinsky-Roa, G.; Caroca, V.; Vallejo, J. Informe sobre la situación de los Relaves Mineros en Chile para ser presentado en el cuarto informe periódico de Chile para el Comité de Derechos Económicos, Sociales y Culturales, perteneciente al consejo Económico Social de la Naciones Unidas. Providencia. 2015. Available online: https://tbinternet.ohchr.org/Treaties/CESCR/SharedDocuments/CHL/INT_CESCR_CSS_CHL_20605_S.pdf (accessed on 7 April 2019).

16. Xenidis, A.; Papassiopi, N.; Komnitsas, K. Carbonate-rich mining tailings in Lavrion: Risk assessment and proposed rehabilitation schemes. *Adv. Environ. Res.* **2003**, *7*, 207–222. [CrossRef]

17. Komnitsas, K.; Kontopoulos, A.; Lazar, I.; Cambridge, M. Risk assessment and proposed remedial actions in coastal tailings disposal sites in Romania. *Miner. Eng.* **1998**, *11*, 1179–1190. [CrossRef]

18. SERNAGEOMIN. *Anuario de la Minería de Chile 2017*; SERNAGEOMIN: San Diego, Chile, 2017.

19. ISA. *Polymetallic Nodule Mining Technology: Current Trends and Challenges Ahead*; ISA: Chennai, India, 2008; p. 276.

20. Douglas, C. *Montgomery: Design and Analysis of Experiments*, 8th ed.; John Wiley & Sons: New York, NY, USA, 2012.

21. Ghosh, M.K.; Barik, S.P.; Anand, S. Sulphuric acid leaching of polymetallic nodules using paper as a reductant. *Trans. Indian Inst. Met* **2008**, *61*, 477–481. [CrossRef]

22. Mitić, M.; Tošić, S.; Pavlović, A.; Mašković, P.; Kostić, D.; Mitić, J.; Stevanović, V. Optimization of the extraction process of minerals from Salvia officinalis L. using factorial design methodology. *Microchem. J.* **2019**, *145*, 1224–1230. [CrossRef]

23. Mathews, P.G.; William, A. *Design of Experiments with MINITAB*; William A. Tony: Milwaukee, WI, USA, 2005.

24. Kanungo, S.B. Rate process of the reduction leaching of manganese nodules in dilute HCl in presence of pyrite. Part I. Dissolution behaviour of iron and sulphur species during leaching. *Hydrometallurgy* **1999**, *52*, 313–330. [CrossRef]

25. Senanayake, G. Acid leaching of metals from deep-sea manganese nodules—A critical review of fundamentals and applications. *Miner. Eng.* **2011**, *24*, 1379–1396. [CrossRef]

26. Komnitsas, K.; Bazdanis, G.; Bartzas, G.; Sahinkaya, E.; Zaharaki, D. Removal of heavy metals from leachates using organic/inorganic permeable reactive barriers. *Desalin. Water Treat.* **2013**, *51*, 3052–3059. [CrossRef]

minerals

MDPI

Article

Consideration of Influential Factors on Bioleaching of Gold Ore Using Iodide-Oxidizing Bacteria

San Yee Khaing [1], Yuichi Sugai [2,*], Kyuro Sasaki [2] and Myo Min Tun [3]

[1] Department of Earth Resources Engineering, Graduate School of Engineering, Kyushu University, Fukuoka 8190395, Japan; poeoo.34@mine.kyushu-u.ac.jp
[2] Department of Earth Resources Engineering, Faculty of Engineering, Kyushu University, Fukuoka 8190395, Japan; krsasaki@mine.kuyushu-u.ac.jp
[3] Department of Geology, Yadanabon University, Mandalay 05063, Myanmar; mintunmyo@gmail.com
*   Correspondence: sugai@mine.kyushu-u.ac.jp

Received: 11 April 2019; Accepted: 30 April 2019; Published: 2 May 2019

Abstract: Iodide-oxidizing bacteria (IOB) oxidize iodide into iodine and triiodide which can be utilized for gold dissolution. IOB can be therefore useful for gold leaching. This study examined the impact of incubation conditions such as concentration of the nutrient and iodide, initial bacterial cell number, incubation temperature, and shaking condition on the performance of the gold dissolution through the experiments incubating IOB in the culture medium containing the marine broth, potassium iodide and gold ore. The minimum necessary concentration of marine broth and potassium iodide for the complete gold dissolution were determined to be 18.7 g/L and 10.9 g/L respectively. The initial bacterial cell number had no effect on gold dissolution when it was 1×10^4 cells/mL or higher. Gold leaching with IOB should be operated under a temperature range of 30–35 °C, which was the optimal temperature range for IOB. The bacterial growth rate under shaking conditions was three times faster than that under static conditions. Shaking incubation effectively shortened the contact time compared to the static incubation. According to the pH and redox potential of the culture solution, the stable gold complex in the culture solution of this study could be designated as gold (I) diiodide.

Keywords: iodide-oxidizing bacteria; bioleaching; gold; iodide; iodine; triiodide; gold diiodide

---

1. Introduction

Recovery of gold from ores and concentrates traditionally relies on cyanide leaching. Gold recovery with cyanidation process generally involves crushing, grinding, leaching, activated carbon adsorption, desorption, and electrowinning. However, cyanide leaching creates serious environmental risks due to its severe toxicity. Several cyanide substitutes have been suggested by some workers [1–4]. The iodine–iodide gold leaching is well known as an effective and environmental-friendly method for gold leaching without toxic chemicals such as cyanide, aqua regia and mercury [3,5]. According to Gos and Rubo (2001) [6], acute toxicity and ecotoxicity of cyanide are much higher than iodine. Cyanide is classified as a strong water contaminant however iodine is designated as slight water contaminant. The triiodide ($I_3{}^-$) is generated in the mixture of iodine (I_2) and iodide (I^-) by the chemical reaction as shown by the Equation (1) [7].

$$I^- + I_2 \rightarrow I_3{}^- \tag{1}$$

Gold can be dissolved in the mixture of iodide and triiodide, forming gold (I) diiodide and/or gold (III) tetraiodide as shown by the Equations (2) and (3) respectively [7].

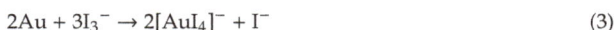

$$2Au + I_3{}^- + I^- \rightarrow 2[AuI_2]^- \tag{2}$$

$$2Au + 3I_3{}^- \rightarrow 2[AuI_4]^- + I^- \tag{3}$$

Iodine–iodide gold leaching also has an advantage that the leaching system could be operated in a wide pH range (pH 2 to 10) [8], whereas the cyanide leaching should be operated under relatively restricted alkaline conditions (pH 10 to 11). Iodine–iodide gold leaching is however disadvantageous in the high cost for both iodine and iodide [9].

The application of microorganisms in metal solubilization processes has become more attractive than conventional pyrometallurgical and hydrometallurgical processes due to its low operating cost and eco-friendly nature [10]. Microorganisms have been applied in the commercial extraction of precious and base metals from primary ores and concentrate through biooxidation and bioleaching. Biooxidation has been utilized as a pre-treatment to dissolve sulfide minerals from refractory gold ores prior to conventional cyanide leaching [11–15]. Bioleaching refers to microbially catalyzed solubilization of metals from solid materials [16]. Iron- and sulfur-oxidizing acidophilic bacteria, such as *Acidithiobacillus ferrooxidans* and *Acidithiobacillus thiooxidans*, oxidize certain sulfidic ores which contain encapsulated particles of elemental gold, resulting in improved accessibility of gold to complexation by leaching agents such as cyanide [13]. In contrast, some microorganisms such as cyanogenic or amino acid-excreting microorganisms solubilize the elemental gold by the formation of gold-complexing metabolic products acting as biogenic lixiviants [13]. Bioleaching of gold by cyanogenic bacteria has been reported [17–20].

Kaksonen et al. [21] proposed the utilization of biogenic iodine–iodide lixiviant solution which was generated by iodide-oxidizing bacteria (IOB) for gold leaching. Khaing et al. [22] isolated IOB strains from environmental samples and demonstrated the gold dissolution from gold ore using the biogenic iodine–iodide lixiviant solution which was generated by IOB strains through the beaker-scale experiments. Iodide can be oxidized into iodine by IOB as shown by the microbial Equation (4) [23].

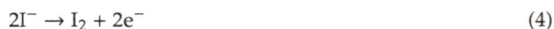

$$2I^- \rightarrow I_2 + 2e^- \tag{4}$$

The underground brine water in natural gas fields in Japan contains iodide at 120 ppm, which is approximately 2000 times that of iodide concentration in sea water. Because of such high iodide concentration in this environment, many kinds of IOB inhabit the brine waters [24]. We therefore collected the brine samples from a natural gas field in Japan and successfully isolated eight bacterial strains of IOB from the brine samples in our previous study [22]. Those strains were then incubated in a liquid culture medium containing nutrients, iodide and gold ore whose gold grade and pulp density were 0.26 wt% and 3.3 *w/v*% respectively under 30 °C. All the strains oxidized iodide into iodine and triiodide in the culture solution, resulting in gold dissolution from the ore as gold diiodide [22]. The best IOB strain dissolved gold from the ore completely within 5 days of incubation. However, the contact time of the gold bioleaching with IOB and iodide may not be as fast as that of gold leaching using cyanide, which was reported to be approximately 24–72 h for gold ore [7]. In addition, the costs of the nutrient and iodide may be expensive, which may be a significant cost factor to consider for the operation of the gold bioleaching with IOB and iodide. The influential factors for the bioleaching with IOB and iodide therefore should be carefully examined in order to improve its performance.

This study focused on determining the impact of incubation conditions such as the concentration of nutrient and iodide, incubation temperature, initial bacterial cell concentration and the shaking incubation on the performance of the gold dissolution from an ore sample through the incubation experiments which were carried out under variable conditions.

2. Materials and Methods

2.1. Characteristics of the Ore

The gold ore used in this study was sourced from the Modi Taung Gold Mine in Central Myanmar. The ore sample was crushed and ground to powder with an average particle diameter less than 75 μm. This powdered ore was used in the experiments of the present study. The chemical composition of the ore, as shown in Table 1, was determined by X-Ray Fluorescence (XRF, ZSX Primus II, Rigaku

Corporation, Tokyo, Japan) analysis. The gold content in the ore was 0.26 wt%. The ore microscopy and X-Ray Diffraction (XRD) analysis using Rigaku RINT-2100 Diffractometer (Rigaku Corporation, Tokyo, Japan), showed the presence of gold, galena, pyrite as the main phases, and sphalerite and chalcopyrite as the minor phase. Free visible gold (>2mm) generally occurred as disseminated grains. The principal gangue mineral was quartz. Mitchell et al. [25] described that high gold values are clearly associated with high pyrite content in the veins.

Table 1. The chemical composition of the ground ore determined by XRF analysis.

Oxides & Elements	SiO_2	SO_3	Pb	Al_2O_3	Fe_2O_3	CaO	K_2O	ZnO	Au	MgO	Ag	Cr_2O_3	LOI
Content (wt%)	78.6	4.57	3.37	3.47	3.12	2.92	1.64	0.82	0.26	0.24	0.20	0.16	0.61

2.2. Characteristics of the IOB Strain

The IOB strain which was selected as the most competent IOB strain through our previous study [22] was used in this study. This IOB strain was identified as *Roseovarius tolerans* by analyzing the sequences of its 16S rRNA gene and named a-1 strain. *R. tolerans* is a heterotrophic, gram-negative bacterium belonging to α-subclass of Proteobacteria that can oxidize iodide to iodine, and iodide-oxidizing reaction was mediated by extracellular oxidase that requires oxygen [24]. It was shown that the a-1 strain had an ability to complete the gold dissolution from the ore within 5 days of incubation in our previous study as described above.

2.3. Culture Medium Preparation

We examined the impact of incubation conditions on the growth of a-1 strain and the gold dissolution in this study. The reference conditions for the incubation experiments were as follows. Difco™ Marine Broth 2216 (Becton, Dickinson and Company, Franklin Lakes, NJ, USA) was used as a nutrient for a-1 strain in this study. The culture medium for the reference contained the marine broth and potassium iodide at the concentration of 37.4 g/L and 21.8 g/L respectively. 15 mL of culture medium was poured into an 80 mL glass tube and lid with a silicone sponge plug. The culture medium was steam-sterilized under 121 °C for 20 min. 0.5 g of ore powder which had been dry-sterilized under 140 °C for 4 h was put into the culture medium (pulp density was 3.3 *w/v*%).

The prescribed amount of preculture solution of a-1 strain was inoculated into the culture medium so that the initial bacterial cell number was set to 3.0×10^6 cells/mL which was the reference initial bacterial cell concentration of this study. The preculture solution was obtained by incubating a-1 strain in the culture medium containing only the marine broth at the concentration of 37.4 g/L at the temperature of 30 °C under aerobic condition for a few days. Subsequently, the a-1 strain in the culture medium was statically incubated in an incubator whose internal temperature was maintained at 30 °C which was the reference incubation temperature of this study. The incubation time of this study was set to 10 days because it had been shown that the gold dissolution in the culture solution of a-1 strain was completed within 5 days by our previous study [22].

The a-1 strain was incubated in the culture medium in which only the concentration of the marine broth was changed to 28.1, 18.7, 9.4 and 4.7 g/L in order to examine the impact of the concentration of the marine broth on the bacterial growth and the gold dissolution. The a-1 strain was also incubated in the culture medium in which only the concentration of potassium iodide was changed to 16.4, 10.9, 5.5 and 2.2 g/L. The initial bacterial cell concentration of a-1 strain was changed to 1.0×10^4 and 1.0×10^5 cells/mL respectively by changing the amount of inoculum whose bacterial cell number was known in order to examine the impact of initial bacterial cell concentration on the bacterial growth and the gold dissolution. The a-1 strain was incubated in the reference culture medium under 20, 35 and 40 °C respectively in order to examine the impact of temperature on the bacterial growth and the gold dissolution. In those experiments, the a-1 strain was incubated in two glass tubes under the same condition to confirm the duplicability of results of the experiments. Lastly, the a-1 strain was incubated in the reference culture medium with the reference initial bacterial cell concentration under the reference temperature while being shaken. The shaking speed was 100 rpm. In this experiment, twenty-two glass tubes including two tubes which were non-inoculated (control) were prepared in order to understand the behavior of the growth of a-1 strain and the gold dissolution under shaking conditions. Both solid samples and liquid samples were extracted from two tubes to confirm the duplicability of results of the experiments every day and those samples were subjected to various analyses as described below. The static incubation experiment was also carried out in parallel with the shaking incubation experiment.

2.4. Analytical Procedures

After the incubation, the bacterial cell number in the culture solution was directly counted with a Petroff–Hausser counting chamber using a phase-contrast microscope (EVOS XL Core Cell Imaging System, Thermo Fisher Scientific Inc., Waltham, MA, USA). After counting the bacterial cell number in the culture solution, the solid phase and liquid phase in the culture solution were separated by the filtration using a membrane filter with 0.2 µm pore size. The solid sample on the membrane filter was washed with pure water and dried in an oven whose internal temperature was 50 °C for a few days. The dried solid sample was subjected to XRF analysis to evaluate the gold content in itself. The leaching yield (%) was calculated based on the XRF analysis using the following equation.

$$LY = (OM_{Au} - FM_{Au}) \times 100/OM_{Au} \qquad (5)$$

where LY is the leaching yield (%), OM_{Au} is the original mass of gold in the ore sample (g/g-ore) and FM_{Au} is the final mass of gold in the ore sample (g/g-ore). The leaching yield is usually calculated based on the mass of soluble gold in the leachate which is measured using ICP-MS/AES/OES or Atomic Absorption Spectrometry (AAS). However, it was calculated based on the mass of gold remaining in the solid phase in this study. Both soluble gold and gold remaining in the solid phase had been measured using ICP-MS and XRF respectively, and good balance between them had been obtained in our previous study [22]. The mass of residual gold which was measured using XRF can be, therefore, used for the evaluation of the leaching yield in this study.

The pH and redox potential of the filtrate were measured at room temperature using a handheld ion/pH meter (IM-32P, DKK-TOA Corporation, Tokyo, Japan). The reference electrode was a silver–silver chloride electrode in 3.3 mol/L solution of potassium chloride.

3. Results and Discussion

3.1. Impact of the Concentration of the Marine Broth and Potassium Iodide on the Bacterial Growth and the Leaching Yield

Figure 1 shows the bacterial cell number (Figure 1a) and the leaching yield (Figure 1b) which were obtained in the incubation experiments under different concentration of the marine broth and potassium iodide. The bacterial cell number became higher as the concentration of the marine broth increased as shown in Figure 1a. In particular, the bacterial cell number increased up to 1×10^8 cells/mL or more in the culture solution whose initial concentration of the marine broth was 37.4 g/L. On the other hand, the bacterial cell number was little different with the variation in the concentration of potassium iodide. The concentration of the marine broth therefore has a significant influence on the growth of a-1 strain, whereas the concentration of potassium iodide has little influence on that.

The leaching yield was influenced by both concentrations as shown in Figure 1b. Gold was completely dissolved from the ore in the culture solution whose initial concentration of the marine broth and potassium iodide was higher than 18.7 g/L and 10.9 g/L respectively. The minimum necessary concentration of the marine broth and potassium iodide can be therefore assumed to be 18.7 g/L and 10.9 g/L respectively under the incubation conditions of this study. It was assumed that the leaching yield did not reach 100% in the other cases because iodine was not generated enough for the complete gold dissolution in those cases. When the concentration of the marine broth was 9.4 g/L or less, the growth of a-1 strain was not so vigorous and the generation of iodine was also not so active. On the other hand, the amount of iodide which was the source for iodine was not enough for complete gold dissolution when the concentration of potassium iodide was 5.5 g/L or less.

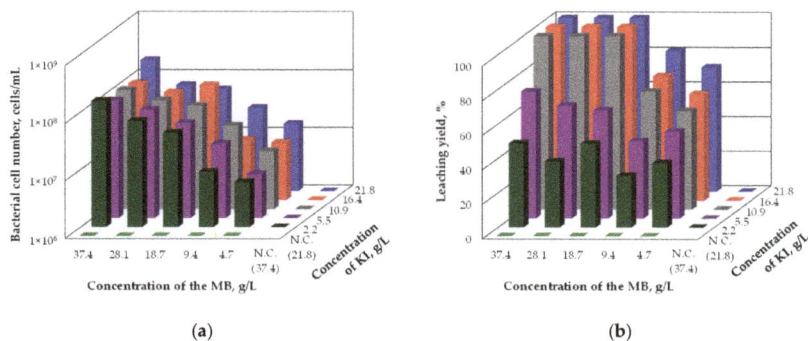

(a)   (b)

Figure 1. Effects of variable concentrations of marine broth (MB) and potassium iodide (KI) on (**a**) bacterial cell number and (**b**) gold leaching yield during 10 days of incubation experiment. Green bar: MB (37.4, 28.1, 18.7, 9.4 and 4.7 g/L) and KI (2.2 g/L), Purple bar: MB (37.4, 28.1, 18.7, 9.4 and 4.7 g/L) and KI (5.5 g/L), Gray bar: MB (37.4, 28.1, 18.7, 9.4 and 4.7 g/L) and KI (10.9 g/L), Red bar: MB (37.4, 28.1, 18.7, 9.4 and 4.7 g/L) and KI (16.4 g/L), Blue bar: MB (37.4, 28.1, 18.7, 9.4 and 4.7 g/L) and KI (21.8 g/L). The pulp density was 3.3 *w/v*%. N.C. is the negative control without inoculation of bacteria.

The mass of gold in a glass tube can be calculated as 0.0013 g from gold grade and the pulp density of the ore used in this study. Assuming that the minimum necessary concentration of the marine broth and potassium iodide is 18.7 g/L and 10.9 g/L respectively, it is calculated that 216 g of the marine broth and 126 g of potassium iodide are necessary for dissolving 1 g of gold from the ore. According to our market price investigation, the price of the marine broth and potassium iodide for industrial use is approximately 45,000 USD per 500 kg and 5400 USD per 50 kg respectively. The cost of the minimum necessary amount of those chemicals for dissolving 1 g of gold from the ore can be calculated as 19.4 USD and 13.6 USD respectively. This result indicates that the cost performance of this method should be improved. In particular, the marine broth which was used in this study is a rich nutrient source containing 16 components such as peptone, yeast extract and inorganic substances and more costly compared to potassium iodide. It is therefore necessary to specify the effective components for IOB among the components in the marine broth and prepare the original nutrient source whose cost is cheaper. Moreover, it is also necessary to provide cheaper effective nutrient sources for IOB as an alternative to the marine broth.

3.2. Impact of the Initial Bacterial Cell Number on the Bacterial Growth and the Leaching Yield

The results of incubation experiments which were started with three different initial bacterial cell numbers are shown in Figure 2. Although the initial bacterial cell number was low such as 1.0×10^4 and 1.0×10^5 cells/mL, a-1 strain grew and the bacterial cell number increased to 4.2×10^7 and 5.4×10^7 cells/mL after 10 days incubation, which were approximately close to 1.2×10^8 cells/mL which was the bacterial cell number in the reference condition. Gold was almost completely dissolved in all cases. Those results indicate that gold can be dissolved completely from the ore regardless of the initial bacterial cell number. In terms of practically of this method, it is an advantage to be able to start the incubation of a-1 strain with low initial bacterial cell concentration.

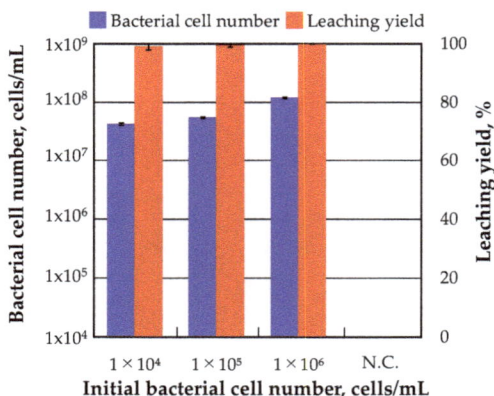

Figure 2. Effects of different initial bacterial cell numbers on the gold extraction during 10 days of incubation experiment. Initial bacterial cell numbers are set to 1×10^4 cells/mL, 1×10^5 cells/mL and 1×10^6 cells/mL respectively. Bacterial cell numbers (cells/mL) is shown by the left vertical axis whereas the leaching yield (%) is indicated by the right vertical axis. N.C. is the negative control without inoculation of bacteria.

3.3. Impact of Temperature on the Bacterial Growth and the Leaching Yield

Figure 3 shows the results of the incubation experiments which were carried out under different temperature conditions. The highest bacterial cell number was observed in the culture solution which was incubated at 30 °C. The growth of a-1 strain was also excellent at 35 °C, whereas it was poor at 25 °C and 40 °C. In particular, the growth of a-1 strain was significantly affected at 40 °C and the bacterial cell number after 10 days of incubation was less than 1.0×10^6 cells/mL. The same tendency was seen as the bacterial cell number with respect to the relation between the leaching yield and temperature. All gold contained in the ore was almost completely dissolved in the culture solution in which a-1 strain was incubated at 30 °C and 35 °C and more than 1×10^8 cells/mL of bacterial cell number was observed. The leaching yield was 79% and 39% in the culture solution which was incubated at 20 °C and 40 °C respectively.

These results suggested that the gold dissolution depends on the activities of a-1 strain in this study although the dissolution of elements from ore is generally promoted with the increase in temperature [26]. It is therefore important to operate the gold bioleaching with IOB and iodide under the optimal temperature for the activities of IOB. The bacterial activities are usually sensitive to temperature change. The optimal growth temperature of the species *R. tolerans* was reported as 33.5 °C by [27]. The optimal temperature for the bacterial growth and the gold dissolution was also 30 °C under the incubation conditions of this study. In particular, a significant difference of the bacterial growth and the gold dissolution was observed between the incubations at 35 °C and 40 °C. The gold bioleaching with IOB and iodide should be therefore operated under the temperature between 30–35 °C.

3.4. Impact of Shaking Condition on the Bacterial Growth and the Leaching Yield

Figure 4 shows the temporal changes of the bacterial cell number and the leaching yield which were obtained by shaking and static incubation experiments. The bacterial cell number increased to 3.3×10^7 cells/mL after 1 day of shaking incubation. The growth of a-1 strain reached stationary phase after 3 days of shaking incubation and the maximum bacterial cell number reached 5.1×10^8 cells/mL at that time. The bacterial cell number was gradually decreased after that and became 1.5×10^8 cells/mL after 10 days of shaking incubation. The leaching yield increased following the bacterial growth. The leaching yield began to increase after starting the shaking incubation and it increased to 7%, 43% and 78% after 1 days, 2 days and 3 days of shaking incubation respectively.

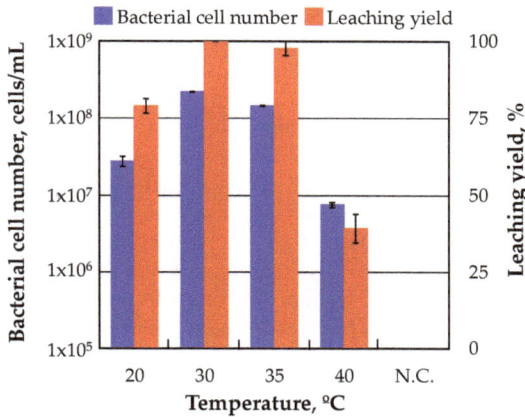

Figure 3. Effect of difference temperature conditions on the bacterial cell numbers and leaching yield during 10 days of incubation experiment. Bacterial cell numbers (cells/mL) is indicated by the left vertical axis with a logarithmic scale and error bars. The leaching yield (%) is indicated by the right vertical axis with a linear scale and errors bars. N.C. is the negative control without inoculation of bacteria. The pulp density of the culture solution was 3.3 *w/v*%. The leaching yield was calculated by the Equation (5) by using the results of XRF analysis of the solid residue collected from culture solution after 10 days of incubation.

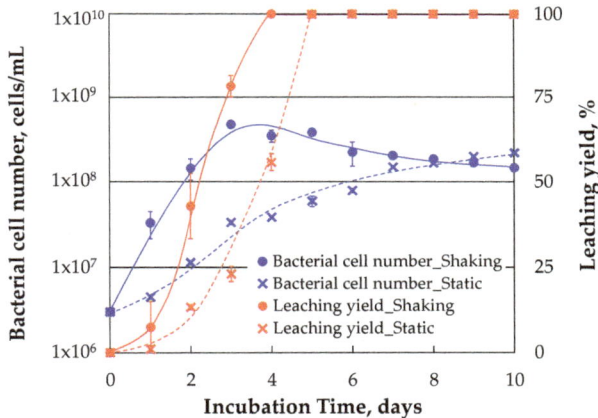

Figure 4. Temporal changes of bacterial cell numbers and Au leaching yield in the ore sample during shaking and static incubation experiments. Dynamic changes of bacterial cell number (blue circle) and Au leaching yield (red circle) from the ore sample in the culture solution with shaking effect. Bacterial cell number (blue cross) and Au leaching yield (red cross) from the ore sample in the culture solution without shaking effect. The bacterial cell number (cells/mL) is indicated in the left vertical axis with logarithmic scale. The Au leaching yield (%) is shown by the right vertical axis with a linear scale.

On the other hand, both bacterial growth and leaching yield observed in the static incubation experiments were inferior to those observed in the shaking incubation experiments. The bacterial cell number of static incubation reached to the level equal to that of shaking incubation after 8 days of incubation. Also, the leaching yield reached to 100% after 5 days of static incubation, which was one day later than the time when the leaching yield reached to 100% in the shaking incubation.

The impact of shaking condition on the bacterial growth and the leaching yield was quantitatively evaluated based on the growth rate and leaching rate which were calculated from the experimental results. Both growth rate and leaching rate were calculated using the bacterial cell number and the leaching yield obtained during the exponential growth phase because the incubation experiments were performed in closed system. Accordingly, the growth rates under shaking and static conditions were calculated using the bacterial cell numbers obtained during 0–3 days and 0–7 days respectively. Also, the leaching rates under shaking and static conditions were calculated using the leaching yields obtained during 1–4 days and 1–5 days respectively. The growth rates of a-1 strain under shaking and static conditions were 1.67 days^{-1} and 0.56 days^{-1}, respectively. The growth rate under shaking condition was three times faster than that under static condition. The leaching rates of gold under shaking and static conditions were 0.41 mg/day and 0.31 mg/day, respectively. The difference between both leaching rates was not so large in comparison with the difference between both growth rates. In particular, the time before the growth of a-1 strain had been started was shortened by shaking incubation and the gold dissolution was also started earlier than the start of the gold dissolution in static incubation. Shaking incubation can be effective to shorten the contact time of the gold bioleaching with IOB and iodide.

3.5. Consideration of the Type of Dissolved Gold by the Measurement of pH and Redox Potential

According to Angelidis et al. [7] and Baghalha [28], gold can be dissolved and stable in the iodine–iodide solution as gold (I) diiodide at the pH range of 0 to 13 and the redox potential range of 400 to 600 mV respectively. When the redox potential of the solution is higher than 600 mV, gold is stable as gold (III) tetraiodide in the solution. The pH and redox potential of the culture solution were measured in order to understand the type of gold dissolved in the culture solution in this study.

The pH and redox potential of the culture solution which was incubated under the reference conditions for 10 days were 8.0 to 8.3 and 522 mV to 547 mV respectively, whereas those of the non-inoculated culture medium (control) was 7.1 and 173 mV respectively. The type of gold in the culture solution could be therefore designated as gold (I) diiodide.

The pH and redox potential of the culture solution in which the complete gold dissolution was observed in the experiments changing the concentration of the marine broth (18.7 g/L to 37.4 g/L) and potassium iodide (10.9 g/L to 21.8 g/L) were within a range of 7.8 to 8.2 and 512 mV to 547 mV respectively. Those of the culture solution which was incubated with initial bacterial cell number of 1×10^5 cells/mL and 1×10^4 cells/mL were within a range of 7.7 to 8.2 and 472 mV to 546 mV respectively. Those of the culture solution which was incubated at 30 °C to 35 °C, which were the optimal temperature for the bacterial growth and the gold dissolution, were within a range of 8.0 to 8.1 and 501 mV to 543 mV respectively. Those of the culture solution which was incubated under shaking condition were 8.4 and 540 mV respectively.

Thus, the pH and redox potential of the culture solution in which gold was completely dissolved from the ore were within a range of 7.7 to 8.4 and 472 mV to 547 mV. The stable gold complex in the culture solution of the present study could be designated as gold (I) diiodide based on the pH and the redox potential of the culture solution.

3.6. Comparison of the Present Study with Other Bioleaching Studies

The present study and other bioleaching studies carried out by the application of some other microorganisms (*Chromobacterium violaceum*, *Pseudomonas aeruginosa*, *Pseudomonas fluorescens*, *Acidithiobacillus* sp., *Aspergillus niger*, *Streptomyces setonii*) were compared and shown in Table 2. Based on the result of the present study, gold can be solubilized completely from high-grade free milling ore having average particle size of 75 μm within 10 days incubation experiment using the IOB-generated iodine–iodide lixiviant. The IOB method is promising due to high gold leaching yields obtained, and the possibility to have cyanide-free process.

Table 2. Comparison of the gold bioleaching efficiency and process conditions from the present study with other gold bioleaching studies.

Bacteria	Particle Size of Ore Sample, μm	Gold Source, Pretreatment	Pulp Density, w/o%	Nutrient, Nutrient Salts	Leaching Efficiency, %	Temp., °C	Initial Bacterial Cell Number, Cells/mL	pH	Reference
Roseovarius tolerans	<75	Sulfide ore with free milling gold without pretreatment	3.3	18.7 g/L Marine broth, 10.9 g/L KI	100	30–35	1×10^4	7.7–8.4	This study
Aspergillus niger	10,000 × 10,000 (rock block)	Printed circuit boards (PCBs) without pretreatment	3.8×10^{-3}	50.0 g/L glucose, 0.10 g/L CaCl$_2$, 0.50 g/L KH$_2$PO$_4$, 1.5 g/L NH$_4$Cl, 0.025 g/L MgSO$_4$·7H$_2$O	87.0	28 ± 2	-	4.4–6.6	[29]
Chromobacterium violaceum	74–400	PCBs biooxidized by At. ferrooxidans	0.5	0.98 g/L MgSO$_4$·7H$_2$O	70.6	30	2.3×10^{14}	11.0	[20]
Chromobacterium violaceum	37–149	Electronic waste biooxidized by At. ferrooxidans	1.0	-	69.3	30	-	7.2–8.8	[19]
Chromobacterium violaceum, Pseudomonas aeruginosa	37–149	Electronic waste Biooxidized by At. ferrooxidans	1.0	-	73.2	30	-	7.2–9.2	[19]
Chromobacterium violaceum, Pseudomonas fluorescens	37–149	Electronic waste biooxidized by At. ferrooxidans	1.0	-	63.1	30	-	7.2–8.8	[19]
Chromobacterium violaceum	1000 × 1000 (rock block)	Waste of mobile phone PCBs without pretreatment	1.5	1.0 g/L MgSO$_4$·7H$_2$O	10.9	30	-	8.0–11.0	[18]
Streptomyces setonii	75	Refractory ore biooxidized by At. thiooxidan, At. ferrooxidans, and L. ferrooxidans	5.0	5.0 g/L tryptone, 3.0 g/L yeast extract, 0.5 g/L KCl, MgSO$_4$·7H$_2$O,	94.7	23–37	-	10.5–11.0	[11]
Acidithiobacillus sp.	<74	Pyrrhotite refractory ore without pretreatment	50	10,100 g/t Ca(OH)$_2$, 16,000 g/t NaCN	19.7	25	-	11.0	[30]
Acidithiobacillus sp.	<74	Pyrrhotite refractory ore biooxidized by At. ferrooxidans	50	19,100 g/t Ca(OH)$_2$, 22,000 g/t NaCN	91.0	25	-	11.0	[30]
Acidithiobacillus ferrooxidans	20–125	Arsenical refractory sulfide concentrate biooxidized by At. ferrooxidans	5.0	160,000 g/t NaCN	85.2	35	-	10.5–12.0	[15]

Note that "-" denotes "absence" of the items or "the literature did not mention".

4. Conclusions

This study focused on the bioleaching of gold from a gold ore by using IOB. The incubation experiments of an IOB strain were carried out using the culture medium containing the marine broth, potassium iodide and a gold ore under various conditions in order to obtain useful information about the influential factors for improving the performance of the bioleaching with IOB and iodide. Specifically, the impact of concentration of the marine broth and potassium iodide, initial bacterial cell number, incubation temperature, and shaking conditions on the growth of the IOB strain and the gold dissolution was evaluated through the incubation experiments in this study. The results obtained are summarized below.

The concentration of the marine broth and potassium iodide should be higher than 18.7 g/L and 10.9 g/L respectively in order to dissolve gold completely from the ore used in this study. From these results, it was calculated that 216 g of the marine broth and 126 g of potassium iodide are necessary for dissolving 1 g of gold from the ore. The cost of those chemicals required for dissolving 1 g of gold from the ore was calculated as 33 USD on the basis of their price for industrial use. Thus, the cost performance of this method should be further improved by screening the effective components from among the components in the marine broth and/or searching for cheaper effective nutrient sources as alternatives to the marine broth.

When the initial bacterial cell number was 1×10^4 cells/mL or higher, the initial bacterial cell number had no significant impact on the growth of the IOB strain and the gold dissolution. The operation of bioleaching with the IOB strain and iodide can be started with low bacterial cell numbers.

Gold contained in the ore was almost completely dissolved in the culture solution incubated at 30 °C and 35 °C, therefore, the optimal temperature for the growth of the IOB strain and the gold dissolution was within a range of those temperatures. The operation of the bioleaching with IOB and iodide should be operated within temperature range between 30 °C and 35°C.

The bacterial growth of the IOB strain was promoted under shaking condition. The growth rate of the IOB strain under shaking condition was three times higher than that under static condition. Accordingly, the gold dissolution was also promoted under shaking condition.

The pH and redox potential of the culture solution in which complete gold dissolution was observed were within a range of 7.7 to 8.4 and 472 mV to 547 mV. The stable gold complex in the culture solution of this study could be designated as gold (I) diiodide.

The only competitor of IOB to obtain iodide–triiodide for gold leaching is the chemical iodine–iodide leaching. Previous study [22] and the present study proved that gold-bioleaching is as effective as chemical iodine leaching. Gold can be recovered completely after 5 days of incubation. However, bacterial leaching of gold using IOB compares unfavorably with chemical iodine–iodide leaching because it is a slower process and the leaching (contact) time may not be as fast as that of direct chemical leaching. In practice, the costs of the reagents iodine and iodide are quite high. Even though gold leaching rate or dissolution in iodine–iodide solution is proved to be much faster than in conventional cyanidation process [2,31], iodine–iodide processing of gold is under-utilized and insufficiently explored mainly due to its high cost [2]. However, one advantage is that, in bioleaching operation using IOB, iodine can possibly be recovered at the end of leaching processes. Only the cost of nutrients (marine broth) and KI must be taken into account. To the best of our knowledge, there are no other ways to produce iodide–triiodide.

Author Contributions: S.Y.K. and Y.S. designed the study and carried out all the experiments. Y.S., K.S. and M.M.T. helped with sample collection, sample analyses and valuable discussions. S.Y.K. and Y.S. wrote and completed the manuscript considering the review comments provided by all authors.

Funding: This research was funded by a grant entitled "Engineering Research for Pioneering of a New Field" provided by the Faculty of Engineering, Kyushu University, Japan. This work was also supported by JSPS KAKENHI Grant Number JP19K05352.

Acknowledgments: We thank U Tin Tun and U Myint Thein Htay, the geologists of Modi Taung Gold Mine (Myanmar) for their supports during the geological excursion for our ore sample collection.

Minerals **2019**, *9*, 274

Conflicts of Interest: The authors declare no conflict of interest.

References

1. Mc Nulty, T. Cyanide substitutes. *Min. Mag.* **2001**, *184*, 256–261.
2. Hilson, G.; Monhemius, A.J. Alternatives to cyanide in the gold mining industry: What prospects for the future? *J. Clean. Prod.* **2006**, *14*, 1158–1167. [CrossRef]
3. Gökelma, M.; Birich, A.; Stopic, S.; Friedrich, B. A review on alternative gold recovery reagents to cyanide. *J. Mater. Sci. Chem. Eng.* **2016**, *4*, 8–17. [CrossRef]
4. Aylmore, M. Alternative lixiviants to cyanide for leaching gold ores. In *Gold Ore Processing: Project Development and Operations*; Adams, M.D., Ed.; Elsevier: Amsterdam, The Netherland, 2016; pp. 447–484.
5. Lucheva, B.; Iliev, P.; Kolev, D. Recovery of Gold from Electronic Waste by Iodine–iodide Leaching. *J. Chem. Tech. Metall.* **2017**, *52*, 326–332.
6. Gos, S.; Rubo, A. The Relevance of Alternative Lixiviants with Regard to Technical Aspects, Work Safety and Environmental Safety. *Cyplus. Degussa AG, Hanau, Germany* **2001**. Available online: http://citeseerx.ist.psu.edu/viewdoc/download?doi=10.1.1.596.4536&rep=rep1&type=pdf (accessed on 1 May 2019).
7. Angelidis, T.N.; Kydros, K.A.; Matis, K.A. A fundamental rotating disk study of gold dissolution in iodine–iodide solutions. *Hydrometallurgy* **1993**, *34*, 49–64. [CrossRef]
8. Qi, P.H.; Hiskey, J.B. Dissolution kinetics of gold in iodide solution. *Hydrometallurgy* **1991**, *27*, 47–62. [CrossRef]
9. Yannopoulos, J.C. *The Extractive Metallurgy of Gold*; Van Nostrand Reinhold: New York, NY, USA, 1991.
10. Johnson, D.B. Biodiversity and interactions of acidophiles: Key to understanding and optimizing microbial processing of ores and concentrates. *Trans. Nonferrous Met. Soc. China* **2008**, *18*, 1367–1373. [CrossRef]
11. Amankwah, R.K.; Yen, W.T.; Ramsay, J.A. A two-stage bacterial pretreatment process for double refractory gold ores. *Miner. Eng.* **2005**, *18*, 103–108. [CrossRef]
12. Lindstrom, E.B.; Gunneriusson, E.; Tuovinen, O.H. Bacterial Oxidation of Refractory Sulfide Ores for Gold Recovery. *Crit. Rev. Biotechnology* **2008**, *12*, 133–155. [CrossRef]
13. Olson, G.J. Microbial oxidation of gold ores and gold bioleaching. *FEMS Microbiol. Lett.* **1994**, *119*, 1–6.
14. Komnitsas, C.; Pooley, F.D. Mineralogical characteristics and treatment of refractory gold ores. *Miner. Eng.* **1989**, *2*, 449–457. [CrossRef]
15. Komnitsas, C.; Pooley, F.D. Bacterial oxidation of an arsenical gold sulphide concentrate from Olympias, Greece. *Miner. Eng.* **1990**, *3*, 295–306. [CrossRef]
16. Watling, H. Review microbiological advances in biohydrometallurgy. *Minerals* **2016**, *6*, 49. [CrossRef]
17. Reith, F.; Lengke, M.F.; Falconer, D.; Craw, D.; Southam, G. The geo-microbiology of gold. *ISME J.* **2007**, *1*, 567–584. [CrossRef]
18. Tran, C.D.; Lee, J.C.; Pandey, B.D.; Jeong, K.K.; Jinki, Y. Bioleaching of gold and copper from waste mobile phone PCBs by using a cyanogenic bacterium. *Miner. Eng.* **2011**, *24*, 1219–1222.
19. Pradhan, J.K.; Kumar, S. Metals bioleaching from electronic waste by *chromobacterium violaceum* and *Pseudomonads* sp. *Waste Manag. Res.* **2012**, *30*, 1151–1159. [CrossRef]
20. Li, J.; Liang, C.; Ma, C. Bioleaching of gold from waste printed circuit boards by *Chromobacterium violaceum*. *J. Mater. Cycles Waste Manag.* **2015**, *17*, 529–539. [CrossRef]
21. Kaksonen, A.H.; Mudunuru, B.M.; Hackl, R. The role of microorganisms in gold processing and recovery—A review. *Hydrometallurgy* **2014**, *142*, 70–83. [CrossRef]
22. Khaing, S.Y.; Sugai, Y.; Sasaki, K. Gold Dissolution from Ore with Iodide-Oxidising Bacteria. *Sci. Rep.* **2019**, *9*, 4178. [CrossRef]
23. Effenberger, J.A. Reactions of iodine and iodide ions in the presence and absence of polysaccharides. *Retrosp. Theses Diss.* **1961**. [CrossRef]
24. Amachi, S.; Muramatsu, Y.; Akiyama, Y.; Miyazaki, K.; Yoshiki, S.; Hanada, S.; Kamagata, Y.; Ban-nai, T.; Shinoyama, H.; Fujii, T. Isolation of iodide-oxidizing bacteria from iodide-rich natural gas brines and seawaters. *Microb. Ecol.* **2005**, *49*, 547–557. [CrossRef] [PubMed]
25. Mitchell, A.H.G.; Ausaa, C.A.; Deiparinea, L.; Hlaingb, T.; Htayc, N.; Khinec, A. The Modi Taung–Nankwe gold district, Slate belt, central Myanmar: Mesothermal veins in a Mesozoic orogeny. *J. Asian Earth Sci.* **2004**, *23*, 321–341. [CrossRef]

26. Deveci, H.; Akcil, A.; Alp, I. Parameters for control and optimization of bioleaching of sulfide minerals. In *Process Control and Optimization in Ferrous and Non Ferrous Industry, Material Science & Techology Symposium*; Kongoli, F., Thomas, B., Sawamiphakdi, K., Eds.; TMS: Pittsburgh, PA, USA, 2003; pp. 77–90.

27. Labrenz, M.; Collins, M.D.; Lawson, P.A.; Tindall, B.J.; Schumann, P.; Hirsch, P. *Roseovarius tolerans* gen. nov., sp. nov., a budding bacterium with variable bacteriochlorophyll a production from hypersaline Ekho Lake. *Int. J. Syst. Bacteriol.* **1999**, *49*, 137–147. [CrossRef] [PubMed]

28. Baghalha, M. The leaching kinetics of an oxide gold ore with iodide/iodine solution. *Hydrometallurgy* **2012**, *113–114*, 42–50. [CrossRef]

29. Madrigal-Arias, J.E.; Argumedo-Delira, R.; Alarcón, A.; Mendoza-López, M.R.; García-Barradas, O.; Cruz-Sánchez, J.S.; Ferrera-Cerrato, R.; Jiménez-Fernández, M. Bioleaching of gold, copper and nickel from waste cellular phone PCBs and computer goldfinger motherboards by two *Aspergillus niger* strains. *Brazil. J. Microbio.* **2015**, *46*, 707–713. [CrossRef]

30. Ubaldini, S.; Veglio, F.; Beolchini, F.; Toro, L.; Abbruzzese, C. Gold recovery from a refractory pyrrhotite ore by biooxidation. *Int. J. Miner. Process* **2000**, *60*, 247–262. [CrossRef]

31. Wang, H.X.; Sun, C.B.; Li, S.V.; Fu, P.F.; Song, Y.G.; Li, L.; Xie, W.Q. Study on gold concentrate leaching by iodine–iodide. *Int. J. Miner. Metall. Mater.* **2013**, *20*, 323–328. [CrossRef]

Article

Sequential Bioleaching of Phosphorus and Uranium

Jarno Mäkinen *, Laura Wendling, Tiina Lavonen and Päivi Kinnunen

VTT Technical Research Centre of Finland Ltd, P.O. Box 1000, FI-02044 Espoo, Finland;
laura.wendling@vtt.fi (L.W.); tiina.lavonen@vtt.fi (T.L.); paivi.kinnunen@vtt.fi (P.K.)
* Correspondence: jarno.makinen@vtt.fi; Tel.: +358-20-722-4703

Received: 31 March 2019; Accepted: 24 May 2019; Published: 28 May 2019

Abstract: Phosphorus and uranium are both vital elements for society. In recent decades, fears have arisen about the future availability of low-cost phosphorus and uranium. This has resulted in pressure to de-centralize production of both elements by utilizing lower-grade or complex deposits. The research presented here focused on phosphorus-containing apatite ores with uranium impurities; in order to separate uranium by selective and sequential bioleaching before phosphorus leaching. This would create an alternative process route for solvent-extraction, used to remove/recover uranium from the phosphorus acid product of apatite H_2SO_4 wet process. In this work, it was seen that the used fluorapatite ore required 24 h leaching at pH 1 by H_2SO_4 to result in 100% leaching yield for phosphorus. As this ore did not contain much uranium, an artificial fluorapatite-uranium ore was prepared by mixing standard uranium ore and fluorapatite. The research with this ore showed that 89% of uranium dissolved in 3 days at pH > 2 and leaching was improved by applying Fe^{3+} oxidant. In these conditions only 4% of phosphorus was leached. By prolonged (28 days) leaching 95% uranium yield was reached. According to the experiments, the iron in the uranium leach solution would be mainly Fe^{3+}, which allows the use of H_2O_2 for uranium recovery and then direct use of spent leachate for another uranium leaching cycle. After the dissolution of uranium, 90% of phosphorus was dissolved by decreasing the pH to 1.3. This was done by bioleaching, by utilizing biogenic sulfur oxidation to sulfuric acid.

Keywords: bioleaching; phosphorus; fluorapatite; uranium

---

1. Introduction

The critical importance of phosphorus is due to its use as a fertilizer: in agriculture, there are no substitutes to this essential element [1–3]. The annual consumption of phosphorus is predicted to increase from approximately 45 kton (2000) to 55–95 kton P_2O_5 (2050), mainly due to the expected population increase and elevated future usage in Africa, South America and China [2]. At present, phosphorus is mainly obtained from non-renewable phosphate rock deposits [2–4]. In 2016, the global mining of phosphate rock reached 255 Mt, while global reserves were estimated as 70,000 Mt [1]. Therefore, with the current mining rate, the phosphate rock reserves would be depleted in 275 years, and with a doubled consumption rate (maximum consumption increase by 2050 [2,3]), in 137 years. However, in 2009 it was estimated that reserves that can be treated with less than 40 US Dollar (USD)/t were only 15,000 Mt [4,5]. Consequently, the highest-grade phosphate rock reserves would be depleted more rapidly, in 59 years with current consumption (or in 29 years with doubled consumption). Estimating the ore reserve volume, quality and processing costs is challenging, due to different methodology and criteria used with global ore deposits [2,4]. While it can be summarized that total phosphorus depletion is unlikely in the near future, depletion of high-grade ores is possible, having impacts on phosphorus price [2].

Major phosphorus rock deposits are unevenly distributed, occurring mainly in Morocco and North Africa, China and the Middle East [1,4,6]. After the depletion of high-grade reserves of major producers, namely United States (but possibly also China), production is likely to concentrate more ton North Africa (especially Morocco) and Middle East [2,4,6]. This may raise the prices, decrease the supply security and generate geopolitical issues [2,6]. This creates pressure to recover phosphorus from low-grade and complex deposits for those countries that do not have high-grade ores. This may cause process challenges, as seen in the historic phosphate rock mines of Florida due to increased Fe and Al concentrations in feed [4]. In addition to Fe and Al, other typical impurities of phosphorus ores are Mg, Na, U, As, Cd and Cr [7].

A vast majority of phosphate rock ores are apatite minerals from which the phosphorus is usually extracted by a wet process [4,7,8]. In this process, apatite is treated with sulfuric acid, resulting in the dissolution of solid mineral phosphorus to soluble phosphoric acid. Simultaneously, insoluble calcium sulfate, called phosphogypsum waste, is generated and stacked in the vicinities of phosphoric acid plants [7,8]. If not properly managed, it can cause negative environmental effects by releasing contaminated run-off and seepage waters [9–11]. The fate of impurities in the wet processes vary. For example, a great majority of uranium ends up to phosphoric acid product and further to fertilizer production [7,12]. It is stated that the uranium enrichment ratio to phosphoric acid is 150% compared to its original concentration in the ore [12]. Removal technologies for uranium from phosphoric acid have been studied in detail; research began in the 1950s [7]. In the late 1970s, solvent extraction technology had reached industrial operations for removing, but also recovering uranium from phosphoric acid plants; however many solvent extraction processes have been shut down due to unfavorable economics [7,13].

Uranium concentration in phosphate rocks varies between 11 and 220 mg/kg [7,14]. According to the estimations, over 10,000 t U per annum is included in phosphoric acid streams, originating from phosphate rock processes [7,13]. This is a remarkable amount compared to total uranium production, reaching 55,975 t U in 2015 [15]. Simultaneously, the highest-grade uranium reserves, i.e., deposits with production costs less than 40 USD/kg U, were only 646,900 t [15], signaling availability for less than 12 years. If new high-grade uranium ore bodies are not found, mining has to focus on poor ores, leading to increased uranium prices. The phosphate rocks are already included in uranium reserves as low-grade/complex ores [15], but with remarkably higher production costs: 1300–6300 USD/kg U [14]. By utilizing processes that recover both phosphorus and uranium from phosphate rock, economic savings would be found, as both elements require similar mining and pre-treatment actions [7]. Estimations of recovering uranium as a by-product from phosphate streams have varied between 60–200 USD/kg U, by using solvent extraction technology [14].

The objective of this research was to study sequential (bio)leaching of uranium and phosphorus from apatite ores. By utilizing a selective uranium pre-leaching step and producing uranium leachate prior to the phosphoric acid step, new process routes may be introduced. For this treatment possibility, it is important to understand dissolution chemistry for the respective elements. Uranium occurs in minerals in tetravalent (U^{4+}) and hexavalent (U^{6+}) oxidation states, uraninite being the most important mineral in the mining industry [16–18]. Tetravalent uranium, found in uraninite, has low solubility in mild acids and requires an oxidizing agent for an effective process [16,17]. A typical solution for dissolution is utilizing sulfuric acid and ferric iron oxidant, resulting in oxidation of insoluble U^{4+} to soluble UO_2^{2+}, as shown in Equation (1) [17]. The produced UO_2^{2+} complexes with sulfate in sulfuric acid media, as shown in Equation (2) [17,18]. This complexation is important for the process, as without sulfate, uranium tends to precipitate due to hydrolysis [18]. In the leaching process, ferric iron is reduced to ferrous iron (Equation (1)) and is therefore oxidized prior to next leaching round with sodium chlorate ($NaClO_3$), pyrolusite (MnO_2), hydrogen peroxide (H_2O_2) or other strong oxidants [16,18]. Use of these chemicals is expensive, may have negative environmental impacts and may increase corrosion and abrasion to the used hydrometallurgical equipment [16].

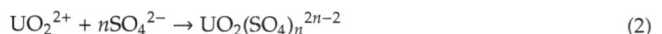

$$UO_2 + 2Fe^{3+} \rightarrow UO_2^{2+} + 2Fe^{2+} \tag{1}$$

$$UO_2^{2+} + nSO_4^{2-} \rightarrow UO_2(SO_4)_n^{2n-2} \tag{2}$$

Another industrially applied hydrometallurgical process for uranium is bioleaching [18–22]. An advantage of bioleaching is that iron-oxidizing bacteria can regenerate Fe^{3+} from Fe^{2+} using only oxygen (air) and protons (Equation (3)) [19,20,22]. The proton needed for the biogenic ferric iron regeneration can be obtained by adding commercial sulfuric acid to the oxidation process, or by biogenic oxidation of elemental sulfur by sulfur-oxidizing bacteria (Equation (4)) [19,20,22]. Both iron and protons can be obtained also from pyrite (FeS_2) by ferric iron attack (Equations (3) and (4)), originating from the activity of iron-oxidizing bacteria (Equation (3)) [19,23,24]. A pyrite content of >1% has been considered suitable for uranium bioleaching [25].

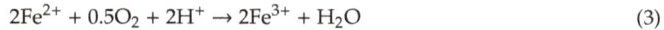

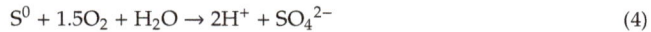

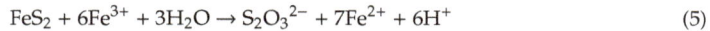

$$2Fe^{2+} + 0.5O_2 + 2H^+ \rightarrow 2Fe^{3+} + H_2O \tag{3}$$

$$S^0 + 1.5O_2 + H_2O \rightarrow 2H^+ + SO_4^{2-} \tag{4}$$

$$FeS_2 + 6Fe^{3+} + 3H_2O \rightarrow S_2O_3^{2-} + 7Fe^{2+} + 6H^+ \tag{5}$$

$$S_2O_3^{2-} + 8Fe^{3+} + 5\,H_2O \rightarrow 2SO_4^{2-} + 8Fe^{2+} + 10H^+ \tag{6}$$

According to the literature, the pH for pitchblende/uraninite leaching should take place at pH 1–2, with ferric iron concentrations between 0.5–3.0 g/L; according to the Pourbaix diagrams UO_2^{2+} occurs at a pH of 0.5–3.5 and at Eh > 300 mV [18]. For the apatite, leaching conditions differ from uraninite, signaling that selective pre-leaching of uranium is theoretically possible. Apatite is leached by using high sulfuric acid concentrations [8]. From Equation (7) it is seen that apatite leaching is not dependent of the oxidant [26,27].

$$Ca_5(PO_4)_3F + 5H_2SO_4 + 10H_2O \rightarrow 3H_3PO_4 + 5CaSO_4 \cdot 2H_2O + HF \tag{7}$$

In addition to utilizing bioleaching for preliminary leaching of uranium, bioleaching was also applied for later leaching of phosphorus from low-grade fluorapatite ore. This was done by utilizing elemental sulfur for biogenic sulfuric acid production and subsequent fluorapatite dissolution (Equation (4); Equation (7)). Bioleaching of apatite has not been applied on an industrial scale but studied on a laboratory scale [26–29].

2. Materials and Methods

2.1. Studied Ores

The ore for experiments was <500 μm particle sized low-grade fluorapatite ore (FA) dominated by magnesioriebeckite, fluorapatite and talc, with 3.6 wt% phosphorus content. The uranium content in FA was very low (7 mg/kg). The other studied ores were obtained by mixing reference uranium ore (RGU-1) prepared by the Canadian Certified Reference Materials Project (CCRMP). The RGU-1 has been prepared by mixing BL-5 standard uranium ore with silica sand to reach a final U content of 400 ± 2.1 mg/kg. The BL-5 ore (and therefore RGU-1) contained plagioclase feldspars, hematite, quartz, calcite and dolomite, chlorite and muscovite; the main uranium-bearing mineral was uraninite. The particle size of RGU-1 was <104 μm [30,31]. A mixture ore of uranium and pyrite (RGU-1-PYR) was prepared by using a pyrite concentrate (particle size of <500 μm, pyrite and pyrrhotite content of 97.4 wt% and 2.5 wt%, respectively). RGU-1-PYR was prepared by mixing RGU-1 (97.5 wt%) and pyrite concentrate (2.5 wt%). This pyrite addition was selected according to previous observations of pyrite requirement [18,25]. A mixture ore of fluorapatite and uranium (RGU-1-FA) was prepared by mixing FA (26 wt%) and RGU-1 (74 wt%) to obtain an artificial ore with a reasonable U and P content, approximately 300 and 9500 mg/kg, respectively. This artificial ore of RGU-1-FA was prepared as the original FA ore and had a very low uranium content. The studied ores and their mixtures are presented in Table 1.

Table 1. Used samples and content of their main elements (wt%). FA: fluorapatite ore, RGU-1: reference uranium ore, RGU-1-PYR: mixed pyrite and reference uranium ore, RGU-1-FA: mixed fluorapatite and reference uranium ore.

Element (wt%)	FA	RGU-1	RGU-1-PYR	RGU-1-FA
P	3.6	-	-	0.95
U	0.0007	0.040	0.039	0.029
Fe	14.0	0.030	1.2	3.7
S	-	0.0020	1.3	0.0015
Si	14.0	46.4	45.3	37.9
Ca	8.7	0.030	0.029	2.3
Mg	7.0	0.010	0.0098	1.8
Al	4.4	0.10	0.10	1.2
K	1.1	0.0020	0.0020	0.29
F	0.71	-	-	0.19

2.2. Leaching Experiments

H_2SO_4 leaching for FA was performed to understand phosphorus leaching in acidic conditions. Experiments were carried in accordance with the European standard leaching method [32]. Separate subsamples were leached at a fixed solid-to-liquid (S/L) ratio of 10. Subsamples (15.0 ± 0.04 g) of FA were suspended in 135 mL ultrapure water and stirred for 30 min to equilibrate prior to testing initiation. H_2SO_4 was added to each suspension via automatic titration (Radiometer TIM 845 TitraLab titration workstations and Radiometer pHC 2005-8 electrodes). The maximum rate of acid addition was limited to prevent a temperature increase. Leaching experiments were conducted for 12, 24, and 48 h with the pH fixed at 1.0, 2.0 or 3.0. The H_2SO_4 titrant concentration was 6.0 M (mol/L), 6.0 M and 3.5 M for pH 1.0, 2.0 and 3.0 experiments, respectively. Solution temperatures remained at 22–25 °C, within the required tolerance of 20 ± 5 °C. After leaching, leachates were vacuum-filtrated (0.45 μm) and analyzed for phosphorus and uranium with ICP-OES (by accredited analytical laboratory Labtium Ltd., Espoo, Finland).

Bioleaching and chemical control tests were executed for RGU-1, RGU-1-PYR and RGU-1-FA ores. The used mixed acidophilic and mesophilic bacterial culture, originally enriched from a sulphide ore mine site [33], contained *At. ferrooxidans, At. thiooxidans, At. caldus, L. ferrooxidans* and *Sulfobacillus thermotolerans*. All experiments were conducted in 250 mL shake flasks with 100 mL working volume. The used nutrient media contained 3 g/L $(NH_4)2SO_4$, 0.5 g/L K_2HPO_4, 0.5 g/L $MgSO_4 \cdot 7H_2O$, 0.1 g/L KCl, 0.014 g/L $Ca(NO_3)_2 \cdot 4H_2O$ [34]. For bioleaching experiments, 10 vol% of inoculum was added to the flasks, while for chemical controls inoculum was replaced by media. Then, iron and sulfur sources were supplemented, where applied. The RGU-1 test was supplemented with 14.9 g/L $FeSO_4 \cdot 7H_2O$; no sulfur source was introduced. The RGU-1-PYR test was not supplemented with any iron or sulfur source (except the pyrite from the sample itself). The RGU-1-FA test was supplemented with 14.9 g/L $FeSO_4 \cdot 7H_2O$ and 10 g/L elemental sulfur. Finally, artificial ores, 100 g/L (RGU-1 and RGU-1-FA) and 102.5 g/L (RGU-1-PYR), were added, followed by a pH adjustment to the pH 2.0 with 95% H_2SO_4. The pH adjustment was continued manually throughout the experiments. The summary of bioleaching (BL) and chemical control (CC) experiments is given in Table 2. During the tests, shake flasks were incubated in a rotary shaker (150 rpm, 30 °C) for 28 days. Sampling was conducted from shake flasks on days 0, 3, 7, 14, 21, and 28, by pipetting 4 mL of leach solution (removed leach solution was compensated by adding 4 mL media), followed by filtration with a syringe and 0.45 μm filter unit (Whatman FP30/0.45 CA-S). The pH and Eh were measured, and phosphate, sulphate and Fe^{2+} were analyzed spectrophotometrically with Hach-Lange LCK349, LCK353 and LCK320 kits, respectively. Elemental analyses were performed from diluted sample solutions with a High-Resolution Sector Field Inductively Coupled Plasma-Mass Spectrometer (HR ICP-MS, Element 2, Thermo Fisher Scientific, Waltham, MA, USA). Calibration curve and control samples were diluted from ICP-MS Multi-Element Solutions 2 and 4 by SPEX and the control sample was diluted from AccuTrace™ Semi-qualitative

Standard (SQS-01-1) and from SPEX Laboratory performance check (LPC-1) standard solutions. Indium was used as an internal standard in all samples, background, calibration and control samples.

Table 2. Summary of bioleaching (BL) and chemical control (CC) experiments.

Ore and Test	Nutrient Media	Inoculum	Iron Source	Sulfur Source	pH Control
RGU-1 (CC)	Yes	No	Yes	No	Yes
RGU-1 (BL)	Yes	Yes	Yes	No	Yes
RGU-1-PYR (CC)	Yes	No	No	No	Yes
RGU-1-PYR (BL)	Yes	Yes	No	No	Yes
RGU-1-FA (CC)	Yes	No	Yes	Yes	Yes
RGU-1-FA (BL)	Yes	Yes	Yes	Yes	Yes

3. Results and Discussion

H_2SO_4 leaching of FA in different pH values and leaching durations are shown in Figure 1. Complete phosphorus leaching was observed at pH 1.0 when leaching duration was 24 h or longer. Uranium leaching yield varied from 66% to 80% at pH 1.0, depending on the leaching duration. At higher pH 2, leaching yields for phosphorus and uranium were approximately 20%, however, phosphorus required at least 24 h leaching duration to reach this level, while for uranium leaching yields were dramatically decreased when leaching duration prolonged to 48 h. The reason for this was not understood. At the highest tested pH (pH 3), phosphorus leaching yield was only 3%–5%, and for uranium 0%. The acid consumption expressed as kg 95% H_2SO_4 per one ton of ore, was 490, 133 and 24 kg/t for pH 1, 2 and 3, respectively (24 h test). Therefore, the FA leaching was considered as an acid intensive method, as discussed, in more detailed examinations of the sulfuric acid wet process for apatite [8,35]. The sulfuric acid leaching of FA had no selectivity between phosphorus and uranium.

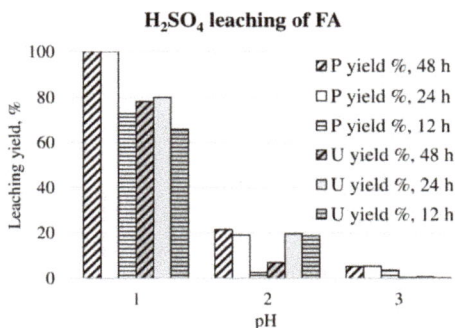

Figure 1. P and U leaching yields during H_2SO_4 leaching of FA in different pH values for 12, 24 and 48 h, at 22 °C temperature.

Bioleaching experiments were started with RGU-1 to understand the effect of ferric iron in uranium leaching, and with RGU-1-PYR to understand if pyrite can serve as iron and sulfur source in the bioleaching system. The bioleaching and chemical control test results are shown in Figure 2. In the inoculated bioleaching experiments an increase in Eh was observed, reaching over the +800 mV level on day 14, while for the chemical control tests Eh remained lower at +500–600 mV. The chemical control experiments resulted in solutions without Fe^{3+}, while in bioleaching experiments with pyrite (RGU-1-PYR) and added $FeSO_4 \cdot 7H_2O$ (RGU-1), Fe^{3+} concentrations rose to 0.5 g/L and 3.0 g/L, respectively. The uranium leaching yield was increasing by the increase of Fe^{3+} concentration, being 79%–87%, 92% and 98% with chemical control experiments, RGU-1-PYR bioleaching and RGU-1 bioleaching, respectively. In the literature, an Fe^{3+} concentration of 0.5 g/L has been considered as a minimum for the process [18] and leaching with only sulfuric acid is considered ineffective in the literature [16–18]. Controversially, in the experiments conducted here, uranium extraction was considered effective with chemical control experiments without the presence of Fe^{3+} ions. It has been reported that in uraninite ores U^{4+} can oxidize to U^{6+}, which is an acid soluble form [17,18]. U^{6+} can reach up to 60% concentration [18], which would explain the majority of the dissolution in RGU-1 chemical control experiment. The acid consumption expressed as kg 95% H_2SO_4 per one ton of ore, was 22.1 and 12.9 kg/t for RGU-1 bioleaching and chemical leaching, respectively. It is emphasized that the bioleaching experiment resulted in higher acid consumption due to biogenic oxidation of Fe^{2+} (supplemented with $FeSO_4 \cdot 7H_2O$) to Fe^{3+}, according to Equation (3) [19,20]. This reaction consumed protons and did not occur in the abiotic chemical control experiment. The acid consumption for RGU-1-PYR bioleaching and chemical leaching was 7.2 and 12.6 kg/t, respectively. In this case, bioleaching resulted in a lower acid consumption, as biological pyrite oxidation produces acidity, as explained in Equations (5) and (6) [19,23]. In this research, it was shown that pyrite can be used in the bioleaching process as a source for iron and sulfur. However, pyrite dissolution in the reported bioleaching experiment was incomplete. This may be due to large particle size (<500 μm) or possible surface passivation of pyrite, as the used concentrate was not re-ground.

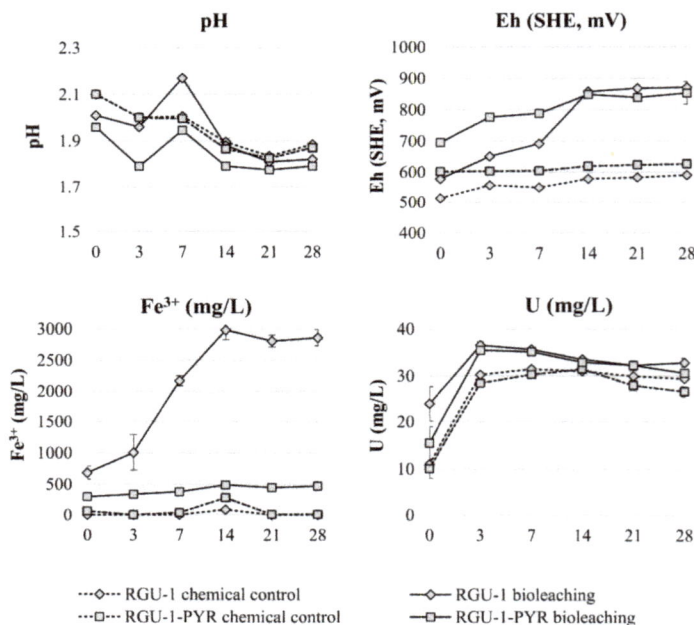

Figure 2. Evolution of pH, Eh and Fe and U concentrations during bioleaching and chemical leaching of RGU-1 and RGU-1-PYR ores. The x-axis represents the leaching duration (days).

Bioleaching experiments with RGU-1-FA were done to understand sequential bioleaching possibility and the efficiency of uranium and phosphorus leaching from the ore. With bioleaching, pH 1.2 was reached by biogenic production of sulfuric acid, while in the chemical control the pH decreased to pH 1.7 (Figure 3). In the bioleaching experiment, the Eh rose to +800 mV after 7 days, and all iron was present as Fe^{3+} (Figures 3 and 4). However, later the Eh and Fe^{3+}/Fe_{tot} ratio started to decrease. This occurred when the pH decreased from 1.9 to 1.4, possibly signaling that on later days the pH was lower than the optimal range for iron-oxidizing bacteria [21]. In the chemical control experiment, Eh remained rather stable at +500–+600 mV (Figure 3) and Fe^{3+}/Fe_{tot} ratio remained extremely small, indicating that Fe^{2+} was the dominating iron species (Figure 4). Major uranium leaching was observed already during the first three days, in both bioleaching and chemical control experiment; in bioleaching, uranium extraction was 10% higher than in chemical control experiment (Figure 5). Phosphorus started to dissolve much later than uranium, in bioleaching test on day 7 (at pH 1.9), reaching a plateau on day 21 (at pH 1.3) (Figure 4). In the chemical control test, phosphorus extraction was very slow. After three days, 89% of uranium was leached (pH 2.2, Eh +650 mV, Fe^{3+} 1.3 g/L), while the simultaneous phosphorus leaching yield was only 4%, illustrating that the selective leaching of uranium is possible. The final phosphorus leaching yields on day 28 were 90% and 24% for bioleaching and chemical leaching, respectively. The phosphorus bioleaching yield was in accordance with other studies: P yields of 97% from low-grade fluorapatite [28] and 70% from phosphate rock were achieved [27]. However, also lower P yields have been reported, like 28% for P concentrate [28], 20–30% [29] and 12% for phosphate rock [36]. The studies presenting lower phosphorus yields utilized pH ≥ 2.0, which may be the reason for lower dissolution rates. However, mineralogy is also expected to play an important role. In this study, the acid consumption, expressed as kg 95% H_2SO_4 per one ton of ore, was 14.7 and 36.8 kg/t for RGU-1-FA bioleaching and chemical leaching, respectively. Therefore, it can be considered that supplemented elemental sulfur resulted in biogenic sulfuric acid production and decreased the total acid consumption remarkably.

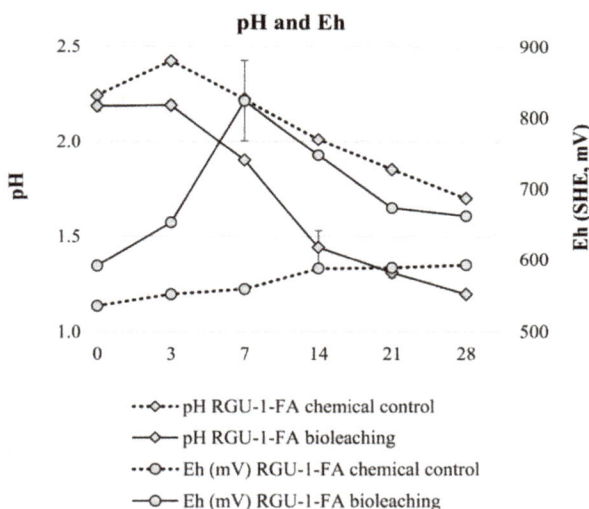

Figure 3. Evolution of pH and Eg during bioleaching and chemical leaching of RGU-1-FA ore; pH and Eh (mV). The x-axis represents the leaching duration (days).

The results shown here illustrate that uranium and phosphorus can be selectively bioleached from ores that contain both uraninite and fluorapatite. A suitable reagent for precipitation/recovery of uranium to concentrate would be hydrogen peroxide, as it can be applied in sulfate-rich solutions at low pH [37]. We consider that the process does not require iron removal after bioleaching since Fe^{3+} was the predominant form of iron in leachate, and therefore excess H_2O_2 is not consumed to oxidize iron. Therefore, after uranium recovery, no regeneration of oxidant (iron) would be needed but spent leachate can be used directly for new uranium leach cycle. Moreover, uranium precipitation would regenerate sulfuric acid as explained in Equation (8) [37].

$$UO_2SO_4 + H_2O_2 + 2H_2O \rightarrow UO_4 \cdot 2H_2O + H_2SO_4 \tag{8}$$

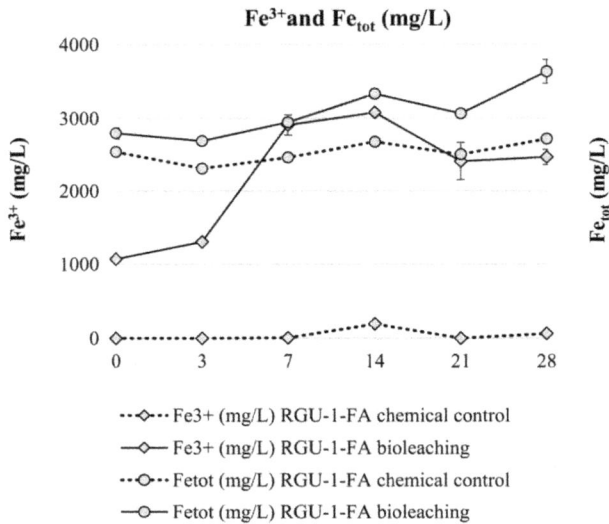

Figure 4. Evolution of Fe^{3+} and total Fe concentrations during bioleaching and chemical leaching of RGU-1-FA ore; iron concentrations (mg/L). The x-axis represents the leaching duration (days).

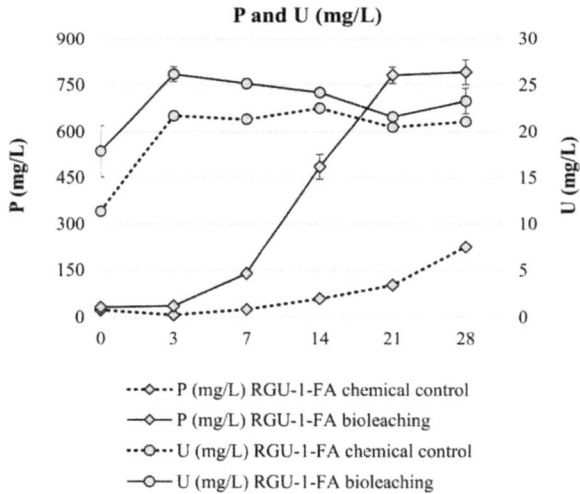

Figure 5. Evolution of P and U concentrations during bioleaching and chemical leaching of RGU-1-FA ore; phosphorus and uranium concentrations (mg/L). The x-axis represents the leaching duration (days).

A schematic process flowsheet has been prepared for the sequential bioleaching of phosphorus and uranium (Figure 6). It is highly noteworthy that the proposed process has been tested only with an artificial ore, by combining fluorapatite ore and standard reference uranium ore. For native ores of various types, the amenability and process parameters must be validated separately.

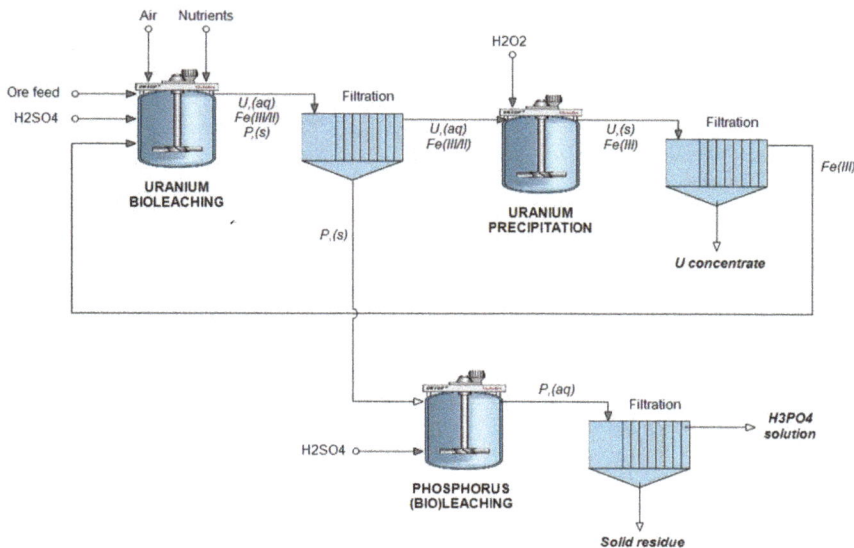

Figure 6. Proposed flowsheet of the sequential bioleaching of phosphorus and uranium.

4. Conclusions

In this research, phosphorus and uranium leaching was studied with bioleaching and chemical leaching, with the final objective to separate current leaching process with two-step sequential leaching. The first leaching step aimed to remove uranium to its own pregnant leach solution for recovery and considered to operate at pH ≥2, Eh +650 mV and Fe^{3+} concentration of ≥1.0 g/L. After uranium extraction and solid-liquid separation, the second leaching step aimed to recover phosphorus from the solid leach residue and was considered to operate at pH ≤ 1.5. Despite not tested here in practice, we consider H_2O_2 precipitation for uranium recovery from the pregnant leach solution as it allows direct reuse of spent leach solution for uranium extraction. It is noteworthy, that phosphorus bioleaching has not been applied on an industrial scale, and therefore the viability of the process has not been tested. In these experiments, despite the high leaching yield for phosphorus, the duration of the process was rather long.

Author Contributions: Conceptualization, J.M., L.W. and P.K.; methodology, J.M., L.W. and P.K.; validation, formal analysis, investigation, resources, data curation and writing the article, J.M., L.W., T.L. and P.K.; visualization, J.M.

Funding: The authors greatly acknowledge the Academy of Finland, Key project funding Forging ahead with research (Decision number 306079, EcoTail project) for funding the research.

Conflicts of Interest: The authors declare no conflict of interest.

References

1. U.S. Geological Survey. *Mineral Commodity Summaries*; U.S. Geological Survey: Reston, VA, USA, 2018; pp. 122–123.
2. Van Vuuren, D.P.; Bouwman, A.F.; Beusen, A.H.W. Phosphorus demand for the 1970–2100 period: A scenario analysis of resource depletion. *Glob. Environ. Chang.* **2010**, *20*, 428–439. [CrossRef]
3. Cordell, D.; Drangert, J.-O.; White, S. The story of phosphorus: Global food security and food for thought. *Glob. Environ. Chang.* **2009**, *19*, 292–305. [CrossRef]
4. Van Kauwenbergh, S.J. *World Phosphate Rock Reserves and Resources*; International Fertilizer Development Center (IFDC): Muscle Shoals, AL, USA, 2010; 50p.

5. U.S. Geological Survey. *Mineral Commodity Summaries*; U.S. Geological Survey: Washington, DC, USA, 2009; 2195p.

6. Cooper, J.; Lombardi, R.; Boardman, D.; Carliell-Marquet, C. The distribution and production of global phosphate rock reserves. *Resour. Conserv. Recy.* **2011**, *57*, 78–86. [CrossRef]

7. Beltrami, D.; Cote, G.; Mokhtari, H.; Courtaud, B.; Moyer, B.A.; Chagnes, A. Recovery of Uranium from Wet Phosphoric Acid by Solvent Extraction Processes. *Chem. Rev.* **2014**, *114*, 12002–12023. [CrossRef] [PubMed]

8. Abu-Eishah, S.I.; Abu-Jabal, N.M. Parametric study on the production of phosphoric acid by the dehydrate process. *Chem. Eng. J.* **2001**, *81*, 231–250. [CrossRef]

9. Komnitsas, K.; Kontopoulus, A.; Lazar, I.; Cambridge, M. Risk Assessment and Proposed Remedial Actions in Coastal Tailings Disposal Sites in Romania. *Miner. Eng.* **1998**, *11*, 1179–1190. [CrossRef]

10. Komnitsas, K.; Lazar, I.; Petrisor, I.G. Application of a Vegetative Cover on Phosphogypsum Stacks. *Miner. Eng.* **1999**, *12*, 175–185. [CrossRef]

11. Salo, M.; Mäkinen, J.; Yang, J.; Kurhila, M.; Koukkari, P. Continuous biological sulfate reduction from phosphogypsum waste leachate. *Hydrometallurgy* **2018**, *180*, 1–6. [CrossRef]

12. International Atomic Energy Agency. *Extent of Environmental Contamination by Naturally Occurring Radioactive Material (NORM) and Technological Options for Mitigation*; International Atomic Energy Agency: Vienna, Austria, 2003; 198p.

13. International Atomic Energy Agency (IAEA). The Recovery of Uranium from Phosphoric Acid. Report of An Advisory Group Meeting Organized by the International Atomic Energy Agency and Held in Vienna, 16–19 March 1987. Available online: https://www-pub.iaea.org/MTCD/Publications/PDF/te_0533.pdf (accessed on 24 January 2019).

14. Gabriel, S.; Baschwitz, A.; Mathonnière, G.; Eleouet, T.; Fizaine, F. A critical assessment of global uranium resources, including uranium in phosphate rocks, and the possible impact of uranium shortages on nuclear power fleets. *Ann. Nucl. Energy* **2013**, *58*, 213–220. [CrossRef]

15. Nuclear Energy Agency. *Uranium 2016: Resources, Production and Demand*; Nuclear Energy Agency: Boulogne-Billancourt, France, 2016; 548p.

16. Venter, R.; Boylett, M. The Evaluation of Various Oxidants Used in Acid Leaching of Uranium. Hydrometallurgy Conference 2009, The Southern African Institute of Mining and Metallurgy, 2009. Available online: https://www.saimm.co.za/Conferences/Hydro2009/445-456_Venter.pdf (accessed on 22 March 2019).

17. Ram, R.; Charalambous, F.A.; McMaster, S.; Pownceby, M.I.; Tardio, J.; Bhargava, S.K. An investigation on the dissolution of natural uraninite ores. *Miner. Eng.* **2013**, *50–51*, 83–92. [CrossRef]

18. Muñoz, J.A.; Gonzáles, F.; Blázquez, M.L.; Ballester, A. A study of the bioleaching of Spanish uranium ore. Part I: A review of the bacterial leaching in the treatment of uranium ores. *Hydrometallurgy* **1995**, *38*, 39–57. [CrossRef]

19. Bosecker, K. Bioleaching: Metal solubilisation by microorganisms. *FEMS Microbiol. Rev.* **1997**, *20*, 591–604. [CrossRef]

20. Rawlings, D.E.; Silver, S. Mining with Microbes. *Biotechnology* **1995**, *13*, 773–778. [CrossRef]

21. Rawlings, D.E. Heavy Metal Mining Using Microbes. *Annu. Rev. Microbiol.* **2002**, *56*, 65–91. [CrossRef]

22. Abhilash, S.S.; Mehta, K.D.; Kumar, V.; Pandey, B.D.; Pandey, V.M. Dissolution of uranium from silicate-apatite ore by *Acidithiobacillus ferrooxidans*. *Hydrometallurgy* **2009**, *95*, 70–75. [CrossRef]

23. Schippers, A.; Sand, W. Bacterial leaching of metal sulfides proceeds by two indirect mechanisms via thiosulfate or via polysulfides and sulfur. *Appl. Environ. Microbiol.* **1999**, *65*, 319–321.

24. Tuovinen, O.H.; Hiltunen, P.; Vuorinen, A. Solubilization of phosphorus, uranium and iron from apatite- and uranium-containing rock samples in synthetic and microbiologically produced acid leach solutions. *Eur. J. Appl. Microbiol. Biotechnol.* **1983**, *17*, 327–333. [CrossRef]

25. Umanskii, A.B.; Klyushnikov, A.M. Bioleaching of low grade uranium ore containing pyrite using *A. ferrooxidans* and *A. thiooxidans*. *J. Radioanal. Nucl. Chem.* **2013**, *295*, 151–156. [CrossRef]

26. Mäkinen, J.; Kinnunen, P.; Arnold, M.; Priha, O.; Sarlin, T. Fluoride Toxicity in Bioleaching of Fluorapatite. *Adv. Mater. Res.* **2015**, *1130*, 406–409. [CrossRef]

27. Bhatti, T.M.; Yawar, W. Bacterial solubilisation of phosphorus from phosphate rock containing sulfur-mud. *Hydrometallurgy* **2010**, *103*, 54–59. [CrossRef]

28. Priha, O.; Sarlin, T.; Blomberg, P.; Wendling, L.; Mäkinen, J.; Arnold, M.; Kinnunen, P. Bioleaching phosphorus from fluorapatites with acidophilic bacteria. *Hydrometallurgy* **2014**, *150*, 269–275. [CrossRef]

29. Xiao, C.Q.; Chi, R.A.; Li, W.S.; Zheng, Y. Biosolubilization of phosphorus from rock phosphate by moderately thermophilic and mesophilic bacteria. *Miner. Eng.* **2011**, *24*, 956–958. [CrossRef]
30. Faye, G.H.; Bowman, W.S.; Sutarno, R. *Uranium ore BL-5—A Certified Reference Material*; CANMET report 1979, 79–4; CANMET, Energy, Mines and Resources Canada: Ottawa, ON, Canada, 1979.
31. IAEA. *Preparation of Gamma-ray Spectrometry Reference Materials RGU-1 RGTh-1 and RGK-1 Report*; International Atomic Energy Agency: Vienna, Austria, 1987.
32. CEN. *Characterization of Waste—Leaching Behaviour Tests—Influence of pH on Leaching with Continuous pH-Control*; CEN/TS 14997; CEN: Brussels, Belgium, 2006.
33. Halinen, A.-K.; Rahunen, N.; Kaksonen, A.H.; Puhakka, J.A. Heap bioleaching of a complex sulfide ore Part I: Effect of pH on metal extraction and microbial composition in pH controlled columns. *Hydrometallurgy* **2009**, *98*, 92–100. [CrossRef]
34. Silverman, M.P.; Lundgren, D.G. Studies on the chemoautotrophic iron bacterium *Ferrobacillus ferrooxidans*. I. An improved medium and a harvesting procedure for securing high cell yields. *J. Bacteriol.* **1959**, *77*, 642–647.
35. Anwar-ul-Haq, M.; Ahmad, I.; Niazi, M.T.; Arif, M.; Mahmood, K. Evaluation of problematic indigenous phosphate deposits for phosphoric acid manufacture. *Fert. Res.* **1990**, *24*, 53–56. [CrossRef]
36. Chi, R.; Xiao, C.; Gao, H. Bioleaching of phosphorus from rock phosphate containing pyrites by *Acidithiobacillus ferrooxidans*. *Miner. Eng.* **2006**, *19*, 979–981. [CrossRef]
37. Gupta, R.; Pandey, V.M.; Pranesh, A.B.; Chakravarty, A.B. Study of an improved technique for precipitation of uranium from eluated solution. *Hydrometallurgy* **2004**, *71*, 429–434. [CrossRef]

Article

Laboratory Scale Investigations on Heap (Bio)leaching of Municipal Solid Waste Incineration Bottom Ash

Jarno Mäkinen [1,*], Marja Salo [1], Jaakko Soini [2] and Päivi Kinnunen [1]

[1] VTT Technical Research Centre of Finland Ltd, P.O. Box 1000, FI-02044 Espoo, Finland;
    Marja.Salo@vtt.fi (M.S.); Paivi.Kinnunen@vtt.fi (P.K.)
[2] Fortum Waste Solutions, Kuulojankatu 1, FI-11120 Riihimäki, Finland; Jaakko.Soini@fortum.com
* Correspondence: jarno.makinen@vtt.fi; Tel.: +358-20-722-4703

Received: 25 March 2019; Accepted: 6 May 2019; Published: 11 May 2019

Abstract: Municipal solid waste incineration bottom ash (MSWI BA) is the main output of the municipal solid waste incineration process, both in mass and volume. It contains some heavy metals that possess market value, but may also limit the utilization of the material. This study illustrates a robust and simple heap leaching method for recovering zinc and copper from MSWI BA. Moreover, the effect of autotrophic and acidophilic bioleaching microorganisms in the system was studied. Leaching yields for zinc and copper varied between 18–53% and 6–44%, respectively. For intensified copper dissolution, aeration and possibly iron oxidizing bacteria caused clear benefits. The MSWI BA was challenging to treat. The main components, iron and aluminum, dissolved easily and unwantedly, decreasing the quality of pregnant leach solution. Moreover, the physical nature and the extreme heterogeneity of the material caused operative requirements for the heap leaching. Nevertheless, with optimized parameters, heap leaching may offer a proper solution for MSWI BA treatment.

Keywords: municipal solid waste incineration; bottom ash; heap leaching; bioleaching

1. Introduction

Incineration has become the typical method for treating municipal solid waste (MSW) in European Union, where 68 million tons of MSW was treated with incineration (MSWI) technology in 2016 [1]. During incineration, the organic content of the MSW is converted to thermal energy that can be utilized in generation of heat and power with simultaneous 90 wt % and 75 vol % reduction of initial waste [2]. The main residue of MSWI is bottom ash (BA), representing approximately 80% of solid incineration rejects [3]. The scrap iron and some other metals of MSWI BA can be separated and utilized by the metal refining industry, followed either by reuse of remaining MSWI BA (e.g., as construction materials) or disposal, if the quality does not allow reuse [2,4,5]. Residual heavy/toxic metals can cause challenges in both reuse and disposal of MSWI BA, but also serve as a secondary source for valuable elements [5–7].

MSWI BA is an extremely heterogeneous residue stream including glass, synthetic ceramics, natural minerals, unburned organic matter and a variety of metals in different forms [3,8–10]. The particle size is heterogeneous varying from a few μm particles to the chunks of several centimeters, effecting also significantly to the elemental composition; in particular, magnetic metals (Fe, contaminated with e.g., Zn and Cu) seem to accumulate to fine fractions, while diamagnetic metals (Al and Cu) are found in all size fractions [3,9]. Moreover, different elements tend to accumulate on different mineralogical fractions. Heavy metals have been reported to concentrate on glass matrix; Cr, Zn and Mn have been incorporated into spinels, while Cu and Pb associated to Fe, Sn and Zn metallic inclusions [9,10]. Despite the incineration process, some metals may be partly in the metallic phase instead of their oxidized form (e.g., copper wires) [3,10]. Also, Cu sulfides may be formed in the process, even though CuO is the dominant form [8].

The above-mentioned characteristics of MSWI BA make the treatment for metal removal/recovery demanding. One proposed solution for the challenge has been hydrometallurgical treatment, however this particular technology has been studied more with other MSWI reject classes (e.g., fly ashes [11–13]), despite the fact that a vast majority of Cu and Zn end up in bottom ash [6]. A great majority of current bottom ash research concentrates mainly on more mild leaching conditions (or water leaching/washing), and is linked often more to the metal release during the aging or after landfilling or utilization as material [14–17]. With MSWI BA, it has been observed that Zn and Cu dissolution start at pH 4–5 and pH 3–4, respectively [14,17,18]. In harsh acidic conditions (acid: 3 M H_2SO_4, temperature: 80 °C, leaching duration: 2 h, solid/liquid ratio: 20%) and applying reactor apparatus, Mo and V recoveries were high (>80%), but for Ni and Cu low (<40%) [7]. Similar leaching yields for Ni and Cu was observed with HNO_3; moreover, Zn leaching yield was found low (<40%) [19]. In addition to low-to-moderate yields of these key metals from MSWI BA, challenges may also be foreseen in high concentrations of iron in the final leachate (treatment costs due to iron removal), as well as in solid-liquid separation due to gel-like formations (filtration costs) [7,19]. Organic acids have also been studied for metals removal from MSWI BA. By using 1 M citric acid and 1% solid/liquid ratio, Cu and Zn leaching yields of >90% have been reached; however, when increasing the solid/liquid ratio to 5%, Cu and Zn leaching yields decreased to <65% [20]. Bioleaching of MSWI BA with iron and sulfur oxidizing microorganisms has been reported to reach high leaching yields for Cu (100%) and Zn (80%) when the system is supplemented with bioleaching microorganisms, elemental sulfur, ferrous iron and 10% (*v/v*) of MSWI BA [21]. The drawback, compared to chemical leaching tests, was the long duration of leaching.

It has been suggested that acid leaching of MSWI BA may not be reasonable, due to requirement of low pH and low solid/liquid-ratio [5]. In this paper, the approach of acid heap leaching process instead of stirred tank reactor leaching is introduced for treating MSWI BA for improving these factors: in acid heap leaching higher solid/liquid ratio can be applied and residual acid circulated back to the heap decreasing the acid consumption. Additional benefits include the lack of expensive reactor vessels, agitation instruments and motors, as well as tolerance for larger particles, removing the need of comminution. The expected negative impacts of acid heap leaching application were expected to be slow leaching kinetics (i.e., long leaching duration) and blockage/channeling of heap due to release of fine particles. Both these may result in poor leaching yields of target elements. It is noteworthy, that temperature increase in MSWI BA leaching has not dramatically increased leaching yields [7], and therefore, restriction of heap leaching to stay on ambient temperature is expected not to jeopardize the process.

For the leaching medium, H_2SO_4 was selected in this work. This enabled two different leaching strategies: the addition of commercial H_2SO_4 to the heap or the production of H_2SO_4 by sulfur oxidizing bacteria from an elemental sulfur source (Equation (1)). The autotrophic sulfur oxidizers, such as *Acidithiobacillus thiooxidans*, are acidophilic microorganisms, which utilize reduced or elemental sulfur as their energy source by using oxygen as an electron acceptor, and fix carbon from the atmosphere as CO_2 (for a review, see [22]).

$$S^0 + 3O_2 + 2H_2O \rightarrow 2SO_4^{2-} + 4H^+ \tag{1}$$

Despite the origin of the acid, it will be consumed to overcome the high buffer capacity of MSWI BA by dissolving carbonates, hydroxides, silicates and oxides, but also to dissolve target metals of Zn and Cu. MSWI BA may also contain plenty of metals that have not oxidized in the incineration process, but remain in the metallic form [7]. Especially metallic Cu is resistant to acid attack, but can be efficiently leached with Fe^{3+} oxidant (Equation (2)), which can be regenerated rapidly by iron oxidizing bacteria (Equation (3)) [23]. Another encountered form of Cu in MWSI BA is Cu_2S [8], which can be dissolved by Fe^{3+} oxidant (Equation (4)), followed by biogenic iron regeneration (Equation (3)) and oxidation of elemental sulfur to sulfuric acid by sulfur oxidizing bacteria (Equation (1)).

$$Cu + 2Fe^{3+} \rightarrow Cu^{2+} + 2Fe^{2+} \tag{2}$$

$$2Fe^{2+} + 0.5O_2 + 2H^+ \rightarrow 2Fe^{3+} + H_2O \tag{3}$$

$$Cu_2S + 4Fe^{3+} \rightarrow 2Cu^{2+} + 4Fe^{2+} + S \tag{4}$$

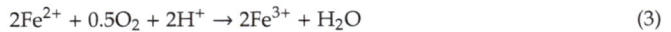

Typical examples of iron oxidizing bacteria are, e.g., *At. ferrooxidans* and *Leptospirillum ferrooxidans*, which are acidophiles and obtain their oxygen and carbon from the atmosphere (for a review, see [22]). The biological oxidation of Fe^{2+} to Fe^{3+} can proceed more than million times faster than the abiotic oxidation by oxygen [24]. Therefore, the biogenic iron regeneration process can be an economically attractive method for obtaining the oxidant in certain cases [25].

The objective of this study was to understand how MSWI BA behaves in acid heap leaching process, with the target of dissolving valuable elements of Zn and Cu. In addition to these elements, Fe and Al dissolution was monitored as their dissolution in the process would have negative impacts for the complete process (higher acid consumption, more complex pregnant leach solution). The behavior and effects of bioleaching microorganisms were studied in heap leaching systems to reveal whether they offer advantages for leaching, and preliminary investigate how they could be applied in the heap leaching process.

2. Materials and Methods

2.1. MSWI BA Sample Material

Fortum Environmental Construction Oy provided a 300 kg MSWI BA sample, which was classified to a particle size of 3–40 mm by Geological Survey of Finland (GTK), and then transported to VTT Technical Research Centre of Finland for leaching studies. The selection of wide particle size range was justified by occurrence of target metals (Zn, Cu) in many fractions, and due to their possible adherence to larger particles [3]. Moreover, objective was to study a robust process that does not require pre-treatment, but allows a simple process for the majority of generated MSWI BA. After the transport, the sample was homogenized by mixing and dividing it to four equal shares on a tarpaulin. The material was very brittle, generating small amount of <1 mm powder already in homogenization. The generated small amount of this finer fraction was not removed, but left mixed with the original material for experiments. This decision was justified by maintaining the representativeness of the sample (sieving increased the generation of fine fractions), and possible loss of Cu and Zn. It is noteworthy that fine material in heap leaching may cause disruption in permeability and solution/gas flows. MSWI BA sample was also extremely heterogeneous, as the ash agglomerates occurred in different sizes and shapes, visibly consisting of a variety of materials in different shares. In addition, metallic pieces (e.g., short copper wires, pieces of stainless-steel cutlery) were observed. The heterogeneity caused difficulties in the determination of chemical composition. As reported, the average chemical composition of MSWI BA differs very slightly [3] and therefore, analysis provided by the material owner was used as an average composition (Table 1).

Table 1. Average composition of municipal solid waste incineration bottom ash (MSWI BA) sample material.

Element	Fe	Ca	Al	Zn	Cu	Pb	Cr	Ni	Sb	As
g/kg	89	62	38	3.2	1.9	1.5	0.16	0.11	0.023	0.019

2.2. Adaptation of Microorganisms to the Material

A mixed culture of iron and sulfur oxidizing microorganisms, enriched from acidic mine waters, was used. The mixed culture contained Marinobacter sp., Acidithiobacillus (such as *At. ferrooxidans*, *At. thiooxidans*, *At. albertensis*, *At. ferrivorans*), Leptospirillum (*L. ferrooxidans*), Cuniculiplasma, Nitrosotenius and Ferroplasma. The culture was adapted to washed MSWI BA in shaken flasks in incubator-shaker (30 °C and 150 rpm) by gradually increasing the solid/liquid ratio up to 10% (*w/v*) in 0 K media (composition presented in the Table 2). Solution pH was regularly adjusted to near 2.0 with 95% H_2SO_4. A separate adaptation with an addition of elemental sulfur was also created (1 g S^0/100 mL); here, the bacteria could produce H_2SO_4 through their own sulfur oxidizing reactions (Equation (1)) and the amount of added H_2SO_4 was reduced. With elemental sulfur addition, the pH remained lower and the RedOx potential higher than in the adaptations without elemental sulfur, resulting in increased heavy metal accumulation in this adaptation.

Table 2. The used 0 K medium.

Chemical Compound	Concentration (g/L)
$(NH_4)_2SO_4$	3.0
KCl	0.1
K_2HPO_4	0.5
$MgSO_4 \cdot 7H_2O$	0.5
$Ca(NO_3)_2 \cdot 4H_2O$	0.14

2.3. Preliminary Tests

Prior to heap leaching experiments, acid consumption and leaching behavior were studied. First, the sample material was washed with 5-fold weight of distilled water compared to the sample to remove the majority of chloride. The washed sample (20 g) was mixed with water (200 g) in 500 mL Erlenmeyer flask. The flasks were placed to the orbital shaker (150 rpm, 30 °C; Stuart SI-500), and pH was maintained at 1.0, 1.5, 2.0, 2.5 or 3.0 for 8 h with 95% H_2SO_4 using manual titration. In addition to the pH static tests, two experiments with Fe^{3+} were conducted to clarify the effect of oxidant for leaching. In these experiments, the bioleaching culture was used to oxidize iron by adding inoculum (25 mL), 0K medium (225 mL) and solid $FeSO_4 \cdot 7H_2O$ (to reach Fe^{2+} concentration of 4.5 or 9.0 g/L). The solution was incubated (150 rpm, 30 °C) until the RedOx potential had risen to +650 mV, illustrating biological oxidation of Fe^{2+} to Fe^{3+} (Equation (3)). The biologically produced ferric solutions (200 mL, with measured Fe^{3+} concentrations of 5.3 and 8.4 g/L), were mixed with the sample material (20 g). The working protocol was the same as with the pH-static tests (150 rpm, 30 °C, pH 2.0, 8 h). After the 8-hour leaching tests, all leachates were filtrated (0.45 µm) and analyzed for dissolved metals with ICP-OES (by external accredited laboratory Metropolilab Oy, Helsinki, Finland). The pH and RedOx were measured with a Consort multi-parameter analyzer C3040, with Van London-pHoenix Co. electrodes (Ag/AgCl in 3 M KCl).

2.4. Column Experiments

The experiments were conducted in four columns (height 31 cm, diameter 10 cm) with approximately 2000 grams of MSWI BA sample material. The material was first washed by filling the column containing the material with distilled water for 15 min, then draining the column, and repeating this process for a total of five times. Four different configurations were chosen for the column tests (see Table 3).

Table 3. Column configurations used in the experiment.

	COL I	COL II	COL III	COL IV
Inoculation of microorganisms	No	Yes	Yes	Yes
Elemental sulfur addition	No	No	No	Yes
Circulation solution volume (L)	2	2	10	2

Three columns were operated as bioleaching heaps (COL II, COL III, COL IV) with an addition of 400 mL of adapted inoculum (COL II and COL III adaptation without elemental sulfur, COL IV adaptation with elemental sulfur), while one column (COL I) was kept as a chemical control. In COL IV, 104 grams of elemental sulfur (5% of column material) was agglomerated to the sample material with an addition of 18.73 g of 0.1 M H_2SO_4 (the same amount of acid was added to the other columns as well). The bottom and the top of the columns contained glass beads (diameter 6 mm) and perforated plates to prevent the washout of the material, as well as to help to distribute the influent evenly to the column. In addition, glass wool was used in the bottom to keep the finer fractions inside the column. The column solutions (0 K media) were circulated through circulation tanks (plastic containers). The solution was not removed from the system, but it was collected and fixed for the pH in the circulation tank, and pumped back to the respective column. Evaporation was compensated by adding distilled water to the circulation tank. The flow rate of solutions was adjusted with peristaltic pumps (Watson Marlow 205S). The pH and RedOx were measured with a Consort multi-parameter analyzer C3040, with Van London-pHoenix Co. electrodes (Ag/AgCl in 3 M KCl) from both column effluents and circulation tanks. Solution samples were taken from circulation tanks, followed by filtration (0.45 µm), and analysis for dissolved metals with ICP-OES (by external accredited laboratories Metropolilab Oy, Helsinki, Finland, and Labtium Oy, Espoo, Finland).

Heap experiments lasted for a total of 139 days with different operational phases. First, the columns were operated as standard irrigation heaps by pumping the solution to the top of the heap (5 $L/m^2 \cdot h$, days 0–54). This operation manner and volume flow is typical for acid heap leaching [26]. The pH of the circulation tanks was adjusted daily to approximately 2.0 with 95% H_2SO_4. After the irrigation trial, the flow direction was reversed on day 54 and the influent was pumped upwards from the bottom, consequently flooding the column (this is a typical operating manner in CEN/TS column tests, often used for MSWI BA research [15]). The solution flow was maintained at 5 $L/m^2 \cdot h$. From day 82 onwards, more intense acid addition was deemed necessary to overcome the acid neutralization potential of the material. The circulation tank pH was adjusted daily to 1.3 and the solution flow was increased to 10 $L/m^2 \cdot h$. From day 119 onwards, the circulation tanks were aerated by pumping air (1.5 L/min adjusted by a rotameter, Kytola Instruments Oy, Muurame, Finland) through a tube and a plastic nozzle to study the effect of iron oxidizing bacteria. Moreover, the solution flow rate was decreased back to 5 $L/m^2 \cdot h$ and COL III supplemented with 400 mL of re-inoculation, to secure the presence of active iron oxidizing bacteria.

3. Results

3.1. Preliminary Experiment

Acid consumption varied from 27 to 556 kg 95% H_2SO_4 per ton of sample material, depending on the selected pH (Figure 1). Acid consumption increased linearly with the decrease of pH from 3.0 to 1.5, but then raised dramatically at pH 1.0. Copper and zinc dissolution increased rather linearly with the pH decrease, while iron dissolution was closer to an exponential rise (Figure 1).

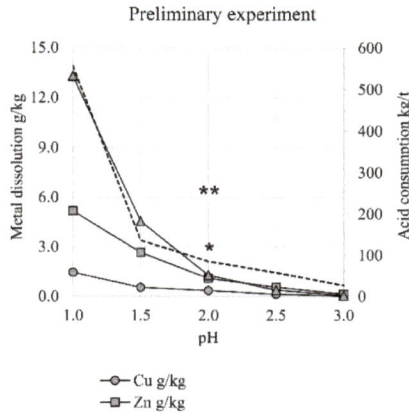

Figure 1. Leaching of the MSWI BA sample material in different pH values in shake flasks (pH 1.0–pH 3.0; symbols "*" and "**" illustrate copper dissolution when Fe^{3+} concentration was 5.3 and 8.4 g/L, respectively). Acid consumption expressed as kg 95% H_2SO_4 per ton of sample material.

Iron dissolution is an unwanted phenomenon in leaching processes, as it consumes extra acid and decreases the quality of produced leachate. The leaching process was estimated to be the most beneficial at pH 2.0, as it resulted in similar zinc and iron concentrations in the leachate and consumed only 85 kg 95% H_2SO_4/t of sample material. The introduction of an oxidizing agent (Fe^{3+}) into the leaching media had drastic effects: in the static pH 2.0, the copper concentrations in the solutions rose from 35 mg/L (no added Fe^{3+}) to 300 and 642 mg/L, when the leaching solutions contained 5.3 and 8.4 g/L of Fe^{3+}, respectively (Figure 1).

3.2. Column Experiments

During the material washing, chloride concentration of the washing solution decreased from 680 to 52 mg/L, but only negligible amounts of metals (Cu, Zn, Fe, Al) were detected. After washing, the heap leaching experiments were started by irrigation tests with the target of pH 2.0, according to the preliminary experiments. Despite the identical pH and flow rate of the column influents, differences in effluents were observed. With all inoculated bioleaching columns effluents decreased to the pH 2–4 level. However, this took much longer in the absence of elemental sulfur (COL II) (Figure 2: days 0–54). The pH in the abiotic column (COL I) remained rather stable at pH 7.5–8.5. The acid consumption was clearly lower in the column with supplemented elemental sulfur (COL IV) compared to the other columns (Figure 2: days 0–54); with all columns, acid consumption was remarkably lower than with preliminary experiments. The RedOx-potential rose rapidly to +450–550 mV level in the inoculated columns; some slower increase was also observed in the abiotic column reaching only +300–400 mV level (Figure 3: days 0–54). Despite the rather low pH of the inoculated columns, iron, zinc and copper dissolutions were negligible during the irrigation trial (Figures 3 and 4: days 0–54). It is noteworthy that the increase in copper dissolution during the early days was caused by the inoculum that contained a remarkable copper impurity that was accumulating to the inoculums during the adaptations. The adaptation culture supplemented with elemental sulfur had the highest copper content, which is clearly seen in COL IV (Figure 4).

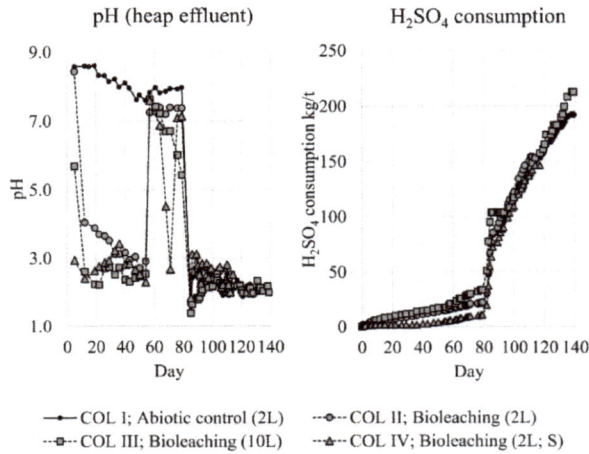

Figure 2. The pH of the column effluents, and cumulative H_2SO_4 consumption (kg pure acid per ton of sample material) during the experiments (H_2SO_4 used to maintain pH of the circulation tank at desired value). Circulation tank volume presented in parentheses; symbol "S" illustrates the column supplemented with elemental sulfur.

The low metals dissolution was most likely caused by uneven distribution and channeling of leaching solution inside the columns. Therefore, the test protocol was changed on day 54 to flooding in order to maximize the contact between solution and sample material. In flooding, the pH rose rapidly in all columns to pH >7 (Figure 2: days 54–79); as the influent parameters were maintained at pH 2.0 and 5 L/m^2·h, all acid was consumed completely inside the column. The RedOx-potential decreased in all columns to +150–350 mV (Figure 3: days 54–79), referring possibly to acid and/or oxygen depletion in columns. Virtually no metals (Cu, Zn, Fe, Al) were leached during this flooding period (Figures 3 and 4: days 54–79). However, an anomaly was observed in pH and RedOx-potential between days 68–71 in the column supplemented with elemental sulfur (COL IV). The pH decreased to pH <3, and the RedOx-potential peaked to +450 mV. The reason for the anomaly was not found, but an explanation could be an oxygen leak in the junction of the column top and effluent tubing. This would cause the oxidation of elemental (or intermediate) sulfur found from the tubings.

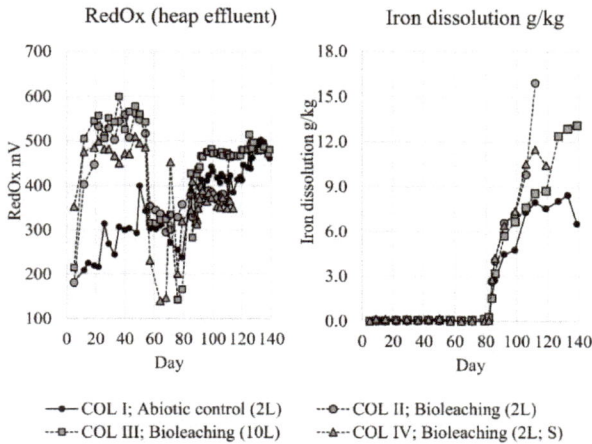

Figure 3. RedOx-potential in column effluents, and iron dissolution from sample material (grams of dissolved iron per kg of sample material) in column experiments. Circulation tank volume presented in parentheses; symbol "S" illustrates the column supplemented with elemental sulfur.

With the original circulation tank pH and flow rate back to the columns there was not enough acid to perform the targeted leaching of metals. Therefore, the flooding trial protocol was changed on day 79, by decreasing the pH in circulation tanks to pH 1.3 and increasing the influent flow rate to 10 L/m^2·h. This caused a drastic drop in the pH with all columns to the level of 1–3 and later stabilizing to pH 2.0–2.5 (Figure 2: days 79–119) with subsequent increase in the acid consumption. The rise of the RedOx-potential was clearly lower with inoculated COL II and COL IV. The highest RedOx-potential was reached with the inoculated column with 10L circulation solution volume (COL III), followed by the abiotic column (COL I) (Figure 3: days 79–119). Iron and zinc dissolution started rapidly when introducing more acid to the columns (Figures 3 and 4: days 79–119). The highest iron dissolution seemed to occur in the inoculated columns COL II and COL IV. However, the iron dissolution in the column supplemented with elemental sulfur (COL IV) collapsed later on day 112, but even intensified without elemental sulfur (COL II). For zinc dissolution, the inoculation of microorganisms had a positive effect, but supplemented elemental sulfur decreased the dissolution. For copper, the dissolution was generally slower and tended to level, especially with the columns that had a lower RedOx-potential (COL II and COL IV). The dissolution of copper increased linearly only with the COL III. Unfortunately, two columns were lost during the intensified flooding trial: the COL IV on day 114 and the COL II on day 119. Column tubings broke during the night, and all column and circulation tank solutions were lost.

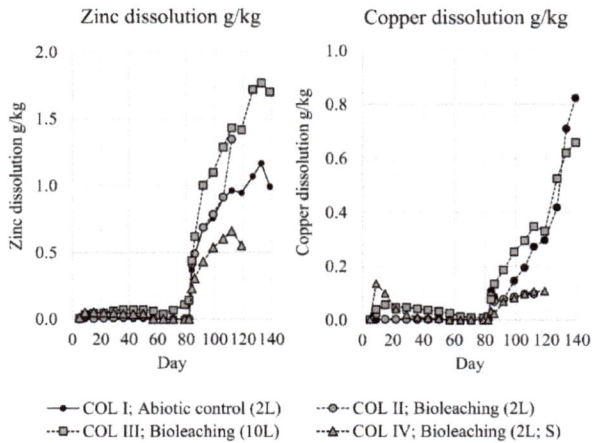

Figure 4. Zinc and copper dissolution from sample material (grams of dissolved iron per kg of sample material) in column experiments. Circulation tank volume presented in parentheses; symbol "S" illustrates the column supplemented with elemental sulfur.

After the loss of two inoculated columns, a final flooding test was done for the remaining columns (COL I and COL III) by adding aeration to the circulation tanks on day 119. Simultaneously, the influent flow rate was decreased back to 5 L/m²·h, and the inoculated column COL III was re-inoculated to secure the presence of active iron oxidizing bacteria. These changes did not have remarkable effects on the pH or acid consumption (Figure 2: days 119–139). The RedOx-potential remained stable with the inoculated column COL III, but increased with the abiotic column COL I, so that both operative columns reached +500 mV level (Figure 3: days 119–139). During the aeration, iron and zinc dissolution tended to level off. However, copper dissolution intensified clearly in both columns (Figures 3 and 4: days 119–139).

In this study, aluminum dissolution was not monitored as often as iron. Nevertheless, with several measurements it was seen that aluminum required stronger acid addition (that occurred from day 79 onwards) for rapid leaching, reaching the final 6–12 g/kg leaching yield level. Therefore, aluminum and iron behaved similarly in terms of leaching. The final leaching yields for zinc, copper, iron and aluminum are presented in Table 4.

Table 4. The final leaching yields in column experiments.

	COL I	COL II	COL III	COL IV
Zn, leaching yield %	32.1	42.9	53.4	17.8
Cu, leaching yield %	44.0	5.8	34.8	5.9
Fe, leaching yield %	7.6	18.1	14.8	12.0
Al, leaching yield %	26.2	30.5	31.0	17.5

4. Discussion

When approaching the challenge of removing heavy metals from MSWI BA, several factors must be stressed. Iron and aluminum were the main metallic components of the sample material, while targeted zinc and copper occurred in lower concentrations. Therefore, certain leaching selectivity must be obtained to avoid extra use of leaching chemicals, but also to produce a solution that can be economically purified to metal products and inert residues. It was seen that all heavy metals required rather strong acid addition and leaching was not possible before the effluent reached pH < 3. This was slightly controversial to some earlier leaching studies [14,17,18]. However, it is noteworthy that refereed studies were done in agitated bottom ash slurry, where leaching may be different, as compared to heap leaching; moreover, these experiments were done in batch-wise while this study examined continuous heap leaching. However, the difference does not possibly link only to the actual dissolution reaction inside the column, but to later hydroxide precipitation of heavy metals when the pH rises, if strongly neutralizing sectors still occur in the heap. For example, the precipitation of $Fe(OH)_3$, $Al(OH)_3$, $Cu(OH)_2$, $Zn(OH)_2$ and $Fe(OH)_2$ occur at pH > 3.5, pH > 4.0, pH > 6.5, pH > 8.0 and pH > 8.5, respectively (the exact pH depends greatly of residual concentration of a studied metal) [27,28]. Therefore, the whole heap must reach and maintain acidic conditions to enable the liberation of heavy metals. In this study, the actual leaching was shown to require a much lower pH than required by the formation of hydroxides (preliminary experiment; Figure 1), indicating that not much metal hydroxides are present in MSWI BA, which has also been shown earlier [8–10]. However, the preliminary experiment also showed that too high an acid addition caused drastic liberation of iron, and that the heap should not be operated below pH 2.0. Later in the column experiments, it was seen that when pH 2.0 was reached in the column effluents, iron and aluminum dissolutions were already tenfold, as compared to zinc and copper. On day 112 (the last recorded acid consumption before two columns were lost), acid consumption in the columns had reached 142–154 kg 95% H_2SO_4/t, compared to 85 kg 95% H_2SO_4/t of preliminary experiment at pH 2.0. Therefore, it can be estimated that the acid addition was too strong and too rapid in columns. When the first parts of the columns reached pH < 2 and started to liberate iron and aluminum, the final parts of columns still possessed some acid neutralization potential and the effluent remained at pH >2. This led to prolonged acid addition and uncontrolled dissolution of unwanted elements. However, we assume that this can be neglected by increasing the influent flow rate but maintaining the influent pH at 2.0–3.0. This secures that enough acid is supplemented to the heap, but the pH gradient in the heap is not too severe. It was also observed that MSWI BA may require a more rapid influent flow rate than typical applications with ores, due to the material characteristics and the risk of channeling effect.

For copper, it was shown in the preliminary and column experiments that dissolution was intensified rapidly by raising RedOx-potential, by aeration and/or iron oxidizing bacteria. This is most likely linked to the presence of metallic copper in the MSWI BA, which requires oxidant, e.g., Fe^{3+}, to dissolve. The role of iron oxidizing bacteria would be intensified Fe^{2+} to Fe^{3+} regeneration for improved copper leaching (the iron regeneration loop presented in Equations (2) and (3)). In addition, iron oxidation may benefit heap leaching by precipitating excess iron as jarosite, schwertmannite or goethite in sulfate rich solutions [29]; these reactions also generate acid that could be reused in heap leaching. For the above-mentioned reasons, it may be justified to add aeration to the heap leaching process of MSWI BA. However, the role of iron oxidizing bacteria is more complex. In the column experiments presented here, no clear benefit was seen between the abiotic aerated (COL I) and the inoculated aerated column (COL III) in copper leaching; both reached high RedOx-potential. However, it is very possible that the abiotic column and/or circulation tank was taken over by microorganisms during the aerated period, as the system was not completely sealed, and the material was not sterilized. Despite the fact that no microorganisms were found in a microscopic examination of the circulation tank solution, the presence of bacteria could not be ruled out. Nevertheless, an industrial heap leaching process cannot be operated in a sterile mode, and most likely, the circulation pond would offer suitable conditions for iron oxidizing bacteria. Another studied biochemical possibility, the oxidation of supplemented elemental sulfur and biogenic sulfuric acid production by sulfur oxidizing bacteria, was also not completely understood in this experiment. During the early days of operation (days 0–54), the inoculated column with supplemented elemental sulfur (COL IV) consumed only 25–35% acid compared to the other columns, which proved the original theory of using elemental sulfur in biological treatment of MSWI BA to decrease the acid consumption. However, as the irrigation approach resulted in slow leaching, the operative mode was changed to flooding to speed up the leaching (day 54 onwards), which caused the dramatic pH increase to the unsuitable levels for, e.g., *At. thiooxidans* [22]. Therefore, it is justified to assume that the sulfur oxidizing bacteria perished, or at least were strongly inhibited by the high pH. Later, the acid consumption of the column with supplemented elemental sulfur was similar to the other columns, showing that the bacteria were no longer active enough, and the true biological acid generation potential remained unclear. However, it was seen that elemental sulfur disrupted at least zinc and copper dissolution, perhaps due to the decreased RedOx-potential.

A simple examination of chemical costs (i.e., H_2SO_4) versus the market value of liberated metals (Cu and Zn) was done (Table 5) to check if the process had any economic viability. The data of COL I was used for these studies (end of experiment on day 139) as it seemed most successful in terms of economic aspects (high Cu and Zn dissolution, lowest Fe dissolution). With the sulfuric acid price of 200 $/t the chemical costs were app. 4.5 times higher than value of liberated metals. As many other costs would be generated by the complete process and also metals would be recovered from the leachate, it is obvious that (1) great improvements must be obtained in the process in terms of leaching yields and acid consumption, (2) an extremely cheap sulfuric acid source must be utilized and (3) added value must be generated for treated MSWI BA.

Table 5. Comparison of sulfuric acid costs to the value of liberated valuable metals (per ton of treated MSWI BA) [30,31].

	Content in MSWI BA	Leaching Yield	Recovered or Consumed per t	Price per t	Revenue per t	Cost per t
Cu	1.9 kg/t	44.0%	0.832 kg/t	6410 $/t [30]	5.36 $/t	
Zn	3.2 kg/t	32.1%	1.02 kg/t	2840 $/t [30]	2.91 $/t	
H_2SO_4			192 kg/t	200 $/t [31]		38.40 $/t

After the heap leaching, valuable metals must be recovered from the pregnant leach solution. This can be done by conventional technologies, such as chemical precipitation, solvent-extraction or ion-exchange [27,28,32]. Moreover, other elements and compounds must be removed from the leachate, to allow water discharge back to environment or circulation back to process. The iron and aluminum removal can be conducted by chemical precipitation with lime, which also results in precipitation of calcium sulfate from the solution [32,33]. Residual elements can also be removed before discharge/recycling by chemical precipitation with lime, if pH is raised to basic area [27,33].

Finally, it is noteworthy that sulfuric acid leaching system has its limitations as some elements are not mobile with this chemistry (e.g., lead forms insoluble $PbSO_4$ in H_2SO_4 leaching) [34]. One possibility is to utilize organic acids that are known to dissolve variety of metals, including Pb [18,20,34]. Organic acids can also be produced by heterotrophic bioleaching method, studied with MSWI fly ashes [35,36]. Currently, it cannot be stated which leaching chemistry is superior for the MSWI BA treatment, and more research is needed regarding acid consumption (and consequent acid cost), leaching duration and leaching yields, obtained leachate composition and the effect of these factors on the down-stream processing possibilities with water circulation. Moreover, sequential leaching, as well as additive chemicals (e.g., pH buffers) may be considered to utilize the strengths of different leaching chemistries and avoid the limitations.

5. Conclusions

MSWI BA is the main output of municipal solid waste incineration process, both in mass and volume. It contains some heavy metals that possess market value, but may also limit the reuse. In this study, we illustrated a robust and simple heap leaching method for recovering zinc and copper from the MSWI BA. Leaching yields for zinc and copper varied between 18–53% and 6–44%, respectively. For intensified copper dissolution, aeration was needed. The main contaminants, iron and aluminum, were easily liberated from the material by sulfuric acid, setting limitations for the industrial utilization of the process. Moreover, the extreme heterogeneity (elemental composition, mineralogy, size, dimension, porosity and wearing fragility) of the material is challenging and seems to cause very different physical heap behavior and requirements compared to the heap leaching of, e.g., ores. With the current results, economics are very challenging for the acid heap leaching process, requiring improvements on leaching efficiency, introducing a cheaper sulfuric acid source, and clear value increase for treated MSWI BA. Despite the challenges, we stress that the heap leaching of MSWI BA should be further studied, with the two main aims of optimizing the leaching chemistry and duration, and physical questions of irrigation.

Author Contributions: Conceptualization, J.M., M.S., J.S. and P.K.; methodology, J.M., M.S. and P.K.; validation, formal analysis and investigation, J.M. and M.S.; resources and data curation M.S. and J.M.; writing—original draft preparation, review and editing, J.M., M.S., J.S. and P.K. visualization, J.M.

Funding: The authors greatly acknowledge the Business Finland Circular Metal Ecosystem (CMEco) project (8161/31/2016) for funding the research.

Conflicts of Interest: The authors declare no conflict of interest.

References

1. Eurostat, Eurostat Database, Municipal Waste. 2018. Available online: https://ec.europa.eu/eurostat/statistics-explained/index.php/Municipal_waste_statistics (accessed on 5 November 2018).
2. Sabbas, T.; Polettini, A.; Pomi, R.; Astrup, T.; Hjelmar, O.; Mostbauer, P.; Cappai, G.; Magel, G.; Salhofer, S.; Speiser, C.; et al. Management of municipal solid waste incineration residues. *Waste Manag.* **2003**, *23*, 61–88. [CrossRef]
3. Chimenos, J.M.; Segarra, M.; Fernández, M.A.; Espiell, F. Characterization of the bottom ash in municipal solid waste incinerator. *J. Hazard. Mater.* **1999**, *64*, 211–222. [CrossRef]
4. Müller, U.; Rübner, K. The microstructure of concrete made with municipal waste incinerato bottom ash as an aggregate component. *Cem. Concr. Res.* **2006**, *36*, 1434–1443. [CrossRef]
5. Todorovic, J.; Ecke, H. Treatment of MSWI Residues for Utilization as Secondary Construction Minerals: A Review of Methods. *Miner. Energy* **2006**, *3*, 45–59. [CrossRef]
6. Kuo, N.-W.; Ma, H.-W.; Yang, Y.-M.; Hsiao, T.-Y.; Huang, C.-M. An investigation on the potential of metal recovery from the municipal waste incinerator in Taiwan. *Waste Manag.* **2007**, *27*, 1673–1679. [CrossRef]
7. Agcasulu, I.; Akcil, A. Metal Recovery from Bottom Ash of An Incineration Plant: Laboratory Reactor Tests. *Miner. Process. Extr. Metall. Rev.* **2017**, *38*, 199–206. [CrossRef]
8. Bayuseno, A.P.; Schmahl, W.W. Understanding the chemical and mineralogical properties of the inorganic portion of MSWI bottom ash. *Waste Manag.* **2010**, *30*, 1509–1520. [CrossRef] [PubMed]
9. Kowalski, P.R.; Kasina, M.; Michalik, M. Metallic Elements Fractionation in Municipal Solid Waste Incineration Residues. *Energy Procedia* **2016**, *97*, 31–36. [CrossRef]
10. Wei, Y.; Shimaoka, T.; Saffarzadeh, A.; Takahashi, F. Mineralogical characterization of municipal waste incineration bottom ash with an emphasis on heavy metal-bearing phases. *J. Hazard. Mater.* **2011**, *187*, 534–543. [CrossRef] [PubMed]
11. Fedje, K.K.; Ekberg, C.; Skarnemark, G.; Steenari, B.-M. Removal of hazardous metals from MSW fly ash—An evaluation of ash leaching methods. *J. Hazard. Mater.* **2010**, *173*, 310–317. [CrossRef]
12. Weibel, G.; Eggenberger, U.; Kulik, D.A.; Hummel, W.; Schlumberger, S.; Klink, W.; Fisch, M.; Mäder, U.K. Extraction of heavy metals from MSWI fly ash using hydrochloric acid and sodium chloride solution. *Waste Manag.* **2018**, *76*, 457–471. [CrossRef] [PubMed]
13. Herck, P.V.; Bruggen, V.; Vogels, G.; Vandecasteele, C. Application of computer modelling to predict the leaching behaviour of heavy metals from MSWI fly ash and comparison with a sequential extraction method. *Waste Manag.* **2000**, *20*, 203–210. [CrossRef]
14. Meima, J.A.; Comans, R.N. The leaching of trace elements from municipal solid waste incinerator bottom ash at different stages of weathering. *Appl. Geochem.* **1999**, *14*, 159–171. [CrossRef]
15. Hyks, J.; Astrup, T.; Christensen, T.H. Leaching from MSWI bottom ash: Evaluation of non-equilibrium in column percolation experiments. *Waste Manag.* **2009**, *29*, 522–529. [CrossRef]
16. Sivula, L.; Sormunen, K.; Rintala, J. Leachate formation and characteristics from gasification and grate incineration bottom ash under landfill conditions. *Waste Manag.* **2012**, *32*, 780–788. [CrossRef]
17. Yao, J.; Li, W.; Xia, F.; Wang, J.; Fang, C.; Shen, D. Investigation of Cu leaching from municipal solid waste incinerator bottom ash with a comprehensive approach. *Front. Energy* **2011**, *5*, 340–348. [CrossRef]

18. Zhang, H.; He, P.-J.; Shao, L.-M.; Li, X.-J. Leaching behavior of heavy metals from municipal solid waste incineration bottom ash and its geochemical modeling. *J. Mater. Cycles. Waste Manag.* **2008**, *10*, 7–13. [CrossRef]

19. Tang, J.; Steenari, B.-M. Leaching optimization of municipal solid waste incineration ash for resource recovery: A case study of Cu, Zn, Pb and Cd. *Waste Manag.* **2016**, *48*, 315–322. [CrossRef]

20. Jadhav, U.U.; Biswal, B.K.; Chen, Z.; Yang, E.-H.; Hocheng, H. Leaching of Metals from Incineration Bottom Ash Using Organic Acids. *J. Sustain. Metall.* **2018**, *4*, 115–125. [CrossRef]

21. Funari, V.; Gomes, H.; Cappelletti, M.; Dinnelli, E.; Fedi, S.; Rogerson, M.; Mayes, W. Bioleaching of fly ash and bottom ash from Municipal Solid Waste Incineration for metal recovery. In Proceedings of the NAXOS 2018 6th International Conference on Sustainable Solid Waste Management, Naxos Island, Greece, 13–16 June 2018.

22. Rawlings, D.E. Heavy Metal Mining Using Microbes. *Annu. Rev. Microbiol.* **2002**, *56*, 65–91. [CrossRef] [PubMed]

23. Mäkinen, J.; Bachér, J.; Kaartinen, T.; Wahlström, M.; Salminen, J. The effect of flotation and parameters for bioleaching of printed circuit boards. *Miner. Eng.* **2015**, *75*, 26–31. [CrossRef]

24. Brierley, C. Microbial mining. *Sci. Am.* **1982**, *247*, 42–50. [CrossRef]

25. Ballor, N.R.; Nesbitt, C.C.; Lueking, D.R. Recovery of scrap iron metal value using biogenerated ferric iron. *Biotechnol. Bioeng.* **2006**, *93*, 1089–1094. [CrossRef] [PubMed]

26. Halinen, A.-K. Heap Bioleaching of Low-Grade Multimetal Sulphidic Ore in Boreal Conditions. Ph.D. Thesis, Tampere University of Technology, Tampere, Finland, 2015.

27. Monhemius, J. Precipitation diagrams for metal hydroxides, sulfides, arsenates and phosphates. *Tran. Inst. Min. Metall. Sect. C* **1977**, *86*, 202–206.

28. Lewis, A.E. Review of metal sulphide precipitation. *Hydrometallurgy* **2010**, *104*, 222–234. [CrossRef]

29. Bigham, J.M.; Schwertmann, U.; Pfab, G. Influence of pH on mineral speciation in a bioreactor simulating acid mine drainage. *Appl. Geochem.* **1996**, *11*, 845–849. [CrossRef]

30. London Metal Exchange. Available online: https://www.lme.com/ (accessed on 19 March 2019).

31. Alibaba. Available online: https://www.alibaba.com/showroom/sulfuric-acid.html (accessed on 19 March 2019).

32. Free, M.L. *Hydrometallurgy: Fundamentals and Applications*, 1st ed.; John Wiley & Sons, Inc.: Hoboken, NJ, USA, 2013; 444p.

33. Kinnunen, P.; Kyllönen, H.; Kaartinen, T.; Mäkinen, J.; Heikkinen, J.; Miettinen, V. Sulphate removal from mine water with chemical, biological and membrane technologies. *Water Sci. Technol.* **2017**, *1*, 194–205. [CrossRef]

34. Huang, K.; Inoue, K.; Harada, H.; Kawakita, H.; Ohto, K. Leaching of heavy metals by citric acid from fly ash generated in municipal waste incineration plants. *J. Mater. Cycles Waste Manag.* **2011**, *13*, 118–126. [CrossRef]

35. Xu, T.-J.; Ting, Y.-P. Fungal bioleaching of incinerator fly ash: Metal extraction and modeling growth kinetics. *Enzyme Microb. Technol.* **2009**, *44*, 323–328. [CrossRef]

36. Wang, Q.; Yang, J.; Wang, Q.; Wu, T. Effects of water-washing pretreatment on bioleaching of heavy metals from municipal solid waste incineration fly ash. *J. Hazard. Mater.* **2009**, *162*, 812–818. [CrossRef]

Article

Bioreductive Dissolution as a Pretreatment for Recalcitrant Rare-Earth Phosphate Minerals Associated with Lateritic Ores

Ivan Nancucheo [1],*, Guilherme Oliveira [2], Manoel Lopes [2] and David Barrie Johnson [3]

[1] Facultad de Ingeniería y Tecnología, Universidad San Sebastián, Lientur 1457, Concepción 4080871, Chile
[2] Instituto Tecnológico Vale, Rua Boaventura da Silva 955, Belém, Pará 66055-090, Brazil;
    Guilherme.Oliveira@itv.org (G.O.); manoeljpl@gmail.com (M.L.)
[3] School of Natural Sciences, Bangor University, Deiniol Road, Bangor LL57 2UW, UK;
    d.b.johnson@bangor.ac.uk
*   Correspondence: inancucheo@gmail.com; Tel.: +56-(41)-2487433

Received: 18 January 2019; Accepted: 21 February 2019; Published: 26 February 2019

Abstract: Recent research has demonstrated the applicability of a biotechnological approach for extracting base metals using acidophilic bacteria that catalyze the reductive dissolution of ferric iron oxides from oxidized ores, using elemental sulfur as an electron donor. In Brazil, lateritic deposits are frequently associated with phosphate minerals such as monazite, which is one of the most abundant rare-earth phosphate minerals. Given the fact that monazite is highly refractory, rare earth elements (REE) extraction is very difficult to achieve and conventionally involves digesting with concentrated sodium hydroxide and/or sulfuric acid at high temperatures; therefore, it has not been considered as a potential resource. This study aimed to determine the effect of the bioreductive dissolution of ferric iron minerals associated with monazite using *Acidithiobacillus* (*A.*) species in pH- and temperature-controlled stirred reactors. Under aerobic conditions, using *A. thiooxidans* at extremely low pH greatly enhanced the solubilization of iron from ferric iron minerals, as well that of phosphate (about 35%), which can be used as an indicator of the dissolution of monazite. The results from this study have demonstrated the potential of using bioreductive mineral dissolution, which can be applied as pretreatment to remove coverings of ferric iron minerals in a process analogous to the bio-oxidation of refractory golds and expand the range of minerals that could be processed using this approach.

Keywords: iron reduction; reductive mineral dissolution; *Acidithiobacillus*; laterites; phosphate mineral; REE

---

1. Introduction

Currently the main supplier of REE to the world market is China, which accounted for 86% of total world production in 2014 and hosts the largest (~42%) proportion of the total global reserves, estimated to be ca. 110 million tons [1,2]. The demand for REE is growing at a rate of approximately 5–10% as these metals are increasingly used in modern technology including the metallurgy, fine chemical, automotive, oil, and renewable energies industries, etc. [3,4]. In recent years, there has been an increasing effort to identify additional potential supplies of REE to limit economic risk in the supply chain, given the fact that cuts in China's exports quota in recent years have provoked uncertainty among the hi-tech markets [4]. The three main rare-earth-bearing minerals are monazite, bastnaesite, and xenotime; these are highly refractory minerals and are not solubilized by conventional chemical treatment, requiring digestion with sulfuric acid and/or concentrated sodium hydroxide at high temperatures [5,6].

Biomining is now well established as an important applied biotechnology in the metal mining sector; it is often perceived as a technology that is less energy consuming and has also been promoted as being a more "environmentally friendly" approach to processing minerals ores and concentrates than conventional practices such as smelting. Biomining operations use far lower temperatures and pressures than conventional extraction processes, require less energy, and produce lower CO_2 emissions, though extraction of metals from primary ores and wastes tend to be far more protracted [7]. Although current commercial-scale biomining operations use the ability of acidophilic microorganisms to extract metals exclusively from reduced (sulfidic) ores, it has also been shown that bioreductive mineral dissolution can be used to process iron oxide ores to facilitate the recovery of metals [8–10]. Some acidophilic prokaryotes can catalyze the dissimilatory reduction of ferric iron; these include several species of *Acidithiobacillus*, which are better known for their ability to catalyze the oxidative dissolution of metal sulfide minerals. *A. ferrooxidans* was shown to couple the oxidation of sulfur to the reduction of ferric iron present in the mineral phase, facilitating the recovery of valuable metals [8]. In addition, it has been well established that heterotrophic acidophilic bacteria also catalyzed dissimilatory ferric iron reduction from ferric iron mineral (i.e., schwertmannite) using organic compounds as the electron donor and carbon source [11].

Many valuable metals occur in nature in oxidized ores and are therefore not amenable to oxidative bioprocessing. There are some reports of solubilization of metals from low-grade lateritic ores using organic acids produced by heterotrophic fungi, though sometimes poor efficiency in metal recovery, high costs of microbial substrates, and the issue of biomass disposal are reasons why this approach has not been exploited commercially [8,12]. Using acidophilic bacteria to enhance the recovery of nickel from laterites was found to be more cost effective in low-pH liquors (where cationic transition metals are far more soluble) and ambient temperatures. Under anaerobic conditions and using a pH- and temperature-controlled bioreactor (pH 1.8 and 30 °C), *A. ferrooxidans* was found to be effective in recovering both nickel (mostly associated with iron oxide phases, such as goethite) and cobalt (mostly associated with manganese oxide minerals, such as asbolane) from limonitic laterite [13]. Elsewhere, oxidized mineral waste from a copper mine from Brazil that contained 0.8% copper was subjected to reductive dissolution using pure and mixed cultures of iron-reducing acidophiles [14]. This study evaluated different operating parameters to increase copper recovery; bacterial numbers, in particular, were noted to have a critical role in copper extraction. More recently, *Acidithiobacillus* spp. were reported to extract cobalt (50%) and nickel (70%) from nickel laterite tailings within 7 days under aerobic conditions at extremely low pH (pH 0.8; [10]). Furthermore, this process was used to leach a laterite overburden in an aerobic bioreactor maintained at pH 0.8 [15]. The major ionic form of iron in this case was ferric iron, which was released from the tailings by the sulfuric acid generated by *A. thiooxidans*, which is more tolerant of extreme acidity than the iron-oxidizing acidithiobacilli [16].

Lateritic deposits that contain phosphate minerals with high grades of REE are important resources found in Brazil (in average ~8% in Goias state; [17]), where monazite is one of the most abundant rare-earth phosphate minerals. Catalão, State of Goiás is reported to have 120 million tons of ore, confirming that is one of the countries with the largest reserves of valuable light REE (mainly La and Ce). So far, this material has not been studied using this radical approach in mineral bioprocessing and also not considered as an important source of recovered REE. We have tested the hypothesis that reductive dissolution of these deposits by *Acidithiobacillus* species, maintained under both aerobic and anaerobic conditions and at between pH 0.9 and 1.8, could mediate the dissolution of ferric iron minerals and facilitate the further processing of monazite.

2. Materials and Methods

2.1. Characteristics of the Ore

The lateritic ore containing monazite was sourced from a phosphate mine in Brazil and subjected to beneficiation. Mineralogical analysis showed that monazite was the main phosphate ore (~90%), and the major iron phases consisted mainly of goethite (~60%) together with magnetite and hematite. The material contained 6.9% REE, 42% Fe_2O_3, 12% SiO_2, 8.1% P_2O_5, 7% TiO_2, and 6.5% Al_2O_3 as the major oxides. Mineralogical and chemical composition analyses of the laterite were carried out as described elsewhere [14]. All tests were performed with monazite-containing lateritic material which was ground and sieved to below 1 mm particle size. Table 1 shows the elemental composition of the lateritic material.

Table 1. Chemical composition of elements of the laterite ore evaluated in this study.

Element	Fe	P	Ce	La	U	Th
Concentration (g/kg)	301	34.9	27	13.2	0.088	0.401

2.2. Bacteria and Cultivation Conditions

The mesophilic acidophiles *A. ferrooxidans* (ATCC 23270[T]) and *A. thiooxidans* (DSM 14557[T]) were used in this study. Bacteria were grown in shake flasks containing 200 mL of liquid medium, comprising basal salts with trace elements [18] and 1 g of elemental sulfur. Cultures were incubated at 30 °C and an initial pH of 3.0 before being used as inocula for the bioreactor experiments.

2.3. Reductive Mineral Dissolution under Anaerobic Conditions

The iron-oxidizing/-reducing acidophile *A. ferrooxidans* was used in these experiments. Prior to inoculation, two stirred bioreactors, each with 2 L working volumes and fitted with pH, temperature, and aeration control (Electrolab, UK), were set up. Fifty grams of elemental sulfur and 1.95 L of basal salts and trace elements [18] adjusted to pH 1.8 were put into each reactor vessel, which were then autoclaved at 110 °C for 40 min [14]. Once cooled, each bioreactor was inoculated with 50 mL of *A. ferrooxidans* culture grown on sulfur. The bioreactors were maintained at 30 °C and a constant pH of 1.8, stirred at 150 rpm, and aerated at 1 L/min with sterile atmospheric air. When numbers of planktonic cells had reached ~10^9/mL (enumerated using a Thoma counting chamber), the air supply to one of them was removed and replaced with oxygen-free nitrogen (OFN) to promote anaerobic conditions while the second bioreactor was maintained with OFN enriched with 10% CO_2. Twenty-five grams of lateritic ore were then added to each of the bioreactors. To assess whether the dissolution of the lateritic ore was catalyzed by the bacteria, a third non-inoculated reactor with basal salts and trace elements was operated abiotically by gassing with OFN at pH 1.8 (by the addition of 0.5 M sulfuric acid) and 30 °C. The bioreactors were operated for up to 30 days and samples were removed at regular intervals for chemical analysis and to determine numbers of viable bacteria by plating on selective solid media [14].

2.4. Reductive Dissolution under Aerobic Conditions

Two 2 L (working volume) bioreactors, each containing 50 g of elemental sulfur and 1.95 L of basal salts and trace elements (pH 1.8), were sterilized as described previously and, when cool, inoculated with *A. thiooxidans*. The bioreactors were aerated at 1 L sterile air/min and stirred at 150 rpm. Once the pH had fallen to either 0.9 (Reactor 1) or 1.2 (Reactor 2) as a consequence of biogenic sulfuric acid production, 25 g of lateritic ore was added to each vessel. Samples were withdrawn as before. As controls, non-inoculated reactors containing lateritic ore and elemental sulfur, together with basal salts and trace elements, were operated at fixed pH values of 0.9 and 1.2 by the addition of 0.5 M sulfuric acid.

2.5. Analytical Techniques

Concentrations of ferrous iron were measured colorimetrically using the Ferrozine assay [19]. Concentrations of total soluble iron were measured using a Dionex ICS-5000 ion chromatography [20] system fitted with an IonPAC CS5A column and an AD25 absorbance detector (Thermo Fisher Scientific

Inc.). Phosphate concentrations were measured using a Dionex ICS-5000 ion chromatograph fitted with an Ion Pac AS-11 equipped with a conductivity detector [21]. Concentrations of lanthanum, cerium, uranium, and thorium in filtered (through 0.2 μm pore-sized polycarbonate filters) mineral leachate samples at the end of the experiments were determined by ICP-MS (NexION 300, PerkinElmer, 2015).

3. Results

The concentrations of total soluble iron in the two bioreactors that were operated under anoxic conditions and fixed pH (1.8) using *A. ferrooxidans* and, the corresponding abiotic control are shown in Figure 1A. The reductive dissolution of ferric iron minerals was greater in the bioreactor that was continuously sparged with the OFN/CO$_2$ gas mix than that which was sparged only with OFN. In both reactors, virtually all the soluble iron was present as ferrous iron, concentrations of which increased rapidly during the first 10 days of the experiment but slowed down thereafter. Dissolution of the laterite sample also occurred under aerobic conditions catalyzed by *A. thiooxidans* (Figure 1B), and this was more effective at pH 0.9 than at pH 1.2. Again, this was a reductive process, as most of the soluble iron was present as ferrous iron throughout the experiment. Concentrations of soluble iron increased at a relatively constant rate in the aerobic pH 1.2 bioreactor, whereas a more biphasic pattern was observed at pH 0.9. The final concentration of soluble iron in the aerobic pH 0.9 bioreactor (985 mg/L) was about twice that in the pH 1.8 anaerobic bioreactor sparged with OFN/CO$_2$ (530 mg/L) and ~3 times greater than that in the pH 1.8 anaerobic bioreactor sparged with OFN (340 mg/L). The greatest extent of total iron minerals extracted from the laterite ore (~30%) occurred in the aerobic pH 0.9 bioreactor. In contrast to the inoculated reactors, most of the iron solubilized from the lateritic ore under abiotic conditions (both aerobic and anoxic bioreactors) was present as ferric iron (data not shown), and both the rates and extents of mineral dissolution were much less than when bacteria were present.

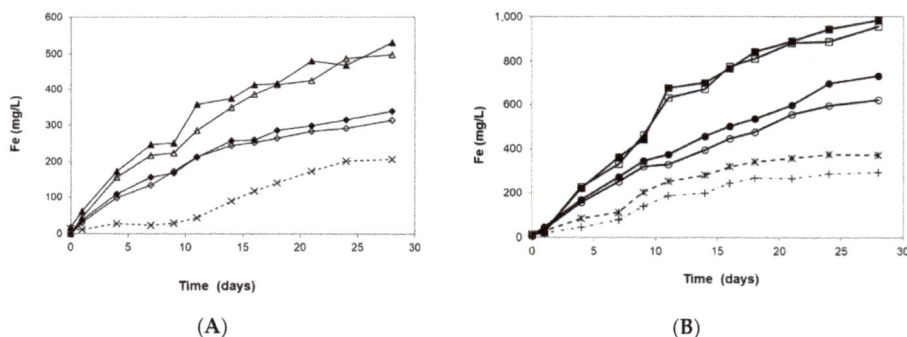

Figure 1. Total soluble iron (solid symbols) and ferrous iron (hollow symbols) leached from the laterite ore (**A**) under anoxic conditions using *A. ferrooxidans* at pH 1.8 and 30 °C, sparged with (i) oxygen-free nitrogen (OFN) (♦, ◊) and (ii) OFN enriched with 10% of CO$_2$ (▲, Δ). (**B**) Under aerobic conditions using *A. thiooxidans* at (i) pH 0.9 (■, □) and (ii) pH 1.2 (•, o). Dashed lines show the concentrations of total soluble iron in the non-inoculated reactors: anoxic, pH 1.8 (x); aerobic, pH 1.2 (+); aerobic, pH 0.9 (✖).

The concentrations of soluble phosphate increased rapidly within the first 2 days of leaching in all reactors (aerobic and anoxic; inoculated and non-inoculated) but solubilization of phosphate minerals slowed down in most cases thereafter (Figure 2). The concentrations of phosphate in both of the inoculated anoxic bioreactors were similar throughout the experiment and were much greater than those in the sterile control (Figure 2A). In the aerobic reactors, more phosphate was both abiotically and microbially leached at pH 0.9 than at pH 1.2, and again, in both cases, the presence of bacteria (*A. thiooxidans* in this case) greatly enhanced the solubilization of phosphate minerals (Figure 2B).

The amount of phosphate released from the lateritic ore using *A. thiooxidans* at pH 0.9 was 35% greater than that by *A. ferrooxidans* sparged continuously with OFN/CO_2 at pH 1.8.

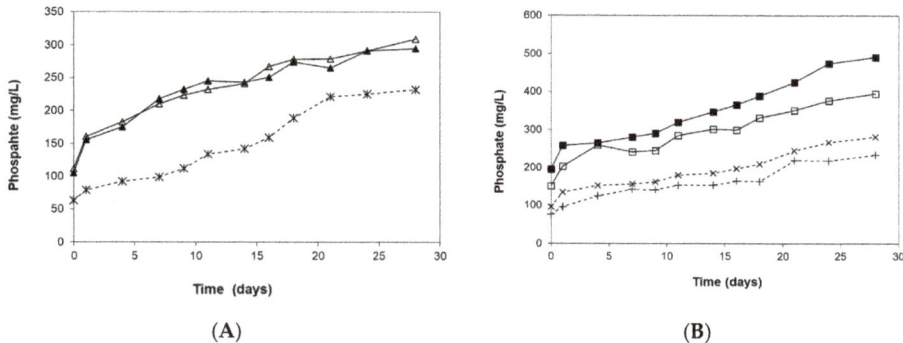

(A)   (B)

Figure 2. Solubilization of phosphate in the bioreactors subjected to reductive dissolution (**A**) under anoxic conditions using *A. ferrooxidans* sparged with (i) OFN (closed triangles) and (ii) OFN enriched with 10% of CO_2 (open triangles). (**B**) Under aerobic conditions using *A. thiooxidans* at (i) pH 0.9 (closed squares) and (ii) pH 1.2 (open squares). Dashed lines show the solubilization of total iron in the non-inoculated reactors at pH 1.8(✖), 1.2 (+), and 0.9 (x).

Analysis of the laterite leachates at the end of the experiments showed that there were similar concentrations of cerium, lanthanum, and uranium present in all three reactors (inoculated and control) that were operated at pH 1.8 and under anoxic conditions (Figure 3A).

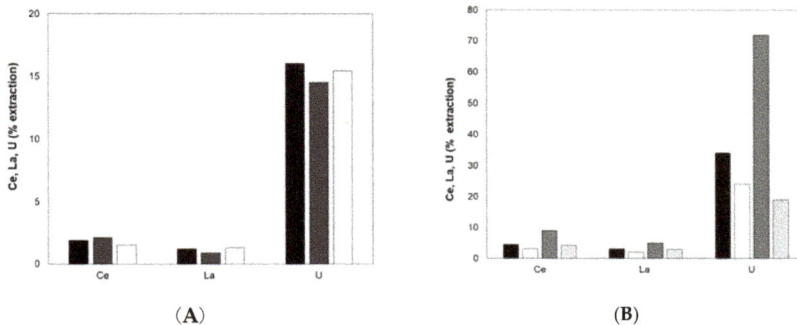

(A)   (B)

Figure 3. Percentages of cerium, lanthanum, and uranium extracted from the laterite ore, determined from the concentrations of these metals present in leachates at the end of the experiments. (**A**) *A. ferrooxidans* at pH 1.8, sparged with OFN (black bars) or with OFN/CO_2 (dark grey bars) and the corresponding abiotic control (white bars); (**B**) *A. thiooxidans* at pH 1.2 (black bars) and the corresponding abiotic control (white bars), and at pH 0.9 (dark grey bars) and the corresponding abiotic control (light grey bars). Note: the axis scales show the differences obtained under the anoxic and aerobic conditions using *Acidithiobacillus* species.

In contrast, the concentrations of all three metals were greater in the bioreactors that were inoculated with *A. thiooxidans* than in the corresponding abiotic controls, particularly at pH 0.9 (Figure 3B). Thorium levels were also determined, and the amount extracted in all three reactors even at pH 0.9 showed similar concentrations (~4%; data not shown).

Although the numbers of total viable planktonic bacteria decreased in all bioreactors during the course of these experiments, the extents to which this occurred were different in the four inoculated reactors (Figure 4). The numbers of planktonic *A. ferrooxidans* grown under anoxic conditions and gassed with OFN/CO_2 remained fairly constant throughout the experiment. More pronounced was the decline in numbers of *A. thiooxidans* in both aerobic bioreactors, though it is worth mentioning that most of the solubilization of iron occurred during the first 15 days of operation (~80%).

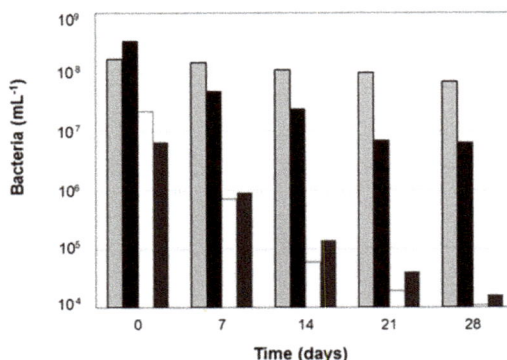

Figure 4. Changes in numbers of viable acidophilic bacteria in the bioreactors: *A. ferrooxidans* sparged with OFN (black bars) or OFN/CO_2 (light grey bars); *A. thiooxidans* pH 0.9 (white bars) and 1.2 (dark grey bars).

4. Discussion

The reductive dissolution of limonitic laterite ores represents a radical development in mineral bioprocessing and contrasts with conventional biomining operations where bioleaching of metal-containing sulfide minerals or the bio-oxidation of refractory gold ores is mediated by microbially catalyzed oxidative dissolution [22]. As the supply of REEs is crucial for many developed countries, it is becoming increasingly necessary to seek new technologies that are more environmentally benign than the existing approaches used to facilitate the processing of REE-bearing ores. The abundance of lateritic ores, particularly in tropical areas, represents an important source of polymetallic metals including nickel, cobalt, manganese, and copper, which have previously been proven amenable to bioreductive dissolution [8,10,14,23]. The results from this work show that the reductive bioprocessing of ferric iron minerals in lateritic ores containing monazite using acidophilic microorganisms can be partially effective in enhancing the solubilization of phosphate minerals that contain REE.

While the percentage of iron leached from the material was relatively small (<30%), this may be perceived as an advantage as the ferrous-iron-rich process liquors produced would require less intensive downstream mitigation [23]. It is worth mentioning that the laterite ore used was rich in goethite but contained no detectable magnesium oxide, therefore delineating it as limonitic [24]. Using *A. thiooxidans* at extremely low pH greatly enhanced the solubilization of iron from ferric iron minerals, as well that of phosphate, which can be used as indicator of the dissolution of monazite. Acid abiotic leaching experiments were partially effective at solubilizing phosphate from the ore, and this was more effective at very low (<1) pH, but in all cases, reductive bioleaching catalyzed by the acidithiobacilli used was superior in this respect. These results suggest that monazite is intimately associated with ferric iron minerals such as goethite present in the ore, and, therefore, that reductive mineral dissolution can be used to remove surface coverings of ferric iron in a process analogous to the bio-oxidation of refractory golds [9].

In addition, the reductive conditions of this bioprocess can partially facilitate the exposure of monazite to attack by acid leaching from the biogenic sulfuric acid produced by the bacteria, enhancing the phosphate solubilization. In contrast, a chemical acid process is not enough by itself to leach phosphate from the matrix, demonstrated with the experiment carried out under abiotic conditions, which is inferior to reductive dissolution. Although ~9% of the cerium and 5% of the lanthanum were leached using *A. thiooxidans* at pH 0.9, the results again suggest that bioreductive dissolution can improve accessibility to monazite, but further studies including analysis of the residue are needed. Recent studies addressing only monazite dissolution using phosphate-solubilizing bacteria have shown different degrees of success with leaching efficiencies for REE recovery (between 0.1% and 25%), though differences in ores and experimental conditions may explain the varying results [5,25,26].

The results of these experiments both confirmed some earlier findings regarding the bioprocessing of lateritic ores. Reductive dissolution of the ore was catalyzed by *A. ferrooxidans* grown at pH 1.8 under anoxic conditions where ferric iron was reduced, and the ferrous iron generated could not be re-oxidized. In addition, aerobic reductive dissolution using bioreactors inoculated with *A. thiooxidans* was previously reported to be effective in recovering nickel from laterite tailings [10,15]. The results from this study confirm that dissolution of ferric iron and phosphate minerals was more rapid and extensive at lower pH values. *A. thiooxidans* was confirmed to mediate reductive mineral dissolution even in well-aerated reactor vessels, and, since this acidophile cannot oxidize iron, the ferrous iron generated remained in the reduced form. Since most of the soluble iron in the abiotic reactors was present as ferric iron, the inference is that, as suggested elsewhere [8], iron was firstly acid-solubilized from goethite and other oxidized minerals and then reduced biologically to ferrous.

However, other data support the notion that *A. thiooxidans* appeared to die off in these bioreactors (viable cell numbers decreased, especially in the pH 0.9 reactor) and little reduction of ferric iron occurred after 15 days. One reason for this could be that *A. thiooxidans* became less viable because of the very low pH values at which the bioreactors were maintained, though this acidophile is known to be more tolerant of extreme acidity than the iron-oxidizing acidithiobacilli [16]. These conundrum observations merit further study, though adding sufficient quantities of exogenously grown cultures could be used as a strategy to maintain the desired reactions. While it has long been known that *A. thiooxidans* can reduce ferric iron [27], later work showed that this acidophile cannot grow by ferric iron respiration [28]. Reduction of ferric iron in aerobically grown cultures has been shown to be widespread amongst the acidithiobacilli [29], and ongoing reduction of soluble ferric iron by *Acidithiobacillus caldus* has recently been shown to sustain the growth of *Leptospirillum ferriphilum* in mixed cultures [30]. The fact that ferric iron can be reduced under aerobic conditions extends the potential for the bioprocessing of minerals such as goethite.

5. Conclusions

This study provides a proof of concept that bacterially catalyzed reductive dissolution of a lateritic ore can be used to remove sufficient ferric iron minerals to facilitate the dissolution of monazite. The process was more effective when carried out at ultra-low (<1) than at higher pH values, where it was mediated by *A. thiooxidans* maintained under aerobic conditions. In addition, the results obtained suggest that reductive dissolution can improve the exposure of monazite to the acidic conditions, though further studies are needed since this approach is a relatively new area of biohydrometallurgy.

Author Contributions: Conceptualization and methodology, I.N.; investigation, I.N., G.O. and M.L.; resources, I.N. and G.O.; writing—original draft preparation, I.N., G.O. and D.B.J.; writing—review and editing, I.N. and D.B.J.; funding acquisition, I.N. and G.O.

Funding: The authors acknowledge the financial support from Conselho Nacional de Desenvolvimento Científico e Tecnológico (CNPq-309312/2012-4) of Guilherme Oliveira and Vale (BioDam). Ivan Nancucheo is supported by Fondecyt, Chile (n° 11150170).

Conflicts of Interest: The authors declare no conflict of interest.

References

1. Fan, H.R.; Yang, K.F.; Hu, F.F.; Liu, S.; Wang, K.Y. The giant Bayan Obo REE-Nb-Fe deposit, China: Controversy and ore genesis. *Geosci. Front.* **2016**, *7*, 335–344. [CrossRef]
2. Wübbeke, J. Rare earth elements in China: Policies and narratives of reinventing an industry. *Resour. Policy* **2013**, *38*, 1–11. [CrossRef]

3. Pourret, O.; Tuduri, J. Continental shelves as potential resource of rare earth elements. *Sci. Rep.* **2017**, *7*, 1–6. [CrossRef] [PubMed]

4. Massari, S.; Ruberti, M. Rare earth elements as critical raw materials: Focus on international markets and future strategies. *Resour. Policy* **2013**, *38*, 36–43. [CrossRef]

5. Shin, D.; Kim, J.; Kim, B.; Jeong, J.; Lee, J. Use of phosphate solubilizing bacteria to leach rare earth elements from monazite-bearing Ore. *Minerals* **2015**, *5*, 189–202. [CrossRef]

6. Maes, S.; Zhuang, W.Q.; Rabaey, K.; Alvarez-Cohen, L.; Hennebel, T. Concomitant leaching and electrochemical extraction of rare rarth elements from monazite. *Environ. Sci. Technol.* **2017**, *51*, 1654–1661. [CrossRef] [PubMed]

7. Johnson, D.B. Biomining-biotechnologies for extracting and recovering metals from ores and waste materials. *Curr. Opin. Biotechnol.* **2014**, *30*, 24–31. [CrossRef] [PubMed]

8. Hallberg, K.B.; Grail, B.M.; Plessis, C.A.D.; Johnson, D.B. Reductive dissolution of ferric iron minerals: A new approach for bio-processing nickel laterites. *Miner. Eng.* **2011**, *24*, 620–624. [CrossRef]

9. Johnson, D.B.; Du Plessis, C.A. Biomining in reverse gear: Using bacteria to extract metals from oxidised ores. *Miner. Eng.* **2015**, *75*, 2–5. [CrossRef]

10. Marrero, J.; Coto, O.; Goldmann, S.; Graupner, T.; Schippers, A. Recovery of nickel and cobalt from laterite tailings by reductive dissolution under aerobic conditions using *Acidithiobacillus* species. *Environ. Sci. Technol.* **2015**, *49*, 6674–6682. [CrossRef] [PubMed]

11. Coupland, K.; Johnson, D.B. Evidence that the potential for dissimilatory ferric iron reduction is widespread among acidophilic heterotrophic bacteria. *FEMS Microbiol. Lett.* **2008**, *279*, 30–35. [CrossRef] [PubMed]

12. Mulligan, C.N.; Kamali, M. Bioleaching of copper and other metals from low-grade oxidized mining ores by *Aspergillus niger*. *J. Chem. Technol. Biotechnol.* **2003**, *78*, 497–503. [CrossRef]

13. Johnson, D.B.; Grail, B.M.; Hallberg, K.B. A new direction for biomining: Extraction of metals by reductive dissolution of oxidized ores. *Minerals* **2013**, *3*, 49–58. [CrossRef]

14. Nancucheo, I.; Grail, B.M.; Hilario, F.; Du Plessis, C.; Johnson, D.B. Extraction of copper from an oxidized (lateritic) ore using bacterially catalysed reductive dissolution. *Appl. Microbiol. Biotechnol.* **2014**, *98*, 6297–6305. [CrossRef] [PubMed]

15. Marrero, J.; Coto, O.; Schippers, A. Anaerobic and aerobic reductive dissolutions of iron-rich nickel laterite overburden by *Acidithiobacillus*. *Hydrometallurgy* **2017**, *168*, 49–55. [CrossRef]

16. Dopson, M. Physiological and phylogenetic diversity of acidophilic bacteria. In *Acidophiles: Life in Extremely Acidic Environments*; Quatrini, R., Johnson, D.B., Eds.; Caistor Academic Press: Haverhill, UK, 2016; pp. 79–92.

17. Tassinari, M.M.L.; Kahn, H.; Ratti, G. Process mineralogy studies of Corrego do Garimpo REE ore, Catalao-I Alkaline complex, Goias, Brazil. *Miner. Eng.* **2001**, *14*, 1609–1617. [CrossRef]

18. Nancucheo, I.; Rowe, O.F.; Hedrich, S.; Johnson, D.B. Solid and liquid media for isolating and cultivating acidophilic and acid-tolerant sulfate-reducing bacteria. *FEMS Microbiol. Lett.* **2016**, *363*, 1–6. [CrossRef] [PubMed]

19. Stookey, L.L. Ferrozine-A new spectrophotometric reagent for iron. *Anal. Chem.* **1970**, *42*, 779–781. [CrossRef]

20. Nancucheo, I.; Johnson, D.B. Selective removal of transition metals from acidic mine waters by novel consortia of acidophilic sulfidogenic bacteria. *Microb. Biotechnol.* **2012**, *5*, 34–44. [CrossRef] [PubMed]

21. Jenke, D.; Sadain, S.; Nunez, K.; Byrne, F. Performance characteristics of an ion chromatographic method for the quantitation of citrate and phosphate in pharmaceutical solutions. *J. Chromatogr. Sci.* **2017**, *45*, 50–56. [CrossRef]

22. Rawlings, D.E.; Johnson, D.B. The microbiology of biomining: Development and optimization of mineral-oxidizing microbial consortia. *Microbiology* **2007**, *153*, 315–324. [CrossRef] [PubMed]

23. Smith, S.L.; Grail, B.M.; Johnson, D.B. Reductive bioprocessing of cobalt-bearing limonitic laterites. *Miner. Eng.* **2017**, *106*, 86–90. [CrossRef]

24. Komnitsas, K.; Petrakis, E.; Pantelaki, O.; Kritikaki, A. Column Leaching of Greek Low-Grade Limonitic Laterites. *Minerals* **2018**, *8*, 377. [CrossRef]

25. Brisson, V.L.; Zhuang, W.Q.; Alvarez-Cohen, L. Bioleaching of rare earth elements from monazite sand. *Biotechnol. Bioeng.* **2016**, *113*, 339–348. [CrossRef] [PubMed]

26. Corbett, M.K.; Eksteen, J.J.; Zhi, X.; Jean, N.; Croue, P.; Watkin, E.L.J. Interactions of phosphate solubilising microorganisms with natural rare-earth phosphate minerals: A study utilizing Western Australian monazite. *Bioprocess Biosyst. Eng.* **2017**, *40*, 929–942. [CrossRef] [PubMed]

27. Brock, T.D.; Gustafson, J. Ferric iron reduction by sulfur- and iron-oxidizing bacteria. *Appl. Environ. Microbiol.* **1976**, *32*, 567–571. [PubMed]

28. Hallberg, K.B.; Thompson, H.E.C.; Boeselt, I.; Johnson, D.B. Aerobic and anaerobic sulfur metabolism by acidophilic bacteria. In *Biohydrometallurgy: Fundamentals, Technology and Sustainable Development*; Process Metallurgy 11A; Ciminelli, V.S.T., Garcia, O., Jr., Eds.; Elsevier: Amsterdam, the Netherlands, 2001; pp. 423–431.

29. Johnson, D.B.; Hedrich, S.; Pakostova, E. Indirect redox transformations of iron, copper, and chromium catalyzed by extremely acidophilic bacteria. *Front. Microbiol.* **2017**, *8*, 1–15. [CrossRef] [PubMed]

30. Smith, S.L.; Johnson, D.B. Growth of *Leptospirillum ferriphilum* in sulfur medium in co-culture with Acidithiobacillus caldus. *Extremophiles* **2018**, *22*, 327–333. [CrossRef] [PubMed]

MDPI
St. Alban-Anlage 66
4052 Basel
Switzerland
Tel. +41 61 683 77 34
Fax +41 61 302 89 18
www.mdpi.com

Minerals Editorial Office
E-mail: minerals@mdpi.com
www.mdpi.com/journal/minerals